CONTROL AND DYNAMIC SYSTEMS

Advances in Theory and Applications

Volume 37

CONTRIBUTORS TO THIS VOLUME

GUY A. DUMONT

XIAOWEN FANG

JANOS J. GERTLER

DAVID M. HIMMELBLAU

D. HROVAT

DANIEL J. INMAN

C. M. KRISHNA

QIANG LUO

L. MILI

CONSTANTINOS MINAS

V. PHANIRAJ

W. F. POWERS

N. LAWRENCE RICKER

P. J. ROUSSEEUW

KANG G. SHIN

K. PRESTON WHITE, JR.

CONTROL AND DYNAMIC SYSTEMS

ADVANCES IN THEORY AND APPLICATIONS

Edited by
C. T. LEONDES

School of Engineering and Applied Science
University of California, Los Angeles
Los Angeles, California
                and
College of Engineering
University of Washington
Seattle, Washington

VOLUME 37: ADVANCES IN INDUSTRIAL SYSTEMS

ACADEMIC PRESS, INC.
Harcourt Brace Jovanovich, Publishers
San Diego   New York   Boston
London   Sydney   Tokyo   Toronto

Academic Press, Inc.
San Diego, California 92101

United Kingdom Edition published by
Academic Press Limited
24-28 Oval Road, London NW1 7DX

Library of Congress Catalog Card Number: 64-8027

ISBN   0-12-012737-7   (alk. paper)

Printed in the United States of America
90 91 92 93   9 8 7 6 5 4 3 2 1

CONTENTS

CONTRIBUTORS

Numbers in parentheses indicate the pages on which the authors' contributions begin.

Guy A. Dumont (65), *Department of Electrical Engineering, Pulp and Paper Centre, University of British Columbia, Vancouver, B.C., Canada V6T1W5*

Xiaowen Fang (159), *George Mason University, School of Information Technology and Engineering, Fairfax, Virginia 22030*

Janos J. Gertler (159), *George Mason University, School of Information Technology and Engineering, Fairfax, Virginia 22030*

David M. Himmelblau (365), *The University of Texas at Austin, Austin, Texas 78712*

D. Hrovat (33), *Ford Motor Company, Dearborn, Michigan 48121*

Daniel J. Inman (327), *Mechanical Systems Laboratory, Department of Mechanical and Aerospace Engineering, University at Buffalo, Buffalo, New York 14260*

C. M. Krishna (1), *Department of Electrical and Computer Engineering, University of Massachusetts, Amherst, Massachusetts 01003*

Qiang Luo (159), *George Mason University, School of Information Technology and Engineering, Fairfax, Virginia 22030*

L. Mili (271), *Virginia Polytechnic Institute and State University, Blacksburg, Virginia 24061*

Constantinos Minas (327), *Mechanical Systems Laboratory, Department of Mechanical and Aerospace Engineering, University at Buffalo, Buffalo, New York 14260*

V. Phaniraj (271), *Virginia Polytechnic Institute and State University, Blacksburg, Virginia 24061*

W. F. Powers (33), *Ford Motor Company, Dearborn, Michigan 48121*

N. Lawrence Ricker (217), *Department of Chemical Engineering, University of Washington, Seattle, Washington 98195*

vii

P. J. Rousseeuw (271), *Vrije Universiteit Brussel, Pleinlaan 2, B-1050 Brussels*

Kang G. Shin (1), *Real-Time Computing Laboratory, Department of Electrical Engineering and Computer Science, The University of Michigan, Ann Arbor, Michigan 48109*

K. Preston White, Jr. (115), *Department of Systems Engineering, University of Virginia, Charlottesville, Virginia 22903*

PREFACE

In the 1940s, techniques for the analysis and synthesis of industrial systems of any degree of complexity were simplistic, at best, and so were the system models. Other essential techniques, such as the detection and diagnosis of industrial plant failures, represented, for all practical purposes, a virtual void in available methods. Fortunately, the industrial systems of the 1940s, were relatively unsophisticated by comparison with the industrial systems of the 1990s. Numerous key advances have occurred in the interim that provide considerable power in the effective control of complex industrial systems. These include advances in computers, electronics, communications, sensor systems, systems techniques, and other areas. The increasing complexity of industrial systems over the past few decades was recognized earlier and manifestly treated at various times in earlier volumes in this series. For example, Volumes 14 and 15 were exclusively devoted to techniques for modeling (the beginning of the industrial system design process) diverse industrial systems, including chemical processes, electric power systems, jet engines, and other systems. Because of the endlessly continuing surge of advances in techniques for industrial systems control, it is now time to revisit this area in this series, and so the theme for this volume is, in fact, "Advances in Industrial Systems Dynamics and Control."

The first contribution, "New Performance Measures for Real-Time Digital Computer Controls and Their Applications, "by Kang G. Shin and C. M. Krishna, deals with the objective measures used to characterize the performance of computers that control critical processes. Such processes are playing an increasing role in industry and require complex algorithms to control them. This is a far cry from the simple transfer-function-based controllers found in the feedback loops of most present-day processes. The choice of a performance measure is a crucial link in the design and development chain of industrial system. There is both the need and the capacity to obtain far more detailed characterizations of the performance of an industrial system control computer than there is for the general purpose computer. There are many essential issues and techniques presented in this chapter on performance measures for real-time digital computer controls, and so this is an appropriate contribution with which to begin this volume.

The next contribution, "Modeling and Control of Automotive Power Trains," by D. Hrovat and W. F. Powers, presents many significant results and techniques for one of the most important major industrial systems areas, namely the automotive industry. Present-day automotive engine control systems contain many inputs and outputs. The unique aspect of the automotive control problem is the requirement to develop systems that are relatively low in cost, will be applied to several hundred thousand systems in the field (indeed, overall, millions of systems), must work on relatively low cost automobiles with their inherent manufacturing variability, will not have scheduled maintenance, and will be used by a spectrum of human operators. Aircraft/spacecraft control problems, for which most sophisticated control techniques have been developed, have nearly an opposite set of conditions. In any event, because the subject area of the automotive industry is one of the major industrial systems areas, this is a most welcome contribution to this volume.

The next contribution, "Control Techniques in the Pulp and Paper Industry," by Guy A. Dumont, is a remarkably comprehensive treatment of the control techniques in this industry. The pulp and paper industry was one of the pioneers of computer process control some 25 years ago. Some of the major breakthroughs in process control, such as minimum-variance and self-tuning controllers, were first tested on paper machines. Today the vast majority of paper machines in the world are computer controlled. Because this is, of course, a major industrial area, this contribution is also a most welcome addition to this volume.

Production scheduling is "the allocation of resources over time for the manufacture of goods," and it thus constitutes a major area of fundamental importance in industrial systems. Thus the next contribution, "Advances in the Theory and Practice of Production Scheduling," by K. Preston White, Jr., constitutes a significant element in this volume. The objective of production scheduling is to find an efficient and effective way to assign and sequence the use of shared resources such that production constraints are satisfied and production costs minimized. The sheer diversity and momentum of activity has made developments in production scheduling difficult to track and assimilate. In the face of this tremendous activity, Professor White presents a remarkably comprehensive treatment of this broadly complex area of great significance, and thus he has contributed another significant element of this volume.

In the next contribution, "Detection and Diagnosis of Plant Failures: The Orthogonal Parity Equation Approach," by Janos J. Gertler, X. Fang, and Q. Luo, a general model-based failure detection and diagnosis methodology is presented. This methodology is based on the techniques of parity equations for the generation of the residuals between the actual plant measurements and the mathematical model of the plant. Heretofore, the emphasis on plant failure determination techniques has been placed on Kalman filters or observers to generate these residuals.

However, these approaches lead to relatively complex algorithms and may not support failure isolation in a very natural way. The detection and isolation (diagnosis) of failures, whether they occur in the basic technological equipment or in the measurement and control system, has always been one of the major tasks of computers supervising complex industrial plants. Further, with the proliferation of the microprocessor, similar systems have appeared more recently on airplanes, automobiles, and even on household appliances. In any event, Professor Gertler and his coauthors have provided an essential element of this volume on industrial system dynamics and control.

The historical approach to the design of complex processes in the chemical industry has been to partition these processes into subsystems without adequate coordination for their overall control. The advent of computer based data acquisition and retrieval systems has made it feasible to implement plant wide control methods that can manage a number of distributed subsystems with due consideration given to their interconnections or interactions. The next contribution, "Model-Predictive Control of Processes with Many Inputs and Outputs," by N. Lawrence Ricker, treats this problem of repeated optimization of high-order dynamic systems with many input and output variables by the methods of Model-Predictive Control (MPC). A key aspect of the optimization problem is the presence of physical constraints on the manipulated variables and, in some cases, safety or product-quality constraints on the output variables. This contribution is a most important and welcome element of this volume.

The next contribution, "Robust Estimation Theory for Bad Data Diagnostics in Electric Power Systems," by L. Mili, V. Phaniraj, and P. J. Rousseew, presents significant techniques for dealing with the vitally essential problem of providing a complete, coherent, and reliable data base from a collection of switch and breaker status data and measurements. Electric power systems consist of large interconnected centralized synchronous generators and decentralized loads. Since only a small amount of electric energy can be stored, the energy production must meet the continuous random fluctuations of energy demand. This is an extremely complex hierarchical multilevel control system whose role is to minimize the cost of generation of electric power while maintaining the system in a normal and secure operating state. At the heart of this problem is the robust data estimation issue which is treated in depth in this significant contribution. Both the United States and western Europe face the prospect of continuing "brownouts," and, what is worse, the economic viability of the United States is threatened because of the continually growing energy shortage. Thus this contribution deals with a fundamental aspect of achieving the best in performance from electric power systems in the face of this limited energy constraint.

The next contribution, "Matching Analytical Models with Experimental Modal Data in Mechanical Systems," by Daniel J. Inman and Constantinos Minas, presents several systematic approaches to modifying an FEM (finite element model) to agree with experimental modal data. Distributed parameter systems are characteristic of many industrial and aerospace systems. There is a large body of theoretical results on the optimization of distributed parameter systems, but, in all cases, the applications of these theoretical techniques are to simplified objects such as bars and disks. However, industrial and aerospace distributed parameter systems are, in general, complex to very complex structures or systems and, as a result, depend on FEM methods for their analysis and design. Until recently the adjustment of the finite element model has been accomplished on an *ad hoc* basis. Because of the great and growing importance of FEM techniques in many different applications, this contribution is also a key element of this volume.

Professor David M. Himmelblau is unquestionably one of the most important figures on the international scene in the field of optimization techniques in industrial chemical systems. His pioneering efforts on the introduction of some powerful techniques in this field have allowed him to attain this status. In the final contribution to this volume, "Techniques in Industrial Chemical Systems Optimization," Professor Himmelblau presents a rather comprehensive treatment of these issues as related to industries that are presently operating under conditions that are quite different from those of the last decade. These changes have occurred because of several factors, including the concern for energy conservation, environmental constraints, and the continuing evolution of computer capabilities. The changes are, of course, motivated by the essential desire to increase profits.

The authors are all to be commended for their superb contributions. The compilation of these contributions in this volume will provide, for many years to come, a unique and significant reference source for practicing professionals as well as those involved with advancing the state of the art.

New Performance Measures for Real-Time Digital Computer Controls and Their Applications

Kang G. Shin[†] and C. M. Krishna[‡]

[†]Real-Time Computing Laboratory
Department of Electrical Engineering and Computer Science
The University of Michigan
Ann Arbor, MI 48109

[‡]Department of Electrical and Computer Engineering
University of Massachusetts
Amherst, MA 01003

1   Introduction

This chapter is concerned with the objective measures used to characterize the performance of computers which control critical processes. Such processes are beginning to play an increasing role in industry, and require complex algorithms to control them. This is a far cry from the simple transfer-function-based controllers found in the feedback loops of most present-day processes.

The specification and validation of control computers requires the existence of appropriate performance measures. A performance measure is, in effect, the currency of discourse, when performance is being assessed. Many design decisions are made on the basis of how they improve the performance as expressed by these measures. Accordingly, the choice of a performance measure is a crucial link in the design and development chain. If it is not appropriate to the application, then all the expensive optimizations that the designers may carry out with respect to it are worth nothing.

There are two key differences between measuring the performance of traditional computers, used in everyday applications, and the specialized machines used to control life-critical processes. First, the consequences of an error in assessing performance in the latter can lead to catastrophe. Second, the designer has much more information about the operating environment and workload in the latter case than in the former. For example, if the controlled process is an aircraft, a model of how much turbulence it is supposed to withstand, and how much it is likely to be subjected to, are both known (albeit stochastically) in advance. Also, the workloads — the jobs that a control computer runs — are all programmed in advance, with the specific application in mind. The run-time characteristics of this software are usually well-known.

As a result, there is both the need and the capacity to obtain far more detailed characterizations of the performance of a control computer than there is in the general-purpose computer. Any measure that is chosen to represent the performance of such a computer must therefore be able to take advantage of this wealth of information and to put out a more specific and appropriate assessment of performance than do such traditional measures as throughput or reliability.

In the rest of this chapter, we consider some application-specific measures of the performance of computers used in control applications. In Section 2, we look at how the controllers of tomorrow are likely to be far more complex than their counterparts of today. In Section 3, we examine the criteria that a performance measure must be judged by. In Section 4, we argue that because the control computer is crucial to the performance of the controlled process, but is designed and built by computer, and not control, specialists, there must be a clean interface between the controlled process and the computer. This interface is a performance measure that makes sense to both control and computer specialists. In Section 5, we introduce performance measures based on response time and consider how well they meet the criteria discussed in Section 4. This is followed, in Section 6, by a simplified example of the control of the elevator deflection of an aircraft in the final stages of its descent, preparatory to landing. In Section 7, we consider the application of these measures to checkpoint placement. In Section 8, we describe the *number-power* tradeoff. We close with a brief discussion in Section 9.

2   The Changing Role of Controllers

The traditional control system consists of some simple elements — electronic, hydraulic, mechanical, or otherwise — in the feedback loop of the controlled process. These elements can often be characterized by means of simple transfer functions. That is, they are implementations of functions of the form

$$\mathbf{u} = \mathbf{C}\mathbf{x} + \mathbf{d}$$

where $\mathbf{x}$ is the current state of the controlled process, and $\mathbf{C}$ and $\mathbf{d}$ are a given matrix and vector, respectively. $\mathbf{u}$ is then the output vector of signals put out by the controller. $\mathbf{C}$ and $\mathbf{d}$ are typically constants.

Such controllers have the virtue of simplicity. When implemented with a suitable amount of redundancy, they have considerable reliability. It is relatively easy to write specifications for such systems.

Recent developments have, however, required more than simple, non-adaptive systems such as traditional controllers. Two examples will suffice to indicate how requirements have changed: aircraft and automobiles.

The traditional aircraft has a multitude of hydraulic, mechanical and electrical controls, each with a specific, simple, function. There are controls for the various actuators: the rudder, elevator, ailerons, and so on. The system is sufficiently stable that maintaining stability in the face of atmospheric turbulence is not difficult: actuators need to respond with time-constants of no less than about a second. These time-constants are sufficiently long, and the controls themselves sufficiently simple, that even if the automatic stability control mechanisms fail, it is possible for the pilot to take over — as long as the mechanical controls are functional — and keep the aircraft stable.

By contrast, in the highly fuel-efficient aircraft planned for the future, a failure of the automatic stability control mechanisms would result in a crash: the time-constants are likely to be so small and the required control responses so complex that the pilot cannot manually adjust control inputs to maintain stability. This is because altering the airframe for optimal fuel-efficiency turns the aircraft into a reduced-stability vehicle, often requiring high-speed and complex reactions to prevent minor disturbances from snowballing into a catastrophic loss of stability.

The pilot's current capability to act as a low-level controller will therefore vanish with such aircraft. His position will be transformed into one of making policy decisions (e.g., whether to abort a landing or takeoff) which will then be translated into actuator controls by a computer acting as an interface.

This expanded role for control devices raises many troubling questions. There are extra limitations placed on the pilot's ability to fly the aircraft close to the edge of its performance/stability envelope. The pilot's intuition, gained by years of experience, has not much of a role to play: if the computer does not have the preprogrammed ability to save the aircraft in an emergency, it will go down. When this fact is combined with the additional fact that those who write the software and design the computer system are likely to be computer engineers/scientists without much expertise in aerospace engineering, the need for accurate and complete specifications becomes an absolute necessity. Such specifications can only be had if there are appropriate measures in hand to express system performance.

Our second example is the automobile of the future. It is reasonable to expect that in a few years, on-board computers will handle the following functions:

- Electrically actuated steering.

- Automatic active suspensions.

- Power-train controls that deliver the amount of engine-torque required by the driver.

- Traction control, combining power-train control and anti-skid procedures.

- Collision avoidance.

- Advanced transmissions, which allow highly adaptive gear shifting.

- Battery charging.

- Navigation assistance.

Many of these items are critical to safety, and for that reason the computer system must be formally validated, again with the means of appropriate performance measures. They require, in many cases, complex algorithms.

To summarize: as time progresses, ever more complex control techniques will be needed. These will be more critical to safety, and will therefore require formal methods for specification and validation. These in turn depend on how apposite the performance measures are. As we argue in Section 4, such measures should form a clean interface between the controlled process and the controlling computer.

3   The Role of Performance Measures

No single measure yet invented has the power to describe computer characteristics completely. Every measure is, in effect, a partial view of the system since it is either sensitive to only a few dimensions of performance, or so general in scope as to have very poor resolution. For this reason, the choice of a performance measure implies the imposition of a scale of values on the various attributes of a computer. The choice of mean response time, for example, implies that occasional abnormalities in response time are tolerable.

Section 3 is extracted from [7].

The incorrect choice of performance measures can therefore lead to an unsatisfactory characterization of performance. The performance analyst must first consider the appropriateness of his measures before proceeding to use them. Unfortunately, while it is possible to devise quantitative performance measures, it is impossible — at present, anyway — to quantify their appropriateness. Perhaps as a result of their training, most analysts are extremely uncomfortable about admitting to anything about their trade that cannot be quantified. This is surely one reason that there is so little in the literature about the appropriateness of performance measures.

A performance measure must do each of the following if it is to be comprehensive:

- Express the *benefit* gained from a system.

- Express the *cost* expended to receive this benefit.

Preferably, both cost and benefit should be expressed in the same or convertible units, so that the net benefit can be computed.

Benefit relates to the rewards that accrue from the system when it is functional. Benefit may be a vector — which happens when we wish to relate each system state (e.g., number of components surviving) to the reward that accrues when the system is in that state.

The *cost* has three components associated with it. They are as follows:

- Costs that arise when the computer does not function even at its lowest level of acceptability.

- Life-cycle costs — capital, installation, repair, and running costs. These include design and development costs, suitably prorated over the number of units manufactured.

While there is a large literature on life-cycle costing (see, for example, [2, 3, 12]), techniques for the accurate estimation of design and development costs are nowhere in sight. Also, estimates of capital, installation, and other costs can be considerably off target.

Given all of this, it is not surprising that the design, or even the choice, of appropriate performance measures is nontrivial. Indeed, no performance measure that accurately expresses either the benefits or the cost in terms of the attributes of a computer has yet been found.

Lowering our sights a little, we can list requirements that are more easily attained. Performance measures must:

R1. Represent an efficient encoding of the relevant information.

R2. Provide an objective basis for the ranking of candidate controllers for a given application.

R3. Be objective optimization criteria for design.

R4. Represent verifiable facts.

Each of these requirements is discussed below.

R1: One of the problems of dealing with complex systems is the volume of information that is available about them, and their interaction with the environment. Determining relevance of individual pieces of data is impossible unless the data are viewed within a certain context or framework. Such a framework suppresses the irrelevant and highlights the relevant.

To be an efficient encoding of what is relevant about a system, the measures must be congruent to the application. The application is as important to "performance" as the computer itself: while it is the computer that is being assessed, it is the application which dictates the scale of values used to assess it.

If the performance measure is congruent to the application, that is to say, if it is a language natural to the application, then specifications can be written concisely and without contortion. This not only permits one to write specifications economically, but also it is important in the attempt to write — and check for — *correct* specifications. The simpler a set of specifications, the more likely it is in general to be correct and internally consistent.

R2: Performance measures must, by definition, quantify the goodness of computer systems in a given application or class of applications. It follows from this that they should permit the ranking of computers for the same application. It should be emphasized that the ranking must always depend on the application for the reasons given above.

R3: The more complex a system, the more difficult it is to optimize or tune its structure. There are numerous side-effects of even simple actions — of changing the number of buses, for example. So, intuition applied to more and more complex computers becomes less and less dependable as an optimization technique. Multiprocessors are among the most complex computers known today. They provide, due to their complexity, a wealth of configurations of varying quality: this complexity can be used to advantage or ignored with danger.

Multiprocessors that adapt or *reconfigure* themselves — by changing their structure, for example — to their current environment (current job mix, expected time-to-go in a mission-oriented system, etc.) to enhance productivity are likely to become feasible soon. All the impressive sophistication of such a reconfigurable system will come to naught if good, application-sensitive, optimization criteria are unavailable.

R4: A performance measure that is impossible to derive is of no use to anyone. To be acceptable, a performance measure should hold out some prospect of being estimated reasonably accurately. What constitutes "reasonably accurate" depends, naturally enough, on the purpose for which the performance characterization is being carried out. Sometimes, when the requirements are too stringent — extremely low failure probability, for example — to be validated to the required level of confidence, it is difficult to decide which, if any, is to be blamed: the performance measure itself, or the mathematical tools used to determine it.

4   The Need For a Clean Interface

The design of the controlling computer — both hardware and software — must take into detailed account the needs of the process that it is meant to control. Figure 1 indicates the ideal logical design flow. The controlled process dynamics are understood, and appropriate control algorithms proposed. These control algorithms are meant to optimize some performance functional (e.g., fuel or time). Failure and other performance criteria are set. From these, specifications are drawn up for the control computer.

From these specifications, the control computer is designed. This includes the hardware design — how many processors, what kind of interconnection network is suitable, how many power supplies, how the processors are to be synchronized, and so on — and the software design — how to reconfigure the computer as its components fail or

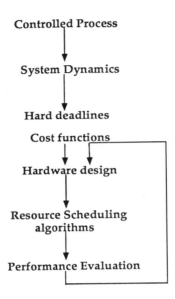

Figure 1: Sequential steps for design of a real-time control computer.

the workload changes, how interrupts are to be handled, how the control algorithms are
to be implemented, what the job priority structure is, what tasks to shed as the computer degrades from perfect functionality to partial collapse, what kind and how much
software redundancy is to be used, and so on. All these decisions will intimately affect
the computer's goodness as a controller, and must be taken in light of the needs of
the particular control application in hand. The scientists and engineers who design the
control computer are not usually specialists in control theory. In fact, they may know
little or nothing about the controlled process. It is therefore vital that the mechanisms
used to translate the needs of the control application be all-encompassing, and that they
be formally expressed in a language that *both* control theorists *and* computer scientists
can understand. That is to say, the interface between the control engineer's domain and
that of the computer engineer must be "clean."

A currency common to the worlds of both the control and the computer engineer is
the *response time* of the control algorithms. The control engineer sees it as the time lag
in the feedback loop of the controlled process. Such a time lag translates into degraded
performance: in fact, given the dynamical equations of the controlled process, this
degradation in performance can be precisely measured in physical terms. For example,
one might be able to say that a time delay of ξ seconds in the feedback loop results
in the expenditure of an extra $f(\xi)$ pounds of fuel. The control engineer is concerned

Task Type	Time-Critical?	Failure-Critical?	Deadline?
Bare-Bones Critical Tasks	Yes	Yes	Yes
Alternate Critical Tasks	No	Yes	Yes
Soft Real-Time Tasks	No	No	Yes
Non-Real-Time Tasks	No	No	No

Table 1. Task classes.

with the *effects* of this delay in the feedback loop on the controlled process: its *cause* is within the province of the computer engineer.

To the computer engineer, the response time is the time consumed between the start (release) of a job and its completion. The response time is determined by (a) the job loading, and (b) the hardware and software decisions made in the design of the computer. The design of the computer involves decisions referred to earlier. Each of these decisions affects the response time, and must therefore be made in the light of how the response time of individual computer jobs affects the performance of the controlled process.

Performance measures which are functions of the job response time thus have the potential to form clean interfaces between the controlled process and the control computer.

5 Performance Measures

We define performance measures, originally introduced in [6], which are application-sensitive and based on task response time.

There are two components to this performance measure: the probability of catastrophic failure over a given period of time, and the additional cost of running the controlled process that is accrued because the computer has a non-zero response time. We consider each of these in turn. Before doing so, however, we need to understand the various kinds of tasks that a control computer will be called upon to run.

In Table 1, we classify the tasks in a real-time control system. A task is said to be *time-critical* if it must be performed correctly at least once every T seconds (for some T) if the system is to survive. It is *failure-critical* if delivery to the actuators or other output devices of an incorrect task output can cause catastrophic failure. It has a *deadline* if it used information that gets dated and is therefore useless if not performed by a certain deadline.

Critical tasks are those which must be performed if the system is to survive. One example would be stability control in an aircraft or rocket. It is *not* true — as is commonly assumed — that each critical task always has an individual and finite *hard deadline*, which, if not met, can cause the controlled process to fail catastrophically. What happens when a task fails to deliver its output is that the corresponding actuator levels are held at their prior values. If, however, a critical task misses a succession of deadlines, the process will fail. We express this by saying that there is a certain amount of critical workload that must be executed successfully (within the deadline) every T seconds, for the system to avoid failure. We *can*, however, still deal with hard deadlines for individual tasks if we are willing to make them contingent on the current state of the controlled process. That is, the controlled process is meant to be within a certain "allowed area" of its state-space if it is to survive. The closer it gets to the edge of the envelope — driven there by a succession of critical tasks failing to meet their deadlines, by an especially hostile operating environment, or any combination thereof — the less the hard deadlines tend to be. Deep within the allowed area, the hard deadlines of the critical tasks will typically be infinite. This is a way of saying that missing one iteration of a critical task when in that region of the state-space does not lead to catastrophe. More formally, we can define the hard deadline associated with a given task as [10]:

$$\tau_{d\alpha|,\sigma}(\mathbf{x}(t_0)) = \inf_{\mathbf{u}\in} \sup \{\tau \ : \ \phi(t_0 + \tau, t_0, \mathbf{x}(t_0), \mathbf{u}) \in \sigma\}$$

where:

- $\tau_{d\alpha}(\mathbf{x}(t_0))$ is the hard deadline associated with task α when the system is in state $\mathbf{x}(t_0)$ (at time t_0), the allowed state-space is a region denoted by σ, and   is the range of control outputs that the actuators can put out, and

- $\phi(t_0 + \tau, t_0, \mathbf{x}(t_0), \mathbf{u})$ denotes the state of the controlled process at time $t_0 + \tau$ given that it was in state $\mathbf{x}(t_0)$ at time t_0, and that controls denoted by $\mathbf{u}$ were applied.

Notice that if the environment is stochastic in character, the hard deadline is a random variable. This is a significant departure from the conventional approach, where a hard deadline is assumed to be a deterministic quantity.

We thus have two ways of defining behavior that can lead to failure of the controlled process. The first is to define a critical volume of workload that must be executed every

T seconds, the second is to define hard deadlines for each task as a function of the state of the controlled process. They are not exactly equivalent because the first approach ignores the impact of the operating environment. However, it can be used without risk provided that suitably pessimistic estimates are made about the operating environment.

We classify critical tasks into two categories: *bare-bones* and *alternate*. By *bare-bones* tasks we mean tasks which just meet the minimum requirements for system safety. They typically require much less execution time to run than their *alternates*, which are their more sophisticated counterparts, which promise better performance (i.e., can calculate control input values that are much closer to the optimal), but not greater immunity from catastrophic failure.

Soft real-time tasks are those which operate upon perishable data, and therefore have deadlines. They do not, however, have an impact on system reliability. Finally, there are *non-real-time* tasks which are do not have deadlines associated with them, and do not affect system reliability.

We now turn to a definition of response-time dependent performance measures.

These measures have two components to them. The first component is *reliability* and the second is *performance*. Reliability is measured by a parameter called the *probability of dynamic failure*, p_{dyn}. The probability of dynamic failure of a control computer over a given period of operation is the probability that it will behave in such a way over that period as to cause catastrophic failure of the controlled process.

A moment's reflection will show that there is an important difference between the traditional reliability of computer systems and the reliability as it is defined above. Traditional reliability focuses on the computer system alone. It identifies certain states of the computer system as *failed states*, and is the probability that the system enters one of these states during the specified period of operation. There is no reference to the current state of the application in the calculation of traditional reliability. By contrast, the probability of dynamic failure explicitly depends on the current state of the controlled process. The same hardware state (e.g., number of processors and buses that are still functional) may be sufficient to keep the controlled process from failing if the latter is in some subset of its state-space, but be insufficient in some other subset. This provides the linkage that is necessary between the controlled process and the control computer.

We now turn to considering the second component of the measure: *performance*. Suppose that some performance functional is available for characterization of the controlled process. Examples are the amount of time it takes for a vehicle to move from one point to another, the amount of fuel consumed, etc. The key point is that it is possible to quantify the performance of the controlled process in units which are physically meaningful, from the point of view of the user. Let $\Omega(\mathbf{x}, t)$ denote the performance of the system when the system is in state $\mathbf{x}$ and the response times of the tasks are given by $\mathbf{t} = (t_1, \ldots, t_n)^T$ for n tasks. Define the cost function

$$g(\mathbf{x}, t) = \Omega(\mathbf{x}, t) - \Omega(\mathbf{x}, \mathbf{0}).$$

That is, the performance measure $g(\mathbf{x}, t)$ represents the extra cost it takes if the controller tasks have response times denoted by the vector $\mathbf{t}$. In many cases, if the tasks are more or less orthogonal, it will be possible to express this performance function as:

$$g(\mathbf{x}, t) = \sum_{i=1}^{n} g_i(\mathbf{x}, t_i)$$

where $g_i(\mathbf{x}, t_i)$ is the task i's contribution to the cost. If all the tasks are not orthogonal to one another, but it is possible to break down the task set into subsets, each set being orthogonal to the others, it would be possible to similarly express the cost function.

The functions $g(\cdot, \cdot)$ are only defined when the controlled process is not driven to failure by the control computer. They are thus always finite.

Given these cost functions and a stochastic model of the operating environment, it is possible to define the *mean cost* over a given period of system operation. This is the expected cost accumulated over the given period. The designer's objective then becomes one of structuring the control computer in such a way as to minimize the mean cost while keeping the probability of dynamic failure above a prespecified threshold.

Note that obtaining the cost functions and hard deadlines is the job of the control engineer. The computer designer takes these quantities as inputs. Each design of the computer system (including the architecture, task schedules, etc.) determines probability density functions for the task execution time. This can be used to determine the mean cost and probability of dynamic failure of the computer when it is controlling the application whose cost functions and hard deadlines are being used.

Feedback Term	Value
b_{11}	-0.600
b_{12}	-0.760
b_{13}	0.003
b_{32}	102.400
b_{33}	-0.400
c_{11}	-2.374

Table 2. Feedback term values.

6   Obtaining the Hard Deadlines and Cost Functions: An Example

The controlled process is an aircraft, in the phase of landing. We focus on the task of computing the elevator deflection. Our purpose is to show how the hard deadlines and finite cost functions can be obtained from a study of the dynamics of the controlled process.

The state-space model and the optimal control solution used are due to Ellert and Merriam [4]. The aircraft dynamics are characterized by the equations:

$$\dot{x}_1(t) = b_{11}x_1(t) + b_{12}x_2(t) + b_{13}x_3(t) + c_{11}m_1(t,\xi)$$
$$\dot{x}_2(t) = x_1(t)$$
$$\dot{x}_3(t) = b_{32}x_2(t) + b_{33}x_3(t)$$
$$\dot{x}_4(t) = x_3(t)$$

where x_1 is the pitch angle rate, x_2 the pitch angle, x_3 the altitude rate, and x_4 the altitude. The constants b_{ij} and c_{11} are given in Table 2. m_1 is the control of the elevator deflection and ξ denotes controller response time. The phase of landing takes about 20 s. Initially, the aircraft is at an altitude of 100 feet, traveling at a horizontal speed of 256 ft/s. This latter velocity is assumed to be held constant over the entire landing interval. The rate of descent at the beginning of this phase is 20 ft/s. The pitch angle is ideally to he held constant at 2^0. Also, the motion of the elevator is restricted by mechanical stops. It is constrained to be between -35^0 and 15^0. For linear operation, the elevator may not operate against the elevator stops for nonzero periods of time during this phase. Saturation effects are not considered. Also not considered are wind gusts and other random environmental effects. The task to calculate the elevator deflection is released every 60 ms.

Section 6 is extracted from [10].

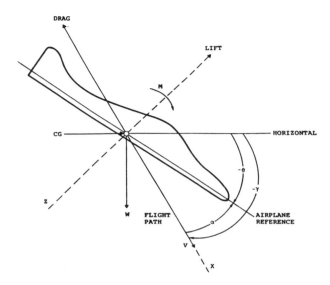

Figure 2: Definition of aircraft angles.

The constraints are as follows. The pitch angle must be held between 0^0 and 10^0 to avoid landing on the nose wheel or on the tail, and the angle of attack (see Figure 2 must be held to less than 18^0 to avoid stalling. The vertical speed with which the aircraft touches down must be less than around 2 ft/s so that the undercarriage can withstand the force of landing.

The desired altitude trajectory (in feet) is given by

$$h_d(t) = \begin{cases} 100e^{-t/5} & 0 \le t \le 15 \\ 20 - t & 15 \le t \le 20 \end{cases}$$

while the desired rate of ascent (in ft/s) is

$$h_d(t) = \begin{cases} -20e^{-t/5} & 0 \le t \le 15 \\ -1 & 15 \le t \le 20. \end{cases}$$

The desired pitch angle is 2^0 and the desired pitch angle rate is 0^0 per second.

The performance index (for the aircraft) chosen by Ellert and Merriam, and suitably adapted here to take account of the nonzero controller response time, ξ, is given by

$$\Theta(\xi) = \int_{t_0}^{t_f} e_m(t, \xi) dt \tag{1}$$

Weighting Factor	Value
$\phi_1(t)$	99.00000
ϕ_{2,t_f}	20.00000
$\phi_3(t)(0 \leq t < 15)$	0.00000
$\phi_3(t)(15 \leq t \leq 20)$	0.00010
ϕ_{3,t_f}	1.00000
$\phi_4(t)$	0.00005
ϕ_{4,t_f}	0.00100

Table 3. Weighting factor values.

where t represents time, and $[t_0, t_f]$ is the interval under consideration, and

$$e_m(t, \xi) = \phi_h(t)[h_d(t) - x_4(t)]^2 + \phi_{\dot{h}}(t)[\dot{h}_d(t) - x_3(t)]^2 +$$
$$\phi_\theta(t)[x_{2d}(t) - x_2(t)]^2 + \phi_{\dot{\theta}}(t)[x_{1d}(t) - x_1(t)]^2 + m_1^2(t, \xi)$$

where the d-subscripts denote the desired (i.e., ideal) trajectory. To ensure that the touchdown conditions are met, the weights ϕ must be impulse-weighted. Thus, we define:

$$\phi_h(t) = \phi_4(t) + \phi_{4,t_f}\delta(20 - t)$$
$$\phi_{\dot{h}}(t) = \phi_3(t) + \phi_{3,t_f}\delta(20 - t)$$
$$\phi_\theta(t) = \phi_{2,t_f}(t)\delta(20 - t)$$
$$\phi_{\dot{\theta}}(t) = \phi_1(t)$$

where the functions ϕ must be given suitable values, and δ denotes the Dirac-delta function. The values of the ϕ are given based on a study of the trajectory that results. The chosen values are listed in Table 3. The control law for the elevator deflection is given by

$$m_1(t, \xi) = \omega_s^2 K_s T_s[k_1(t - \xi) - k_{11}(t - \xi)x_1(t - \xi) - k_{12}(t - \xi)x_2(t - \xi)$$
$$-k_{13}(t - \xi)x_3(t - \xi) - k_{14}(t - \xi)x_4(t - \xi)]$$

where the aircraft parameters are given by $K_s = -0.95 \ s^{-1}$, $T_s = 2.5$ s, $\omega_s = 1$ rad·s^{-1}, and the constants k are the feedback parameters derived (as shown in [4]) by solving the Riccati equation. For these differential equations, we refer the reader to [4].

In a system such as this, it is virtually impossible to find closed-form solutions for the hard deadlines. A numerical solution must therefore be derived. To do this, we divide the allowed state-space down into disjoint *state-subsets* which are aggregates of states in which the system exhibits roughly the same behavior. Even if there do not exist clear boundaries for these state-subsets, one can always force the allowed state-space to be divided into state-subsets so that a sufficient safety margin can be provided. In every such state-subset, each control job has a unique hard deadline (which may be infinity if that job is non-critical in that subset) and finite cost function. Because we obtain them separately for each state-subset, these quantities are functions only of the system response time and not the current controlled-process state.

Our first step is to derive the allowed state-space for the aircraft system. Note that in the idealized model which we are considering, it is implicitly assumed that the constraint on the angle of attack is always honored, so that the only constraints to be considered are the terminal constraints.

The terminal constraints have been given earlier, but are repeated here for convenience. The touchdown speed must be less than 2 ft/s in the vertical direction, and the pitch angle at touchdown must lie between 0^0 and 10^0. To avoid overshooting the runway, touchdown must occur at between 4,864 and 5,120 ft in the horizontal direction from the moment the landing phase begins. The horizontal velocity is assumed to be kept constant throughout the landing phase at 256 ft/s (this would constitute a separate controller job; we do not consider here how that is to be done). Thus, touchdown should occur between 19 and 20 s after the final descent phase begins. The only control we consider is the elevator deflection, which must be held between -35^0 and 15^0.

For representational convenience, we define "component state-spaces" as follows. Any point in the four-dimensional state-space whose individual components each lie in the corresponding component state-spaces is in the overall allowed state space. The component state-spaces are determined by perturbing the state-values around the desired trajectory and determining the maximum perturbation possible under the requirement that no terminal constraint be violated. They were obtained by a search process, and are plotted in Figure 3. It should be stressed that what we plot in Figure 3 is a *subset* of the overall allowed state-space.

The method we employed to obtain the state subsets is the same way that analogue signals are quantized into digital ones. Intervals of hard deadlines and cost function are defined. Then, points are allocated to state-subsets corresponding to these intervals. To take a concrete example, consider a state-space $\mathbf{X} \subset \mathbf{R}^n$ that is to be subdivided on the basis of the hard deadlines. The first step is to define a quantization of the hard deadlines. Let this be Δ. Then, define state-subset S_i as containing all states in which the hard deadline lies in the interval $[(i-1)\Delta, i\Delta)$. Alternatively, one might define

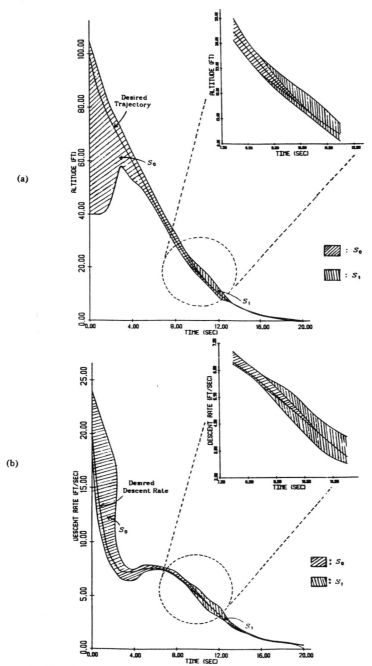

Figure 3: (a) Allowed state-space: altitude. (b) Allowed state-space: descent rate.

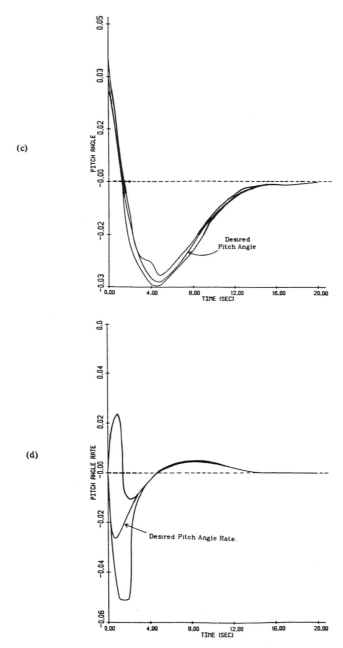

Figure 3: (Continued) (c) Allowed state-space: pitch angle. (d) Allowed state-space: pitch angle rate.

a sequence of numbers $\Delta_1, \Delta_2, \ldots$, and define S_i as containing all states in which the hard deadline lies in $[\Delta_i, \Delta_{i+1})$.

We subdivide the allowed state space into two state-subsets: S_0 and S_1. These correspond to the deadline intervals $[120, \infty)$ and $[60, 120]$, respectively. S_0 is the noncritical region corresponding to the $[120, \infty)$ interval, and S_1 is the critical region corresponding to $[60, 120]$. The hard deadline may conservatively be assumed to be 60 ms in S_1.

The control for various values of fixed controller response time is computed using the control law equation, and is shown in Figure 4. Due to the absence of any random effects, elevator deflections for all the response times considered tend to the same value as the end of the landing phase (20 s) is approached, although much larger controls are needed initially. In the presence of random effects, the divergence between controls needed in the low and high values of controller response time is even more marked. We present an example of this in Figure 5. The random effect considered here is the elevator being stuck at -35^0 for 60 ms 8 s into the landing phase due to a faulty controller order. The controlled process is assumed in Figure 5 to be in the state-subset for which the landing job maps into a noncritical job (i.e., one in which the hard deadline is infinite). The diagrams speak for themselves.

We found that the finite cost function does not vary significantly within the entire allowed state-space. We therefore express this as a function only of the system response time, and not as a function of current airplane state. In Figure 6, we present this function. The cost is in terms of the performance functional of the controlled process. The reader should compare the nature of the cost function to the plots showing elevator deflection and notice the correlation between the marginal increase in cost with increased execution delay and the marginal increase in control needed, also as a function of the execution delay.

7 Application of the Measures: Checkpoint Placement

When a computer fails while carrying out a long computation, it must restart from the beginning unless means have been provided to store intermediate results. Such storage can greatly reduce the mean execution time in database and other general-purpose computer systems. This process is called *checkpointing*. The question is where, in the

---

Section 7 is adapted from [8].

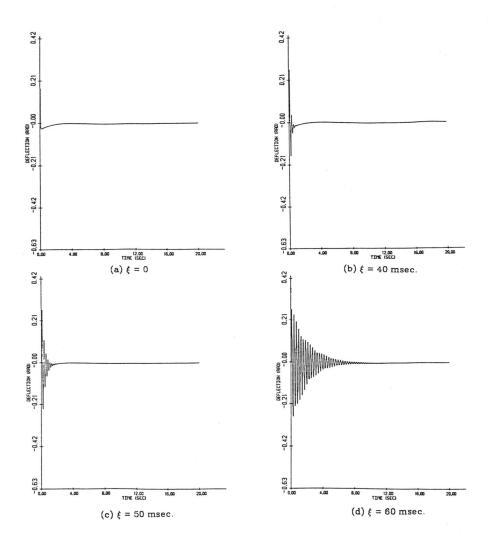

(a) $\xi = 0$

(b) $\xi = 40$ msec.

(c) $\xi = 50$ msec.

(d) $\xi = 60$ msec.

Figure 4: Elevator deflection.

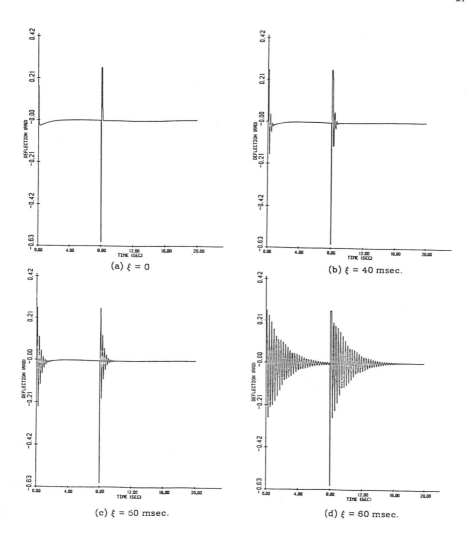

Figure 5: Elevator deflection with abnormality.

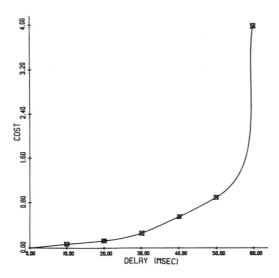

Figure 6: Finite cost function.

time-line of a computation, the checkpoints should be placed.

Optimization of checkpoint placement has long interested researchers [1, 5, 11, 13]. The declared purpose there has been to place the checkpoints in such a way as to minimize the mean execution time.

In real-time systems, on the other hand, the focus is on minimizing the mean cost, while keeping the probability of dynamic failure below a prespecified level. The principal function of the checkpoints is no longer that of affecting the mean costs: instead, it is to reduce the probability of dynamic failure. p_{dyn} is affected by the tail of the task response time distribution. Anything which cuts down on the magnitude of the tail can be expected to improve p_{dyn}. Checkpointing does precisely this.

To provide an example of how to determine the number and placement of the checkpoints, consider the aircraft-landing application that we looked at in the previous section.

Let the Mean Time Between Failure (MTBF) be 10,000 hours, and let errors occur according to a Poisson process with rate $\lambda = 1/\text{MTBF}$. When an error is discovered, the system rolls back to the latest checkpoint. Let t_{roll} denote the time taken (this is a random variable) to roll back. There will be a certain amount of time taken up with restarting the process: let t_s denote this time. Finally, let t_{ov} be the overhead associated with the storage of data which is done when a process encounters a checkpoint.

It is possible that the saved data itself may be incorrect. This can, in many cases, be detected by *acceptance tests*. Let us assume, for simplicity that no faulty data escape these tests. Denote by p_s the probability that the saved data is incorrect. If a checkpoint is incorrect, we have to roll back to the preceding checkpoint, if any, or go back to the very beginning of the computation and start all over again. Let t_{start} denote the extra time taken if this has to be done. Let n denote the number of checkpoints used. Then, the total execution time is given by

$$\xi_t = \xi + nt_{ov} + t_{rec}$$

where

$$t_{rec} = \begin{cases} 0 & \text{if no error occurs} \\ t_s + t_{roll} & \text{if an error occurs and the task} \\ & \text{is recovered by rollback} \\ t_s + t_{start} & \text{if an error occurs and the} \\ & \text{task has to restart.} \end{cases}$$

The ratio of the execution time of any task to the MTBF is typically at least 10^{-7}, and so the probability of a second failure occurring to the same task is negligible. Let the checkpoints be placed at equidistant intervals and let $t_{inv} = \xi/(n+1)$ be the interval between successive checkpoints. The density function of t_{roll} and t_{start} are given by $f_{roll}(t) = \lambda e^{-\lambda t}/(1 - e^{-\lambda t_{inv}})$ for $t \in [0, t_{inv}]$, and $f_{start}(t) = \lambda e^{-\lambda t}/(1 - e^{-\lambda \xi})$ for $t \in [0, \xi]$, respectively. The density function of the total response time can easily be found from these functions.

Three cases are considered for the nominal execution times of the deflection task: 20 ms, 30 ms, 40 ms and 50 ms. For each, the probability of dynamic failure is considered with the hard deadline set at 60 ms. The mean cost that ensues with checkpoints is also computed. To express the marginal benefit accrued (in terms of the reduction of p_{dyn}) against the price paid (in terms of the increased mean cost) we use the following tradeoff ratio:

$$T(n) = \frac{p_{dyn}(n-1) - p_{dyn}(n)}{MC(n) - MC(n-1)}$$

where $p_{dyn}(k)$ and $MC(k)$ are the probability of dynamic failure and mean cost, respectively, if k checkpoints are used. The results are presented in Table 4. When the nominal execution time is 20 ms, all that the checkpoints do is to increase the over-

n	Mean Cost	p_{dyn}	$(Trade\text{-}off\ ratio) \times 10^7$
0	0.12848	$0.3086E - 15$	—
1	0.12909	$0.3086E - 15$	0.0
2	0.12971	$0.3086E - 15$	0.0
3	0.13033	$0.3086E - 15$	0.0
4	0.13095	$0.3086E - 15$	0.0
5	0.13157	$0.3086E - 15$	0.0

(a) Nominal execution time: 20 ms.

n	Mean Cost	p_{dyn}	$(Trade\text{-}off\ ratio) \times 10^7$
0	0.26156	$0.37037E - 07$	—
1	0.26431	$0.37037E - 08$	121.5
2	0.26709	$0.37037E - 08$	0.0
3	0.26991	$0.37037E - 08$	0.0
4	0.27272	$0.37037E - 08$	0.0
5	0.27567	$0.37037E - 08$	0.0

(b) Nominal execution time: 30 ms.

n	Mean Cost	p_{dyn}	$(Trade\text{-}off\ ratio) \times 10^7$
0	0.55352	$0.30555E - 06$	—
1	0.55472	$0.43055E - 07$	2177.0
2	0.55586	$0.30555E - 07$	109.8
3	0.55694	$0.30555E - 07$	0.0
4	0.55795	$0.30555E - 07$	0.0
5	0.55891	$0.30555E - 07$	0.0

(c)Nominal execution time: 40 ms.

n	Mean Cost	p_{dyn}	$(Trade\text{-}off\ ratio) \times 10^7$
0	0.89694	$0.46666E - 06$	—
1	0.90848	$0.35666E - 06$	95.6
2	0.92025	$0.26166E - 06$	80.6
3	0.93231	$0.16666E - 06$	78.7
4	0.94466	$0.71666E - 06$	76.9
5	0.95730	$0.46666E - 06$	19.8

(d) Nominal execution time: 50 ms.

Table 4. Tradeoff ratios.

head, i.e., the mean cost. No discernible drop is noticed in p_{dyn}. The marginal gain in reliability on adding checkpoints is therefore zero.

However, as the nominal execution time increases, checkpointing begins to cause a noticeable decrease in p_{dyn}. This is expressed through a positive tradeoff ratio: $n = 1$ for a nominal execution time of 30 ms; $n = 1, 2$ for 40 ms; and $n = 1, 2, 3, 4, 5$ for 50 ms. These ratios show that a tangible gain in reliability has been made for the indicated number of checkpoints (i.e., the p_{dyn} is reduced by a factor of about 10 in each case). Of course, this has been achieved at the price of a certain increase in the mean cost, which is also reflected in the tradeoff ratio.

Computations such as this can be used by designers to specify the optimal number of checkpoints to use for particular control applications.

8 The Number-Power Tradeoff

Device redundancy has long been used as a means for obtaining systems which are more reliable than their individual components. As devices fail, they can be replaced by their surviving counterparts.

In real-time systems, as we have seen, a massive hardware collapse is not the sole cause of system failure. The computer can also fail if it does not respond quickly enough to the environment. This is nicely illustrated by means of the number-power tradeoff.

The number-power tradeoff is idealized as follows. Suppose we wish to design a system with a given number-power product. A number-power product is the product of the number of the processors and the power of an individual processor. Let this quantity be Π. Then, if we choose to have N processors, each processor will be of power $\pi(N) = \Pi/N$. Power can be expressed in units of throughput or any other suitable measure. We are assuming, in this idealized example, that processors of power $\pi(N)$ are available. The problem is to obtain N so that the system reliability is maximized.

While idealized, this problem neatly illustrates the two components of failure mentioned above: failure due to massive hardware collapse, and due to tasks not being executed before their respective deadlines. If N is very small, there will be few processors, each of considerable power. As a result, the probability of all the processors

---

The figures in Section 8 are quoted from [9].

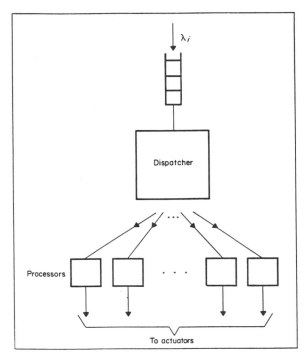

Figure 7: A real-time multiprocessor system.

being wiped out will be considerable. If N is very large, this probability will go to zero (making the usual assumptions of independent failures), but then the individual processors will be so slow as to make it improbable that the deadlines can be met. A happy medium is called for.

To illustrate this, consider the multiprocessor shown in Figure 7. Jobs come into a dispatcher, and wait if necessary, until a processor becomes free, upon which they are dispatched to that processor.

Let us assume that massive hardware collapse happens when there are fewer than n functioning processors. Failure can also happen when there are at least n processors functional, but one or more hard deadlines are missed. Figure 8 models the process as a Markov chain. The failure process is assumed to be Poisson, and the task execution times exponentially distributed. The failure rate of a processor is μ_p, and the distribution time of the execution time is given by $F_{MMp}(t)$ where p is the number of functional processors. $f_d(t)$ is the distribution function of the hard deadlines. (Recall that these

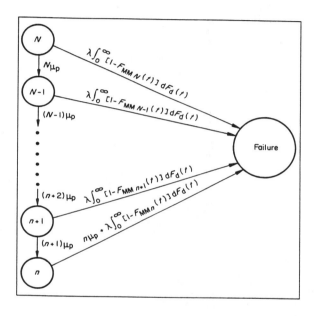

Figure 8: Markov model for number-power tradeoff $(n = [\lambda/\mu] + 1)$.

can be random variables, as well). Here, the state of the functional system is the number of functional processors. Because there is no repair, the system starts in state N and degrades all the way to n. Any failures beyond n processors leads to system failure (i.e., the system enters the trapping state of Failure). However, even if the state is $\ell \geq n$, there is a probability that the system will miss one or more hard deadlines. There is thus a direct transition from each such state ℓ to the Failure state. The rate of this transition is given in the Markov model.

Figure 9 from [9] illustrates the point that we seek to make in this section. Here, we plot, for a given hard fixed deadline, the probability of dynamic failure as a function of the number of processors, N, and the mission lifetime, *ML*. The absolute values of p_{dyn} are unimportant, what is important is the shape. As N is increased from 1, the probability of failure starts to drop. This is because we are providing more redundancy, and this more than makes up for the increased propensity to miss deadlines. Beyond a certain point (the nadir of each curve), the gain in reliability by introducing additional redundancy is more than made up for by the loss in reliability by slowing down the individual processors (recall that the number-power product is fixed), and thus increasing the probability of failure due to missing a hard deadline.

9   Discussion

In this chapter, we have considered performance measures for real-time systems which are based on system response time. We have argued that response time is meaningful both to the control engineer as well as to the computer designer. To the former, it is the cause of degradation of control quality when it appears in the feedback loop of the controlled process. To the latter, it appears as the task response time. The control engineer can quantify the incremental cost (in terms of suboptimal control) that is incurred by having a certain task response time. The computer engineer can control this task response time by suitably choosing the components that go to make up the control computer, and adjusting the operating system (by assigning suitable task priorities, for example). As a result, although the computer and control engineers may have little knowledge of each other's worlds, the response time is a common currency that has meaning to both. Cost functions and hard deadlines thus provide a clean interface between the world of the control and the computer engineer. Such a clean interface is vital if specifications are to be understood by both parties, and objective evaluations conducted of rival computer systems for given applications.

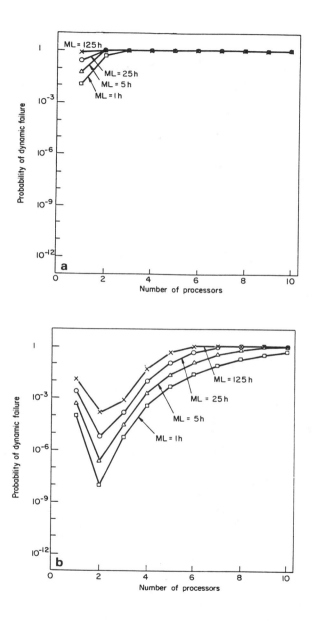

Figure 9: Probability of dynamic failure for a constant number- power product of 1000: (a) $t_d = 0.01$; (b) $t_d = 0.05$.

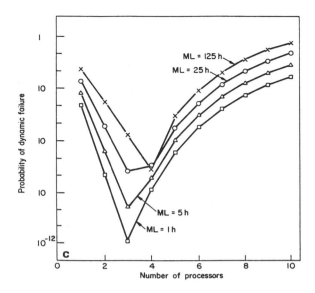

Figure 9: (Continued) Probability of dynamic failure for a constant number-power product of 1000: (c) $t_d = 0.10$.

References

[1] F. Baccelli, "Analysis of a service facility with periodic checkpointing," *Acta Informatica*, vol. 15, no. 1, pp. 67–81, 1979.

[2] R. K. Barasia and T. D. Kiang, "Development of a life-cycle management cost model," *Proc. IEEE Reliability & Maintainability Symp.*, pp. 254–259, 1978.

[3] "Canadian Forces Procedure CFP113 Life Management System – Guidance Manual," Dept. National Defence, Canada.

[4] F. J. Ellert and C. W. Merriam, "Synthesis of feedback controls using optimization theory – An example," *IEEE Trans. Automatic Control*, vol. AC–8, no. 4, pp. 89–103, Apr. 1963.

[5] E. Glenbe and D. Derochette, "Performance of rollback recovery systems under intermittent failures," *Commun. ACM*, vol. 21, no. 6, pp. 493–499, Jun. 1978.

[6] C. M. Krishna and K. G. Shin, "Performance measures for multiprocessor controllers," *Performance '83*, pp. 229–250, 1983.

[7] C. M. Krishna and K. G. Shin, "Performance measures for control computers," *IEEE Trans. Automatic Control*, vol. AC–32, no. 6, pp. 357–366, Jun. 1987.

[8] C. M. Krishna, K. G. Shin and Y.–H. Lee, "Optimization criteria for checkpoint placement," *Commun. ACM*, vol. 27, no. 10, pp. 1008–1012, Oct. 1984.

[9] K. G. Shin and C. M. Krishna, "New performance measures for design and evaluation of real-time multiprocessors," *Int'l J. Computer Science & Engineering*, vol. 1, no. 4, pp. 179–191, Oct. 1986.

[10] K. G. Shin, C. M. Krishna, and Y.–H. Lee, "A unified method for evaluating real-time computer controllers and its application," *IEEE Trans. Automatic Control*, vol. AC-30, no. 4, pp. 357–366, 1985.

[11] A. N. Tantawi and M. Ruschitzka, "Performance analysis of checkpointing strategies," *ACM Trans. Computer Systems*, vol. 2, pp. 123–144, May 1984.

[12] "Life-cycle cost analyzer (LCCA) general information guide," Tech. Analytical Sciences Corp., Engg. Memo E.M. 2184-1, 1981.

[13] J. W. Young, "A first order approximation to the optimum checkpoint interval," *Commun. ACM*, vol. 17, no. 9, pp. 530–531, Sep. 1974.

MODELING AND CONTROL OF AUTOMOTIVE POWER TRAINS

D. HROVAT AND W. F. POWERS

Ford Motor Company
Dearborn, Michigan 48121

I. INTRODUCTION

In the middle 1970s, American automotive manufacturers introduced microprocessor-based engine control systems to meet the conflicting demands of high fuel economy and low emissions. Present-day engine
control systems contain many inputs (e.g., pressures, temperatures,
rotational speeds, exhaust gas characteristics) and outputs (e.g., spark
timing, exhaust gas recirculation, fuel-injector pulse widths). The unique
aspect of the automotive control problem is the requirement to develop
systems that are relatively low in cost, will be applied to several hundred
thousand systems in the field, must work on relatively low-cost automobiles with the inherent manufacturing variability, will not have scheduled maintenance, and will be used by a spectrum of human operators.
(Note that the aircraft/spacecraft control problem, for which most sophisticated control techniques have been developed, has nearly an opposite
set of conditions.)

The software structure of the controllers that have been developed
to date is much like that in other areas (i.e., aircraft controllers, process
controllers) in that there exists an "outer-loop" operational mode structure with "inner-loop" feedback-feedforward-adaptive (learning) modules. The development of a total system involves four major steps: 1)
development of unique, problem-oriented large scale integrated (LSI)
devices; 2) linear digital control system theory preliminary design; 3)

nonlinear simulation/controller design; and 4) hardware-in-the-loop/ real-time simulation capability for identification, calibration, and verification.

To illustrate how control theoretic techniques are employed in the design of power train control systems, the problems of idle speed control and transmission control will be reviewed. The paper concludes with a discussion of current trends in on-board computer control applications.

II. POWER TRAIN CONTROL APPLICATIONS

The design of a power train control system involves tradeoffs among a number of attributes. When viewed in a control theory context, the various attributes are categorized quantitatively as follows:

- Emissions: A set of terminal (final time) inequality constraints.
- Fuel Consumption: A scalar quantity to be minimized over a time interval; usually is the objective function to be minimized.
- Driveability: One or more state variable inequality constraints which must be satisfied at every instant on the time interval.
- Performance: Either part of the objective function or an intermediate point constraint, e.g. achieve a specified 0-60 mph acceleration time.
- Reliability: As part of the emission control system, the components in the computer control system (sensors, actuators, computers) have a 50,000 mile or 5 year warranty. In the design process, reliability usually enters as a sensitivity or robustness condition, e.g., location of roots in the Z-plane.
- Cost: The effects of cost are problem dependent. Typical ways that costs enter the problem quantitatively are increased weights on control variables in quadratic performance indices (which implies relatively lower cost actuators) and output instead of state feedback (which implies fewer sensors but more software).
- Packaging: Networking of computers and/or smart sensors and actuators requires distributed control theory and tradeoffs among data rates, task partitioning and redundancy, among others.
- Electromagnetic Interference: This is mainly a hardware problem which is not treated explicitly in the analytic control design process.
- Tamper-proof: This is one of the reasons for computer control, and

leads to adaptive/self-calibrating systems so that dealer adjustments are not required as the power train ages or changes.

The flexibility of microprocessor based power train control systems allows the designer to deal effectively with the relatively large number of interacting attributes listed above. However, this same flexibility requires a systems oriented discipline to insure that the major attributes are considered continually as the total system design evolves. Control theory can play a major role in such a discipline, and examples of its use in idle speed control and electronic transmission control will be presented in this section. Further details on these applications may be found in references [1-8]. It should be pointed out that the present paper is based on [9], which focuses mainly on work at Ford Motor Company. Representative examples of work outside of Ford can be found in [10-17].

A. Idle Speed Control

In the design of an idle speed control system there exist numerous hardware and strategy possibilities, e.g. electric motor, throttle bypass solenoid, or pneumatic throttle actuator; inclusion/exclusion of throttle position sensor feedback, spark actuation, accessory load sensor (e.g. switch) information, and/or feedforward control, among others. If each of the possible combinations listed above is treated as a separate design study, then considerable human resources and design time are required to determine the "optimal" system for a given application. A more desirable situation is the development a single framework which allows analysis of the various hardware/strategy configurations. Formulation of the problem as an optimal control problem supplies such a framework.

Suppose that $\{x_1,...,x_n\}$ is the set of all state variables and $\{u_1,...,u_m\}$ is the set of all control variables associated with the problem. For a power train control problem such as idle speed control, the full set of state variables is defined by the "essential dynamics" (for control) represented by the blocks in Fig. 1, and the full set of control variables is {throttle rate, spark rate, fuel flow rate, and Exhaust Gas Recirculation (EGR) rate}. (Because of important actuator dynamics, rates of the commonly referred control variables are treated as the "mathematical control variables" in a control theory setting.) During idle, EGR is not in operation so EGR is eliminated from the problem. The resultant simulation model consists of approximately twenty state variables and three control variables. The equations represent a mixture of physical principles and transfer functions which must be calibrated with dynamometer and/or vehicle data.

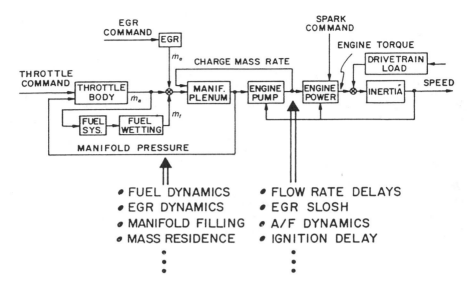

Fig 1. Engine control system block diagram [1,9].

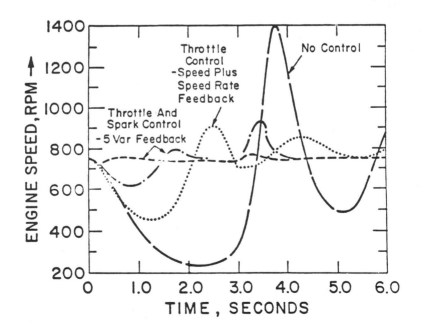

Fig 2. Idle speed control simulation results [1,5,9].

After performing numerous simulations with the twenty-state model to develop insights into the major dynamic interactions, a reduced-order, linear differential equation model consisting of five state and two control variables is formed to initiate the determination of the controller gains. The state vector consists of perturbations on engine speed, manifold pressure, throttle angle, throttle motor rate, and spark advance. The two control variables correspond to throttle rate command and spark advance command perturbation. After linearization about a nominal reference condition a linear quadratic problem (LQP) was defined to determine the controller gains. The performance index emphasized small RPM deviations (for good setpoint control) and small throttle rate deviations (for lower cost throttle actuators).

Figure 2 shows the resultant LQP state feedback controller when simulated with the twenty state simulation program. (In this simulation a load is placed on the engine at t = 0 and removed at t = 3 seconds. Other control policies shown are: no control, throttle-only control, and optimal throttle/spark state feedback with and without feedforward control; the curve with the smallest amplitude oscillation is the feedforward case.) The simulation indicates that coordinated throttle and spark feedback gives a much improved transient response over the standard throttle only control.

After simulating numerous hardware systems, especially various types of throttle actuators, a candidate hardware system was selected. The digital control model structure shown in Fig. 3 was then developed, and the Landau model reference identification technique was employed to obtain the model parameters on an engine dynamometer. The resultant model was then employed with digital control theory to define a coordinated throttle and spark digital controller for vehicle implementation. Comparisons of actual vehicle data and simulation data for the no control, throttle only control, and throttle/spark control cases (all without feedforward) are shown in Fig. 4.

The data shown in Fig. 4 are for a four cylinder engine which means that during each second approximately twenty combustion events occur. Thus, with throttle only control, approximately two seconds are required for recovery to the neighborhood of the setpoint after a major disturbance at time = 0. Note that the throttle only controller is still lightly damped beyond two seconds. Alternatively, throttle-spark control requires approximately one second to return to the setpoint, and the response is relatively well-damped in the neighborhood of the setpoint. Also note that the throttle only case has a speed droop of approximately two hundred RPM while the throttle-spark case droops only one hundred RPM.

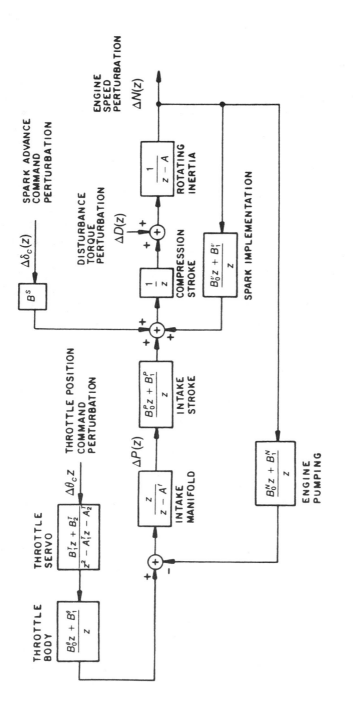

Fig 3. Idle speed control block diagram [5,9].

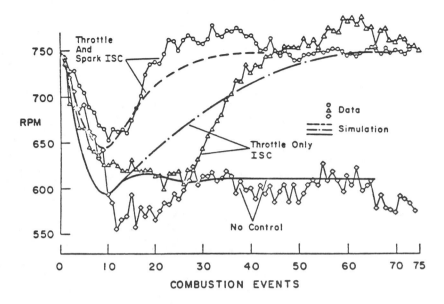

Fig 4. Idle speed control experimental and simulation results [5,9].

Figure 4 indicates that the fidelity of the model is good enough to allow considerable "paper design" before vehicle implementation. The accuracy of the model is probably even better than Figure 4 indicates in that only one vehicle test is shown in the figure. If an average vehicle response was displayed (instead of a single response), then the vehicle and model data would probably be in even closer agreement.   The heuristic reason why the throttle-spark controller is better than the throttle only controller is due to the fact that spark acts much more quickly than the throttle (with its actuator and manifold delays). Qualitatively this is best represented by comparing the root loci of the two cases in the Z-plane. Figure 5 shows the closed-loop poles for zero spark feedback as the throttle feedback gain is increased. The system goes unstable when the magnitude of the throttle gain is equal to 0.4. Figure 6 shows the same system with a fixed, non-zero spark feedback gain as the throttle feedback gain is increased. When the throttle gain magnitude reaches 0.4, the resultant closed-loop poles are well within the unit circle and a stable, relatively insensitive design results.

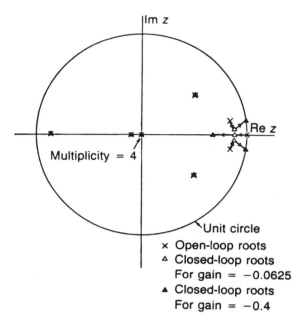

Fig 5. Idle speed control root locus [5,9].

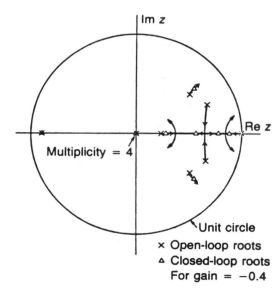

Fig 6. Idle speed control root locus with spark control [5,9].

B. Electronic Transmission Control

The case of Electronically Controlled Transmissions (ECT) is a typical example of power train system control, since it includes almost all major aspects of vehicle longitudinal dynamics described in more detail in [7-8]. In addition to the engine dynamics, the ECT example includes important dynamic effects of torque converter and transmission electrohydraulics. Moreover, unlike in the idle speed control case, the power train can now operate at an almost arbitrary speed/load point during the usually large transmission shift transients, which can amplify any nonlinear effects present in the plant.

ECTs are employed to improve fuel economy, performance, and driveability, the latter being reflected through shift quality as indicated by fore-aft vehicle acceleration. Additional benefits include reduced hardware complexity and packaging requirements, improved functionality through coordination, creation of diagnostics and fault detection capabilities and convenience/communication centers such as the display of present gear information. Perhaps the greatest potential benefit is the flexibility offered by microcomputer software. An example of this can be found in current "adaptive" shift schedules which can be tailored for improved fuel economy, performance or comfort. Potential disadvantages are (at least initially) in the issues of electronic reliability, software complexity and cost.

In general, ECTs are characterized by the fact that many hydraulic functions of conventional automatic transmissions are replaced by electronic and electrohydraulic counterparts. Listed below, in an approximate order of increased complexity, are some of the possible electronically-controlled functions and their typical implementations:

- Torque converter lock-up (open loop; ON/OFF solenoids). The lockup is used to reduce torque converter losses and, thus, improve the fuel economy, particularly while driving in higher gears.
- Lube/clutch cooling with the help of on/off solenoids. Cooling is typically used during the shifts.
- When-to-shift or shift scheduling which can be chosen to enhance performance, fuel economy or driving comfort. This is accomplished with the help of RPM sensors and on/off solenoids, variable force solenoids (VFS), or pulse width modulation (PWM) solenoids for throttle valve pressure control.
- Torque converter bypass clutch slip control, typically implemented in a closed-loop manner.

- Neutral idle: strategies for smooth transition from/to neutral idle that may be used to improve fuel economy.
- Speed ratio (SR) control for continuously variable transmissions (CVTs): closed loop using PWMs or VFSs.
- Line pressure control (open or closed loop with PWM or VFS). The line pressure serves as the main source of actuation for automatic transmission hydraulic system.
- CVT belt load control: open or closed loop via PWM of VFS actuators.
- Drive-away or driving from stop for ECTs without torque converters; open or closed loop; PWM or VFS; possibly augmented by drive-by-wire throttle control.
- How-to-shift or shift execution: open or closed loop; PWMs or VFSs; possibly augmented by spark, fuel, and drive-by-wire throttle control. Here the system design engineer is facing conflicting requirements of good shift quality ("smooth" shifts) versus short shift duration needed for good fuel economy, performance, and clutch/system durability.

As with conventional automatics, electronically controlled transmissions can be divided into two major groups: discrete and CVTs. The work on electronically controlled CVTs is an ongoing research activity in a number of automotive and related companies. Many of the control principles used for discrete ratio ECTs can be used for CVTs as well. However, there are some inherent differences: while the discrete ratio ECTs are characterized by many large transients of short duration that constitute a shift, the CVT control is more of a continuous "process control" nature, and as such it typically results in simpler software.

Discrete ECTs can be classified according to whether only the shift scheduling phase is done electronically or whether both the shift scheduling ("when-to") as well as the shift execution ("how-to") are implemented via electronics. Moreover, the shift execution can be implemented by either open or closed-loop approaches. The main emphasis in the following example will be on discrete ratio ECTs, where both shift scheduling and execution are implemented under computer control, with the shift execution under closed-loop control.

The schematic of an example prototype four-speed transaxle ECT is shown in Fig. 7. The example illustrates some basic shift control principles applicable to many discrete ratio ECTs. The transmission consists of a torque converter, reverse, low, and high clutches, and a single shift actuator - a so-called "dog actuator". The low clutch is constantly

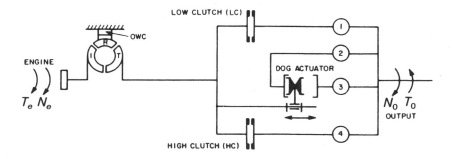

Fig 7. Example schematic for electronic transmission control [9].

engaged in the first gear and the high clutch is constantly engaged in the fourth gear. The dog actuator is used to engage the second gear when displaced to the left in Fig. 7, or the third gear when displaced to the right. Finally, the power is transmitted to the front wheels through the differential and CV joint equipped axles.

The design is characterized by relative simplicity and low part count. In particular, it should be noted that the dog actuators do not include a synchronizer mechanism, which is typically used in manual transmissions. Thus, the dog actuator essentially creates a rigid link-or engagement and as such it tolerates very small speed differentials at the time of the engagement. This is an important constraint for the shift controller, to be discussed later.

As with the conventional automatic transmission, the example ECT has the capability of executing power-on shifts such that positive driving force is supplied throughout the duration of the shift. Manual transmissions, on the other hand, are characterized by a (short) period of zero output torque during the neutral phase of a shift. This large torque "hole" would lead to objectionable shift feel in the case of automatic transmissions. Among the power-on shifts with simultaneous dog actuator engagement, the 1-2 upshift was judged as the most difficult in view of the large torque levels and speed ratio change involved. Consequently, the main emphasis of the modeling and control work was on the 1-2 upshift as described next.

The 1-2 upshift strategy is illustrated in Fig. 8. The shift consists of three phases: torque, inertia, and level holding phases. During the torque phase, the engine combustion torque is transferred from the low to the high clutch as can be seen from the corresponding pressure traces in Fig. 8. This transfer is a necessary first step to be able to reduce the

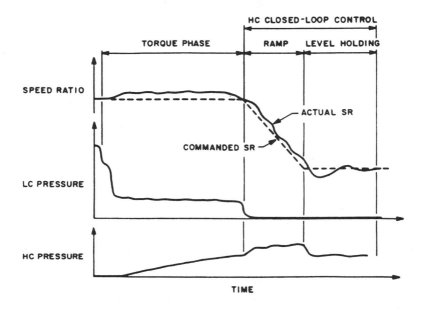

Fig 8. One-two power-on upshift execution phases [9].

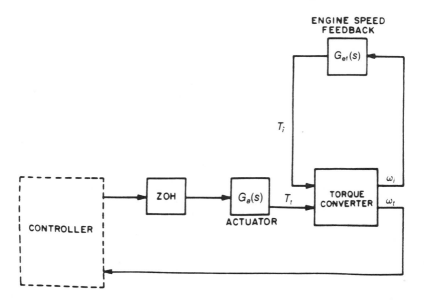

Fig 9. Control system block diagram.

turbine speed to the second gear synchronous level. It should be pointed out that due to the large speed ratio difference between the low and high clutch power paths, the torque phase may result in a relatively large torque "hole". Although the corresponding driveability effects can be minimized through a fast torque transfer, however, there are practical limits in what can be achieved with transmission hydraulics, which is typically limited to 5-10 Hz bandwidth. Once the low clutch has been unloaded, the high clutch controls the turbine speed to the new synchronous level by following the speed ratio ramp as shown in Figure 8. This constitutes the inertia phase of the shift. Subsequent to the inertia phase, the turbine speed is held at the second-gear synchronous level. This level holding phase facilitates the dog actuator engagement which is the most critical phase of the shift for the present ECT, in view of the precise speed ratio control requirements. Once the dog actuator has been engaged, the shift is completed by releasing (or venting) the high clutch.

Closed loop implementation of speed ratio control is essentially mandatory in order to meet the stringent requirements of the inertia and level holding phases. An important starting point in control algorithm design is the development of a suitable power train model. Various details of power train modeling can be found in [18]. A model used for the present ECT control design consists of submodels of an engine, torque converter, and an automatic clutch with its associated electrohydraulic control valve. This model is suitable for the design of controllers which are active during the inertia and level holding phases of a shift, where only one clutch is used as a controlling actuator. Here the torque converter turbine speed is controlled through the clutch pressure modulation, which itself is controlled via duty cycle variations of a PWM solenoid.

The overall model block diagram is shown in Figure 9, and includes the torque converter, engine, control valve-clutch actuator, and a zero-order-hold which reflects the D/A conversion process. The torque converter has been modeled as a two-port device with the turbine and impeller torques as inputs, and the corresponding speeds as outputs. The details of the mathematical model derivation can be found in [7]. The resulting, nonlinear differential equations contain four states (impeller, turbine and reactor speeds, and torus flow) and two inputs - the impeller and turbine torques. It should be mentioned that a simpler model that neglects the torus flow inertia could be used if a desired bandwidth fidelity is below circa 10 Hz.

In cases when the engine control is not used to improve the shift, a relatively simple engine model can often be used [18]. A typical model consists of an inertia term corresponding to the impeller and flywheel

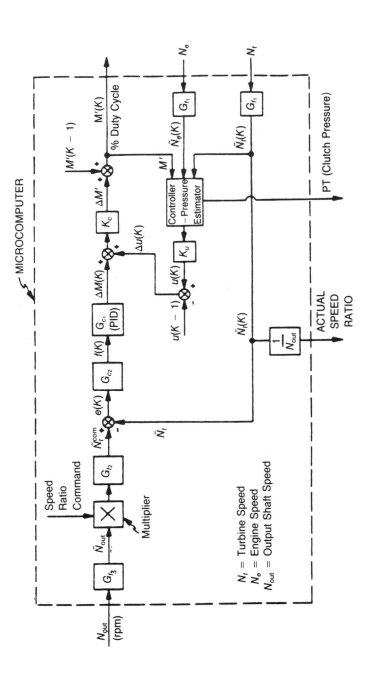

Fig 10. Shift controller block diagram [9].

inertias, and the damper-like term based on so-called cross-sectional curves of engine torque versus speed relation, parameterized by throttle angle position (TAP). At this point, a simple, linearized model of an engine is used to determine the impeller torque. The model consists of an engine inertia term that is appended to impeller inertia, and a friction torque that specifies the impeller torque via the transfer function G_{ef} in Fig. 9. While the impeller torque is influenced by the engine, the turbine torque is assumed to be a static, linear function of clutch pressure only, which itself is determined from electrohydraulic actuator dynamics.

It is important to point out that the electrohydraulic actuator dynamics often determine the dominant (oscillatory) mode in the power train [6]. As such, it must be taken into account in any serious power train control design. In the present case, the clutch actuator transfer function G_a between the clutch pressure and duty cycle in Figure 9 was determined experimentally using a spectral analyzer and resulted in second-order dominant dynamics [6].

The preceding nonlinear model is applicable to detailed studies of power train performance. A linearized and simplified version of the nonlinear model results in a fifth-order system which is used for preliminary control design via root locus and pole placement techniques. The discrete controller update is based on time as the independent variable. As an alternative, a crank-angle based update offers some advantages in the case of engine-torque converter subsystems operating at low speed ratio. This can be seen [19] by introducing the crank angle differential $d\phi$ via $dt = d\phi/\omega_i$ in the expression for torque converter dynamics and engine manifold dynamics. This leads to partial linearization of the two. However, the time-based controller update describes more naturally the dynamics of the electrohydraulic subsystem and structural ("shuffle mode") dynamics of the drive train discussed in more detail in [8]. Since the electrohydraulics constitute the main actuation for the present ECT, and since the duty cycle updates are time based, a constant time sampling is adopted for subsequent control development.

The controller structure is shown in Fig. 10. It consists of a PID block, labeled G_{c1}, and a lead/lag block, labeled G_{c2}. The integral portion of the PID controller is needed to ensure good ramp following and zero steady-state offset during the level holding phase. The lead/lag portion is used for fine tuning of the closed-loop system. In addition, the filters G_{f1} and G_{f3} reduce the measured signal noise, and the filter G_{f2} shapes the commanded signal. Figure 10 contains an additional block for hydraulic pressure estimation and pole placement control, the details of which are described in [6]. (For the above "classical" controller this block is not used

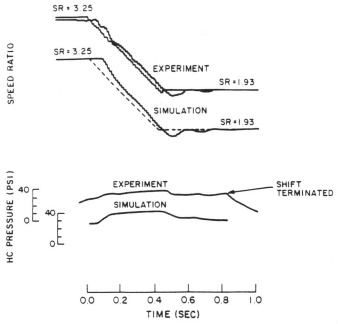

Fig 11. Experimental and simulation results for 1-2 power-on upshift [9].

and is bypassed by setting $K_u = 0$ in Fig. 10.)

Once the preliminary controller parameters, based on linear analysis, were obtained, the controller was further tuned by using a nonlinear power train model. After a satisfactory set of controller parameters was achieved via nonlinear simulations, the controller was programmed in assembly language using the Ford EEC IV microcomputer. The experimental tests were performed next in a dynamometer facility. Typical experimental and simulation results are shown in Figure 11 for the case of a 1-2 power-on upshift. The speed ratio (SR) is commanded to follow a ramp over a 400 ms interval, followed by a level holding phase during which the dog actuator is engaged. It should be pointed out that the initial controller gains, which were based on simulations, resulted in an oscillatory SR response which was quickly corrected by increasing the derivative gain. Subsequent model sensitivity studies revealed that this initial discrepancy can be eliminated by reducing the nominal electrohydraulic actuator bandwidth. With this model modification, the simulation results agreed very well with the experiments, as can be seen from Figure 11. As a further confirmation of model predictions, more detailed, subsequent experimental tests with electrohydraulics demonstrated considerable variability in electrohydraulic bandwidth. This variability was found to depend upon factors difficult to control, such as the amount of entrained air, among others. Thus, in this case, the model predicted which critical hardware areas needed additional experimental and design work.

Figure 11 demonstrates that in addition to good ramp following, the closed-loop SR control also achieves good level holding at the time of dog actuator engagement. The latter occurs about 150-200 msec. after the dog has been set in motion starting at time $t \approx 0.45$ sec. The effectiveness of the controller was further demonstrated through additional experiments, where the ramp time was gradually decreased from 400 to 200 msec. as shown in Fig. 12. It should be noted that all three traces in Fig. 12 were obtained using the same control parameters. All cases were characterized by well-controlled shifts. This example illustrates the flexibility offered by the microcomputer, so that now, unlike with conventional automatics, it is possible to adapt shift execution ("how-to") as well as shift scheduling ("when-to") to different driving conditions. For example, for improved performance and economy, faster shifts may be more appropriate, such as the 200 msec. ramp in Fig. 12. In contrast, for improved comfort the slower 400 msec. ramp may be used.

The above example illustrates typical control design procedures used for next generation power trains. Similar modeling and control system design principles were then applied for design of a number of

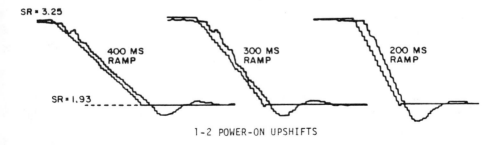

Fig 12. Experimental results for 1-2 upshifts with progressively faster speed ratio ramps [9].

ECTs. This additional work revealed the importance of comprehensive and detailed power train system models, and appropriate digital signal processing which should parallel and complement the control work. The modeling also contributed to better understanding of the new hardware, in particular, the electrohydraulic actuators. This led to modifications which improved the control function.

Thus, it can be concluded that control engineering can contribute not only to software improvements such as more efficient control algorithms, but also to critical hardware improvements that result from improved understanding of the underlying physics and, in particular, system dynamics.

III. PRESENT AND FUTURE TRENDS IN POWER TRAIN CONTROL SYSTEMS

At present, the closed-loop control algorithms are typically only a relatively small portion of an overall power train control strategy. For example, in the case of idle speed control major development efforts and resulting software code are typically devoted to treating various contingencies caused by transients from and to idle speed control mode. Similarly, as shown in the previous section, the automatic transmission shift consists of several transient phases each of relatively short duration. Under these circumstances the usage of open-loop controls can result in good shift performance as demonstrated through many production units.

In conventional production automatics, the open-loop shift control is achieved with the help of ingenious hardware devices such as transmission valve body and one-way clutches. The latter act as mechanical "diodes" zeroing negative torques and thus eliminating the possibility of clutch "fighting" during the torque phase of a shift. Current trends in electronic transmission controls are to implement many of the functions of these open-loop control devices in computer software, thus substantially simplifying the hardware. Moreover, for special cases such as the simple transmission hardware configuration discussed in the previous section, a major burden of precise speed synchronization of different engaging parts is carried by the closed-loop software control. Even in this case the closed-loop control is only a relatively small part of an overall shift strategy that includes many open-loop controlled segments for transient mode operations (e.g., clutch fill, vent, stroking, dog actuation). Because of this, the expert knowledge and thorough understanding of the underlying physics will remain crucial complements to a formal knowledge of closed-loop control techniques.

In view of the above, many current efforts are directed toward obtaining validated and efficient models of power train components and systems. Good, physically based models are key prerequisites for good closed-loop control design, and hardware and overall software modifications. As discussed previously, the latter two have prominent roles in practice and are often more important than a formal utilization of a given control technique per se. For example, strategies that slip the off going clutch during the torque phase of a shift [20-22] offer a number of advantages (e.g., torque hole reduction) over the conventional approaches. Also, the introduction of essentially open-loop spark advance control can significantly reduce the torque transients during the inertia phase of a shift [23]. Other examples include air-fuel (A/F) control where the usage of innovative fuel injection schemes [24] and novel sensors, with simple controls, resulted in major A/F control improvements.

Increased maturity of the power train modeling base will facilitate increased utilization of modern linear and nonlinear control methodologies. Some early attempts in this direction include usage of the Loop Transfer Recovery (LTR) technique [12], H_∞ control theory [25-27], and nonlinear control system techniques such as sliding mode control [28]. As an illustration of possible usage of H_∞ control methodology for power train control system design, the idle speed example from [25-26] will be summarized next. The example includes the choice of an H_∞ performance index, associated weighting functions, and the usage of a CAD package to numerically determine the corresponding optimal controls and their simplifications.

It was shown in [25-26] and supported by experimental data that the ISC model from Section II.A. can be approximated by a structure of Fig. 13. The model parameters were determined as described in [25]. This form of a model is suitable for H_∞ design based on the so-called "Standard Compensation Configuration" from [29], shown in Fig. 14. Here y is a vector of internal outputs-in the ISC example this is just the engine speed; u is the vector of internal inputs - in the present case these are the two control inputs: idle valve setting and ignition timing. Furthermore, e is a vector of external outputs, i.e., the signals to be constrained through the optimization. In the ISC case e is comprised of three signals: the engine speed and the two control inputs. The vector v represents external inputs such as disturbances, noise and reference signals. Although in the present example the main disturbance is created by the torque load in Fig. 13, for computational convenience v is assumed to be additive output

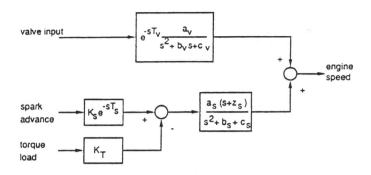

Fig 13. Simplified model structure [26].

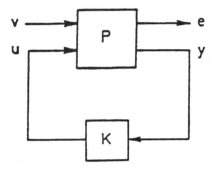

Fig 14. Standard compensation configuration [26].

speed disturbance, which in a sense is a more severe case of plant perturbation.

The design problem now consists of determining the controller **K** for the 3-input, 4-output system **P** shown in Fig. 14. More precisely, the H_∞ design problem is to find a stabilizing **K** that solves the following optimization problem:

$$\min_{K} \left\| \mathbf{W} \cdot \text{LFT}(\mathbf{P},\mathbf{K}) \right\|_\infty = \min_{K} \sup_{\omega} \bar{\sigma} \left[\mathbf{W} \cdot \text{LFT}(\mathbf{P}(j\omega),\mathbf{K}(j\omega)) \right] \quad (1a)$$

or, in component form

$$\min_{K} \left\| \mathbf{W} \cdot \text{LFT}(\mathbf{P},\mathbf{K}) \right\|_\infty = \min_{K} \left\| \begin{matrix} \mathbf{W_1 S} \\ \\ \mathbf{W_2 KS} \end{matrix} \right\|_\infty \quad (1b)$$

where

LFT($\mathbf{P,K}$) designates the linear fractional transformation of $\mathbf{P}(s)$, which is the transfer function between $\mathbf{e}$ and $\mathbf{v}$,

$$\text{LFT}(\mathbf{P,K}) = \mathbf{P}_{11} + \mathbf{P}_{12}\mathbf{K}(\mathbf{I} - \mathbf{P}_{22}\mathbf{K})^{-1}\mathbf{P}_{21}; \tag{2}$$

$\bar{\sigma}(\mathbf{M})$ denotes maximum singular value of matrix $\mathbf{M}$, i.e.

$$\bar{\sigma}(\mathbf{M}) = \max_{\|\mathbf{x}\|_2 = 1}\|\mathbf{M}\,\mathbf{x}\|_2 = \max_{\mathbf{x} \neq 0}\frac{\|\mathbf{M}\,\mathbf{x}\|_2}{\|\mathbf{x}\|_2} = \max_i\sqrt{\lambda_i(\mathbf{M^*M})} \tag{3}$$

with $\mathbf{M^*}$ being the complex conjugate transpose of $\mathbf{M}$;
$\mathbf{S}$ is the output sensitivity matrix, i.e. the transfer fucntion from the disturbance $\mathbf{v}$ to the output speed $\mathbf{y} = \mathbf{e}_1$,

$$\mathbf{S} = (\mathbf{I} + \mathbf{G}\,\mathbf{K})^{-1} \tag{4}$$

where $\mathbf{G}$ is the original plant transfer function from control inputs $\mathbf{u}_1$ and $\mathbf{u}_2$ to the plant output $\mathbf{y}$ (cf. Fig. 13). $\mathbf{KS}$ is the transfer function from disturbance $\mathbf{v}$ to control signals $\mathbf{u}_1$ and $\mathbf{u}_2$, and $\mathbf{W}_1$ and $\mathbf{W}_2$ are weighting transfer functions emphasizing certain frequency regions of $\mathbf{S}$ and $\mathbf{KS}$ as discussed below.

It is instructive to compare the above H_∞ problem with the well-known Linear-Quadratic regulator problem which belongs to the H_2 optimization class. The LQ formulation minimizes the time-averaged, squared values in the performance index, for a limited class of inputs consisting of impulses and white noise signals. On the other hand the H_∞ formulation minimizes the maximal singular value (Eq. 1a), which is the measure of maximal possible disturbance energy amplification through the plant. The disturbance class is now much richer and includes all square-integrable (L_2) functions scaled to a unity bounded norm (cf. Eq. 3). Therefore, the H_∞ control technique is a natural extension of the LQ methodology, and both can have a prominent place in the arsenal of control design and evaluation tools.

For the ISC example the sensitivity function in Eq. (1b) reflects system performance in terms of disturbance attenuation at the output - engine speed. The weighting $\mathbf{W}_1$ is then chosen to provide a small closed-loop sensitivity in the low-frequency range, but is allowed to increase at higher frequencies where good disturbance rejection is not possible due to the modeling errors. Thus the first-order low-pass filter is chosen for the

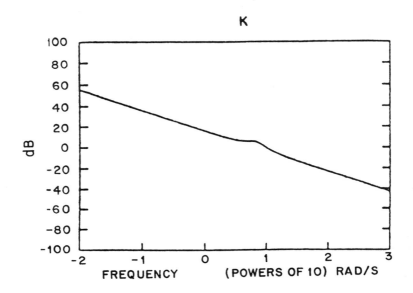

Fig 15a. Singular value : Idle valve controller [26].

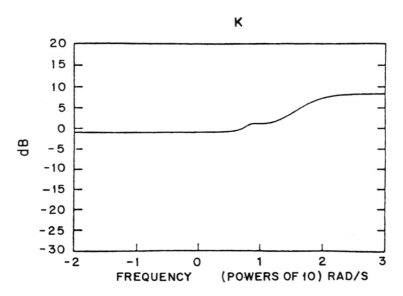

Fig 15b. Singular value : ignition timing controller [26].

sensitivity weighting $\mathbf{W_1}$.

The control energy constraints are embodied through the $\mathbf{KS}$ term in Eq. (1b), which also serves for a robust stability test under additive plant perturbations within $1/\|\mathbf{KS}\|_\infty$ radius. The control weighting functions in $\mathbf{W_2}$ were chosen according to the actuators physical characteristics. The idle valve control loop ($\mathbf{u_1}$) is of a relatively low bandwidth dominated by engine manifold dynamics and induction-to-power delays. Thus the corresponding weighting function is chosen as a high-pass filter penalizing control action at high frequencies. On the other hand, because the ignition timing $\mathbf{u_2}$ provides relatively fast control action, the spark advance (SA) should only be changed in a transient manner, with zero steady-state change, which is necessary to revert to the original spark calibration. An approximate weighting function for $\mathbf{u_2}$ is thus a low-pass filter providing high penalty at low frequencies, but also a relatively small finite penalty at high frequencies. The latter is needed to limit the high frequency gain of the ignition control loop. It should also be noted that although the sensitivity weight $\mathbf{W_1}$ was of low-pass type, the resulting $\mathbf{S}$ will eventually settle to a constant value at high-frequencies where the term $\|\mathbf{W_2 KS}\|_\infty$ dominates in Eq. 1b [29].

To be able to use the CAD program from [29], the original problem was next modified by approximating pure delay terms in Fig. 13 by corresponding first-order lags. Future investigations should address other alternatives such as, for example, Pade approximations of different orders.

At this stage, a number of different control designs were attempted [25, 26] using different weighting functions. In the present context the weighting functions are the "tuning knobs" of H_∞ design, and as such they still imply a need for design iterations depending in good part on control engineers experience and intuition. The final design resulted in disturbance-to-control transfer function characteristics shown in Figs. 15 a and b.

From Fig. 15a, it can be seen that the idle valve control loop acts as a low-frequency, integral-type controller securing good steady-state properties. On the other hand, based on Fig. 15b, the spark controller primarily acts as a transient controller (robustness of this loop could possibly be further enhanced by lowering the SA gain for frequencies above circa 100 rad/sec).

These main characteristics of the two control loops are also visible in the corresponding time responses for a hypothetical unit step in engine speed disturbance shown in Fig. 16. This corresponds to the transient caused by the engine speed error at the moment when the closed-loop idle speed control is turned on. Note that most of the transient control is

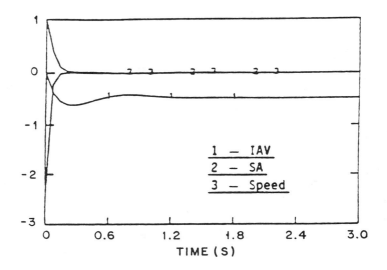

Fig 16. Unit output disturbance response [26].

achieved via the spark ignition timing. The spark control essentially decays to zero at the steady-state, where the disturbance is controlled by a relatively slow idle valve controller.

Final simulation tests were performed using the load torque disturbance. For this purpose the original controller was simplified to result in two 3-state controllers without significant loss of performance. The simplified controllers were next discretized with a 20 ms sample period and applied to the original model of Fig. 13, which included the delays.

Figure 17 shows the closed-loop system response to a 15 Nm torque disturbance step which is representative of a significant air-conditioning load, for example. It should be noted that the speed is measured with respect to an 800 RPM idle operating point, and it drops only 18.3 RPM during the transient. To investigate the closed-loop system robustness the controller was also tested using models that had been identified at different operating points. An example is shown in Fig. 18 where the same 15 Nm load disturbance is applied to the model identified at 1000 RPM, where up to 100% higher open-loop gains were present than in the nominal 800 RPM case. Although the spark now saturates at 20 degrees and the speed drop is greater (37.5 RPM), the overall response is still stable with very little sign of oscillations.

The above H_∞ design was developed in parallel with a classical control approach. The classical controller was similar to the one implemented in Section II.A. It consisted of PID throttle control, and a proportional spark advance control. The throttle controller gains were optimized using [30]. The corresponding performance was initially even better than some early H_∞ designs. This was a clear indication that there was room for improvement of H_∞ results through more judicious choices of weighting functions as presented above.

In return, the optimal H_∞ results (Figs. 15-18) indicated that the classical design could also benefit from a faster SA action. To this end the original proportional SA feedback was expanded to a lead-lag controller structure. The corresponding response for 15 Nm load disturbance applied to the nominal 800 RPM model is shown in Fig. 19. Although the speed drop is now only circa 12% larger than for the H_∞ case of Fig. 17, the classical design lacks the robustness of the H_∞ counterpart. This is demonstrated by simulations in Fig. 20 for the modified, 1000 RPM model responses where an extremely lightly damped mode would most likely not

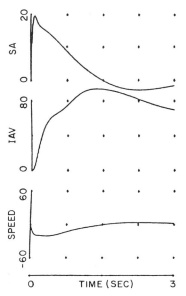

Fig 17. Nominal disturbance
response [26].

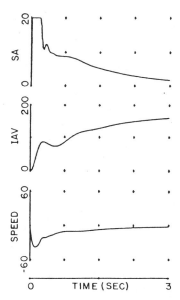

Fig 18. Disturbance response of
perturbed model [26].

be acceptable in practice. Also, comparing Figs. 17 and 19, it can be seen that the H_∞ design requires a much slower idle valve actuator, leading to improved reliability and cost savings.

It is important to point out that the above H_∞ idle speed control design is pending experimental validation planned for future work. Indeed, with exception of the ISC study [12] and SISO H_∞ example [27], there exists no published experimental validations of various new linear and nonlinear control techniques proposed for power train applications. The experimental validation is crucial for eventual successful automotive

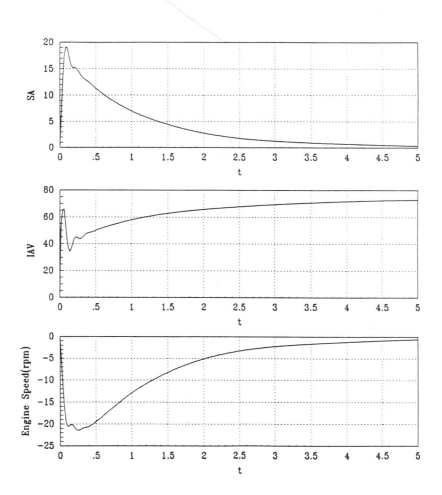

Fig 19. Classical controller : nominal disturbance response

application of any modern control technique. It is important for LTR and H$_\infty$ techniques in view of their implied linear system assumption, and it is particularly important for sliding mode techniques [28] in view of their inherent high-gain characteristics and potential for excitation of unmodeled, high-frequency modes. To secure their place "under the hood" the modern control techniques will have to significantly outperform simple open-loop and classical controls in both performance and reliability, to be proven through numerous vehicle tests under very different driving and environmental conditions.

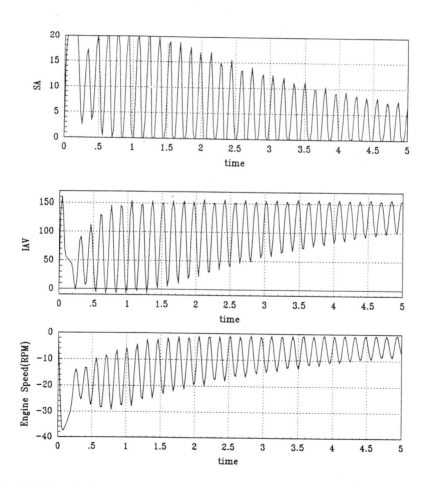

Fig 20. Classical controller : disturbance response of perturbed model.

IV. TRENDS IN AUTOMOTIVE COMPUTER CONTROL SYSTEMS

In addition to the computer control of power trains, other major subsystems on the vehicle are now under computer control. For example, Ford introduced a production computer controlled suspension system in 1984 and a computer controlled anti-lock braking system in 1985. Most control systems to date tend to be input/output oriented as opposed to signal processing oriented. For example, the advantage of Ford's EEC-IV power train control computer is mainly due to the way that it deals with the vast amount of input information and output commands [31, 32]. Another example of this philosophy is the Ford Electronically-controlled Air Suspension (EAS) system described in [32, 33].

The EEC IV and EAS computer control systems emphasize digital information type sensors (e.g., exhaust gas oxygen, revolution-per-minute (RPM), Hall effect devices), digital output devices (solenoids, relays), and mode selection strategies. From a control point of view, the characteristics of such first-decade systems may be summarized as follows:

- Feedback Control: Low-level because feedback sensing devices are switch-oriented; high-level of scheduled and feedforward control.
- Adaptive Control: Low-level, table-oriented; lack of "rich" feedback information and process models limits the application.
- Diagnostics: Concerned mainly with diagnosing faults in the control system and not the process under control; again, lack of rich feedback information and process models limits the application.
- Maintenance-On-Demand: Operation condition calculation oriented; lack of sensors and appropriate feedback information limits the application.
- Communications: Relatively low-speed and limited to noncritical operations; high-speed, critical applications will depend upon a higher-level of multiplexing and/or highly interactive control of major subsystems.

There is considerable world-wide research and development in the areas of automotive dynamic system modeling and sensors. As these areas mature, the role of control theory should increase even more in automotive control systems because the control problems will become more feedback and signal processing oriented (as opposed to the current I/O and mode selection orientation). In fact, four of the major areas mentioned above (feedback, adaptive control, diagnostics, and maintenance- on-demand) should be containable within the same control/signal processing framework. For example, rich feedback signals will be employed for immediate (foreground) system control, and signal processing will be applied in the background to the same signals to adjust controller gains (adaptive), determine system faults (diagnostics), and determine the state of the system with respect to maintenance requirements (maintenance-on-demand).

Present and future efforts in the power train control area are directed toward increased application of coordinated control between the engine, transmission and overall vehicle system. This will intensify the need for validated, efficient, and comprehensive vehicle system models. A strong and mature modeling base should ultimately facilitate full utilization of modern, linear and nonlinear control methodologies.

V. CONCLUDING REMARKS

Applications of control theory to idle speed and transmission control have been presented to illustrate the use of control theory in the computer control of automobiles. In addition, the structure of these first generation systems was analyzed from a control theoretic perspective. Driven by international competition, next generation systems will require a higher level of information processing and coordination. A number of modeling and control engineering techniques have been developed that hold promise for systems with such requirements. In particular, this includes the bond graph [18] and block diagram modeling approaches, and LQG/LTR, H_∞, H_μ [34, 35], and QFT [36, 37] robust control design techniques. When combined with a thorough understanding of the underlying hardware and associated physics, these methodologies will contribute to superior automotive products, both in terms of the software as well as the hardware. Thus, the role of control engineering should increase dramatically in the automotive industry. Since automotive applications involve millions of units and sophisticated microelectronic devices are relatively low-cost in high volume applications, the marriage of microelectronic driven control theory to the automobile represents an exciting frontier in control engineering research and development.

ACKNOWLEDGMENTS

Simulations for Figs. 19 and 20 were efficiently performed by L.F. Chen from Ford Motor Company. The camera-ready version of the paper was formated with the expert help from R. A. Schaefer of Ford Motor Company.

REFERENCES

1. B. K. Powell and W. F. Powers, "Linear Quadratic Control Design for Nonlinear IC Engine Systems," Presented at ISATA Conference, September 1981, Stockholm, Sweden.
2. W. F. Powers, "Internal Combustion Engine Control Research at Ford," *IEEE Proceedings of Conference on Decision and Control*, 1447-1452, Dec. 1981.
3. W. F. Powers, B. K. Powell and G. P. Lawson, "Applications of Optimal Control and Kalman Filtering to Automotive Systems," *International Journal of Vehicle Design*, SP4, 39-53, 1983.
4. R. L. Morris, M. V. Warlick and R. H. Borcherts, "Engine Idle Dynamics and Control: A 5.8L Application," SAE Paper No. 820778, June 1982.
5. R. L. Morris and B. K. Powell, "Modern Control Applications in Idle Speed Control," *Proceedings of the 1983 American Control Conference*, 79-85, June 1983.
6. D. Hrovat, "Powertrain Control Using Observers to Estimate Unmeasured Quantities", *Proceedings of the 1983 ISATA Conference*, Cologne, Federal Republic of Germany, Sept. 1983.
7. D. Hrovat and W. E. Tobler, "Bond Graph Modeling and Computer Simulation of Automotive Torque Converters," *Journal of the Franklin Institute*, Special Issue on Physical Structure in Modelling, Vol. 319, No. 1/2, pp. 93-119, Jan./Feb. 1985.
8. D. Hrovat, W. E. Tobler, and M. C. Tsangarides, "Bond Graph Modeling of Dominant Dynamics of Automotive Power Trains," *Proceedings of the 1985 ASME Winter Annual Meeting*, Miami Beach, Publication DSC-Vol. 1, 293-301, Dec. 1985.
9. D. Hrovat and W. F. Powers, "Computer Control Systems for Automotive Power Trains", *IEEE Control Systems Magazine*, Vol. 8, No. 4, Aug. 1988.
10. D. J. Dobner and R. D. Fruechte, "An Engine Model for Dynamic

Engine Control Development," *Proceedings of the American Control Conference*, Vol. 1, pp. 78-78, June 1983.

11. M. Hubbard, P. D. Dobson, and J. D. Powell, "Closed-Loop Control of Spark Advance Using a Cylinder Pressure Sensor," *ASME Journal of Dynamic Systems, Measurement and Control*, Dec. 1976.

12. C. E. Baumgartner, H. P. Geering, C. H. Onder, and E. Shafal, "Robust Multivariable Idle Speed Control," *Proceedings of the 1986 American Control Conference*, Seattle, June 1986.

13. U. Kiencke, "The Role of Automatic Control in Automotive Systems," *Proceedings of the X IFAC Triennial World Congress*, Munich, West Germany, July 1987.

14. T. Tabe, M. Ohba, E. Kamei, and H. Namba, "On the Application of Modern Control Theory to Automotive Engine Control," *IEEE Transactions on Industrial Electronics*, Vol. 1E-34, No. 1, Feb. 1987.

15. F. J. Winchell and W. D. Route, "Ratio Changing for Passenger Car Automatic Transmission," Paper 311A presented at SAE Congress, Detroit, Jan. 1961.

16. T. Ishihara and R. I. Emori, "Torque Converter as a Vibration Damper and Its Transient Characteristics," SAE Paper No. 660368, *SAE Transactions*, Vol. 75, 1967.

17. T. Tabe, H. Takeuchi, M. Tsujii, "Vehicle Speed Control System Using Modern Control Theory," *Proceedings IECON Conference*, 1986.

18. D. Hrovat and W. E. Tobler, "Bond Graph Modeling of Automotive Power Trains", to appear in *The Journal of the Franklin Institute*, Special Issue on Current Advances in Bond Graph Modeling.

19. D. Hrovat, "Variable Displacement Pump Line Pressure Control," Ford Motor Company Internal Document, Sept. 1984.

20. S. Pierce, P. Jain, and L. T. Brown, "Transmission Clutch Control System and Method", U.S. Patent 4,527,678, July 1985.

21. L. T. Brown and D. Hrovat, "Transmission Clutch Loop Transfer Control", U.S. Patent 4,790,418, Dec. 1988.

22. M. B. Leising, H. Benford, and G. L. Holbrook, "The All-Adaptive Controls for the Chrysler Ultradrive Transaxle", SAE Paper No. 890529.

23. M. Schwab, "Electronic Control of a 4-Speed Automatic Transmission with Lock-Up Clutch", SAE Paper No. 840448.

24. R. Nighiyama, S. Ohkubo and S. Washino, "Effects of Fuel Transport Process on Transient A/F Controls", *JSAE Review*, Vol. 8, No. 2, April 1987.

25. S. J. Williams, D. Maclay, J. W. v.Crevel, D. Hrovat, C. Davey and L.

F. Chen, "An H-Infinity Approach to Idle Speed Control Design", *Proc. 1989 IASTED Conference on Modelling*, Control and Simulation, Grindlewald, Switzerland, Feb. 1989.

26. S. J. Williams, D. Hrovat, C. Davey, D. Maclay, J. W. v.Crevel, and L. F. Chen, "Idle Speed Control Design Using an H-Infinity Approach", *Proc. 1989 American Control Conference*, Pittsburgh, June 1989, pp. 1950-1956.

27. H. Kuraoka, N. Ohka, M. Ohba, S. Hosoe, and F. Zhang, "Application of H_∞ optimal Design to Automotive Fuel Control", *Proc. 1989 American Conference*, Pittsburgh, June 1989, pp. 1957-1962.

28. D. Cho and J. K. Hedrick, "A Nonlinear Controller Design Method for Fuel-Injected Automotive Engines", *Proc. ASME Automotive Engine Technology Symposium*, Dallas, Texas, Feb. 1987.

29. I. Postlethwaite, S. D. O'Young, and D.-W. Gu, "Stable-H User's Guide", Oxford University Report, OUEL 1687/87, 1987.

30. P. V. Kokotovic and D. Rhode, "Sensitivity Guided Design", Reports 1-5, PK Controls, 1986-1987.

31. R. C. Breitzman, "Development of a Custom Microprocessor for Automotive Control," *IEEE Control Systems Magazine*, May 1985.

32. W. F. Powers, "Automotive Computer Control Systems," *Proceedings of the Automotive Microelectronics Advanced Course on Optimal Engine Control*, Capri, Italy, June 1985.

33. E. H. Marquardt and R. J. Sandel, "Development of a control system for an electronic air suspension (EAS) system," *Proceedings of the 1984 American Control Conference*, 1190-1198, June 1984.

34. Doyle, J. C., "Lecture Notes in Advances in Multivariable Control", ONR/Honeywell Workshop, Minneapolis, 1984.

35. M. Morari and E. Zafiriou, *Robust Process Control*, Prentice Hall, Englewood Cliffs, New Jersey, 1989.

36. I. M. Horowitz, "Synthesis of Feedback Systems with Nonlinear Time Varying Uncertain Plants to Satisfy Quantitative Performance Specifications", *IEEE Proceedings*, Vol. 64, 1976, pp. 123-130.

37. S. Jayasuriya and M. A. Franchek, "Loop Shaping for Robust Performance in Systems with Structured Perturbations", ASME Paper 88-WA/DSC-11, Presented at the Winter Annual Meeting, Chicago, Nov. 1988.

CONTROL TECHNIQUES IN THE PULP
AND PAPER INDUSTRY

GUY A. DUMONT

Department of Electrical Engineering
Pulp and Paper Centre
University of British Columbia
Vancouver, B.C., Canada V6T1W5.

I. INTRODUCTION

The pulp and paper industry was one of the pioneers of computer process control some 25 years ago, [1], [2]. Some of the major breakthroughs in process control, such as minimum-variance and self-tuning controllers were first tested on paper machines, [3]-[5]. Today, the vast majority of paper machines in the world are computer controlled. In the past, numerous surveys have been published on various aspects of pulp and paper process control. In [6], Brewster and Bjerring review the status of the field as of 1970. In [7], Church reviews some problems that are specific to the industry. In [8], Perron and Ramaz survey the control strategies available for chemical pulping plant control. Michaelson reviews computer control strategies used for control of thermomechanical pulping (TMP) plants in [9]. Gee and Chamberlain [10] review some of the control algorithms commonly used in the pulp and paper industry. In [11], Dumont surveys the application of advanced control methods in the pulp and paper industry, while in [12] he discusses the application of system identification and adaptive control techniques to papermaking.

Like chemical processing plants, pulp and paper mills are very interactive plants with extensive recycling. They consist of both batch and continuous processes. The nature of these processes is also very varied. For instance pulping processes can be mechanical, chemical or a combination of both. As an illustration of the complexity of a pulp and paper mill, Fig. 1 depicts a schematic diagram of an integrated kraft mill.

The first step is pulping, that is the dissolving of the lignin, the material that binds the fibres together. Pulping is achieved by cooking the wood chips with an alkaline solution, called liquor, under controlled conditions in a digester. There exist two types of digesters, batch and continuous. The latter is most often a Kamyr digester, *i.e.* a vertical plug flow reactor.

Bleaching can be regarded as the continuation of pulping, as it is essentially the removal of residual lignin and colouring compounds. A typical bleach plant consists of five or six chemical reactors with interstage washing. Each stage must be carefully controlled in order to preserve desirable pulp properties and to limit the environmental impact of the plant.

Finally in papermaking, the most important piece of equipment is the paper machine, on which stock is drained and dried to produce a sheet of paper formed into a reel. The paper reel is then sent to the finishing area for rewinding and roll trimming.

The control problem in the pulp and paper industry shares many characteristics of the control problem in the process industries. Because this is an old industry, many pulp and paper mills are old and were designed without much concern about their controllability. Raw materials characteristics vary, especially in the pulping plant and chemical recovery area. Thus a major problem is to produce consistent quality in the face of a highly variable species mix in the feed material. The processes in a pulping plant are poorly known. Despite intense activity in the field, quantitatively accurate kinetic models of delignification are not yet available. Little is known of the mixing and dilution of gas in a fibre suspension, yet it has a direct influence on the bleaching reaction rate. Flow patterns in process vessels are generally poorly modelled.

Processes in a chemical pulping plant are generally slow, with often large dead-times. Typically, the retention time in a Kamyr digester can be 4–5 h, in a chlorination tower 1.5–2 h. Dead-times can vary due to changing mixing conditions, dead zones, channelling, or due to viscosity changes as in black liquor evaporators. Processes in a papermaking plant are faster, with time constants and dead-times that can vary from a few seconds to several minutes. Most processes are non-linear and multivariable.

Very often, sensors are not available to measure key process variables. Although effective alkali sensors for the liquor are used, effective sensors to measure the degree of delignification of a pulp are only starting to appear. Many sensors use inferential measurement, and are affected by factors other than

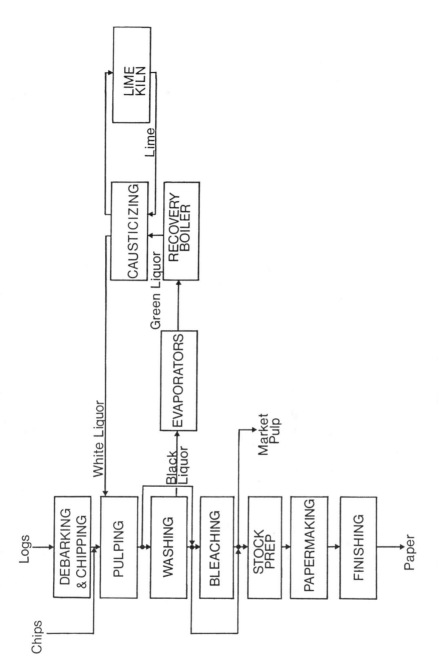

Fig. 1. Integrated Kraft Mill.

the variable of interest, and so require frequent calibration. Sometimes, the key variables are not clearly defined, let alone measurable. For instance, in mechanical pulping, pulp quality is affected by factors like fibre length and its distribution and specific area but is not clearly defined. In general, more sensors have been developed for the paper mill than for the pulp mill.

The applications of computer control systems have centered around the paper mill. In particular, many turnkey control systems and sensors are available for the paper machine. Although fewer suppliers offer them, turnkey systems are also available for most processes in the pulp mill.

The present paper concentrates on four problem areas that, together represent the most challenging control problems in the pulp and paper industry. First, we discuss various techniques that have been or are being developed to solve the problem of control in the presence of a variable dead time. In Section III, we discuss the control of web processes, and in particular of the paper machine. In Section IV, we treat a control problem which is a rather unique in the process industries, the control of wood chip refiner motor load. The uniqueness of this problem is that the process incremental gain can change sign without notice. Finally in Section V, we touch upon the problems due to wood species variations.

II. CONTROL OF VARIABLE DEAD-TIME PROCESSES

A. STANDARD DEAD-TIME COMPENSATION TECHNIQUES

Because significant dead-time is present in many pulp and paper processes, dead-time compensation is routinely used. Of the two major techniques available for dead-time compensation, i.e. the Smith's predictor and the Dahlin regulator, the latter one is most used. The main reason is its simple design and its ease of use.

The Dahlin controller provides implicit dead time compensation and rests on very simple premises. It states that the closed-loop response of the feedback system should be first-order plus dead time, with unit gain and a closed-loop dead time equal to the open-loop one. The closed-loop time constant is a free parameter used for tuning. Like the Smith's predictor, the Dahlin controller is a model-based control design method.

Let the discrete plant be described by

$$y(t) = \frac{B(q^{-1})}{A(q^{-1})} q^{-d} u(t) \tag{1}$$

and the feedback controller be:

$$u(t) = \frac{N(q^{-1})}{D(q^{-1})}(y_r - y(t)) \tag{2}$$

The closed-loop system is then:

$$y(t) = \frac{BNq^{-d}}{AD + BNq^{-d}}y_r(t) \tag{3}$$

Now, to design a Dahlin controller, we wish the closed-loop system to be a first-order plus dead-time

$$y(t) = \frac{1 - p}{1 - pq^{-1}}q^{-d}yr(t) \tag{4}$$

where the closed-loop pole can be used as a tuning parameter, as it is the only free parameter in the design.

Equating the closed-loop transfer function with the desired one, *i.e.*

$$\frac{BNq^{-d}}{AD + BNq^{-d}} = \frac{1 - p}{1 - pq^{-1}}q^{-d} \tag{5}$$

gives the Dahlin controller:

$$\frac{N}{D} = \frac{1 - p}{[1 - pq^{-1} - (1 - p)q^{-d}]}\frac{A}{B} \tag{6}$$

Note that $D(1) = 0$, hence the Dahlin regulator has integral action, unless it is cancelled by an integrator in the process. The Dahlin controller inverts the process dynamics, *i.e.* it cancels all process poles and zeros. Therefore, no process poles or zeros can be allowed outside the unit circle. Also, the presence of stable but poorly damped process zeros will cause ringing in the control signal. Such ringing is extremely bad for actuator wear and has to be eliminated from the regulator. In practice, the Dahlin regulator is mostly used with a first-order model of the plant:

$$y(t) = \frac{bq^{-d}}{1 - aq^{-1}}u(t) \tag{7}$$

then,

$$\frac{N}{D} = \frac{(1 - p)(1 - aq^{-1})}{b[1 - pq^{-1} - (1 - p)q^{-d}]} \tag{8}$$

or

$$u(t) = b_1e(t) + b_2e(t - 1) + b_3u(t - 1) + (1 - b_3)u(t - d) \tag{9}$$

with $b_1 = (1 - p)/b$, $b_2 = -ab$, $b_3 = p$, $e(t) = y_r(t) - y(t)$

This controller can be expressed in incremental, or velocity form:

$$\Delta u(t) = b_1 e(t) + b_2 e(t-1) - (1-b_3) \sum_{i=1}^{d-1} \Delta u(t-i) \tag{10}$$

A reason for the popularity of the Dahlin regulator is that it can be interpreted as the sum of a PI regulator and a dead-time compensator:

$$\Delta u(t) = \underbrace{b_1 e(t) + b_2 e(t-1)}_{PI} \underbrace{-(1-b_3) \sum_{i=1}^{d-1} \Delta u(t-i)}_{\text{dead- time compensation}} \tag{11}$$

For the delay-free case, $i.e.$ $d = 1$, this reduces to a PI regulator.

Although designed for the servo-problem, the Dahlin controller is used most often for the regulation problem. However, many industrial processes may be approximated by a FOPDT process with additive output noise in the form of a first-order integrated moving-average, i.e

$$y(t) = \frac{bq^{-d}}{1 - aq^{-1}} u(t) + (1 - cq^{-1})e(t)/\Delta \tag{12}$$

where $e(t)$ is a zero-mean, white sequence and $\Delta = 1 - q^{-1}$. It is easy to show [13] that if $c > 0$, then with the choice $p = c$, the Dahlin controller is the minimum-variance controller for the system in Eq. (12). This may explain the popularity of the Dahlin controller in the process industries.

Despite all its niceties, the Dahlin controller does have some drawbacks, in particular it is sensitive to time delay mismatch between the model and the plant. This is particularly acute on systems with variable dead-time, which as seen in the introduction are not infrequent. For a stable and inversely stable discrete plant, with perfect model match except for the dead time, the stability of the closed-loop system is defined by the roots of Eq. (13) below, where d and $\hat{d}$ are respectively the true plant delay and the model delay.

$$(q - p) + (q^{-d+1} - q^{-\hat{d}+1})(1 - p) = 0 \tag{13}$$

Using p as a tuning parameter, from Eq. (13), it is possible to provide stability robustness to delay mismatch. Table I summarizes, for a range of d and $\hat{d}$ the lower limit on p that ensures stability. For instance if $d = 5$ and $\hat{d} = 3$ the lower limit is $p = 0.38$. Although this preserves stability, it does not guarantee performance robustness. The latter depends on the desired performance, the open-loop dynamics and the noise dynamics. When robustness cannot be guaranteed without an acceptable loss of performance, then one way to obtain both robustness and performance is to consider the use of adaptive control.

NUMBER OF SAMPLING PERIODS BY WHICH PROCESS TIME-DELAY IS:

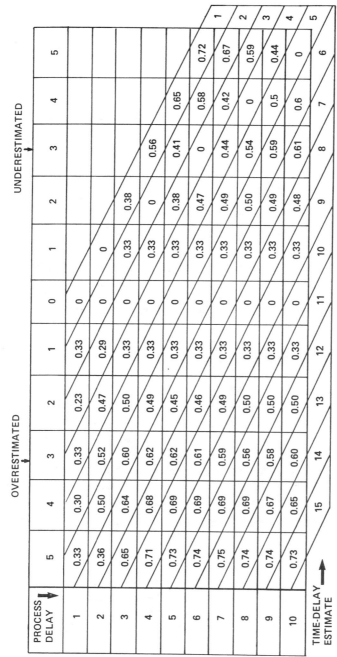

PROCESS DELAY	OVERESTIMATED						UNDERESTIMATED				
	5	4	3	2	1	0	1	2	3	4	5
1	0.33	0.30	0.33	0.23	0.33	0					
2	0.36	0.50	0.52	0.47	0.29	0	0				
3	0.65	0.64	0.60	0.50	0.33	0	0.33	0.38			
4	0.71	0.68	0.62	0.49	0.33	0	0.33	0	0.56		
5	0.73	0.69	0.62	0.45	0.33	0	0.33	0.38	0.41	0.65	
6	0.74	0.69	0.61	0.46	0.33	0	0.33	0.47	0	0.58	0.72
7	0.75	0.69	0.59	0.49	0.33	0	0.33	0.49	0.44	0.42	0.67
8	0.74	0.69	0.56	0.50	0.33	0	0.33	0.50	0.54	0	0.59
9	0.74	0.67	0.58	0.50	0.33	0	0.33	0.49	0.59	0.5	0.44
10	0.73	0.65	0.60	0.50	0.33	0	0.33	0.48	0.61	0.6	0
TIME-DELAY ESTIMATE	6	7	8	9	10	11	12	13	14	15	

Table. 1. Tuning of Dahlin controller in presence of delay mismatch.

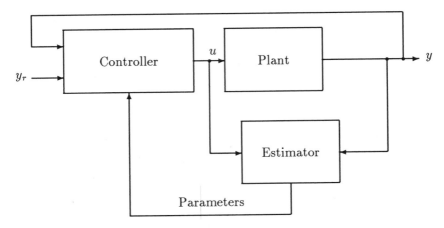

Fig. 2. Block diagram of an adaptive controller.

B. ADAPTIVE CONTROL

Adaptive control is a complex and still somewhat immature field, despite more than thirty years of research which became particularly intense after the seminal 1973 paper by Åström and Wittenmark, [14]. For a comprehensive review of the current state of the art in the field, readers are referred to the book by Åström and Wittenmark, [15].

When a control engineer wants to develop a control scheme for a particular process, one of the first tasks to be accomplished is the identification of the process dynamics. The second major task is then to design the controller that will achieve the desired control objective. Adaptive control can be thought of as an automation of that design procedure, *i.e.* on-line identification and control design. A typical block diagram of such an adaptive controller is shown on Figure 2. A major difference is that the entire procedure is performed on-line, in real-time and without human supervision. Because of this lack of human supervision, extra care has to be taken when implementing such a scheme. One of the current research trends is to develop an emulation of that human supervisor using an expert system overlooking both the implementation and the operation of the adaptive controller. Such a system would greatly improve the reliability of current adaptive control schemes. However, a major drawback with current adaptive control methods is that they require prior knowledge of the structure of the plant dynamics, *i.e* of the order and dead time of the process transfer function. Wrong assumptions may lead to instability, obviously an undesirable feature.We shall discuss some adaptive control techniques applicable to processes with unknown time delay.

1. Extended Polynomial Approach

Commonly used adaptive control techniques assume knowledge of the plant time delay, and are rather sensitive to the presence of a mismatch. As mentioned before, on many processes the time delay is either not known accurately, or is even time-varying. This has prompted the development of adaptive control techniques that can handle such situations. The frequently used method for handling this problem is to use an extended polynomial in the numerator of the transfer function model of the plant as in, [16], [17]:

$$y(t) = \frac{b_1 q^{-1} + \cdots + b_r q^{-r}}{1 + a_1 q^{-1} + \cdots + a_n q^{-n}} q^{-l} u(t) \tag{14}$$

where l is the minimum value of the dead time and r is the possible range for the dead time variations, in sampling intervals. A pole placement controller that preserves the open loop zeros can then be designed in a simple fashion. Although simple, the method has several disadvantages. When using large A and B polynomials, the likelihood of common factors in the estimated model is increasing. This renders the identification more difficult. Also, the increased number of estimated parameters makes the persistency of excitation condition more severe and slows down convergence. Recently, a novel approach known as the variable regression estimator shows some promise for solving these problems. The delay is explicitly estimated, and the number of estimated parameters is not affected by the time delay, see [18], [19] for more details.

2. Generalized Predictive Control

Generalized Predictive Control (GPC) has recently been proposed as a "general-purpose" adaptive control method by Clarke et al. [20], [21]. It is based on the minimization of the following quadratic performance index:

$$J = E\left(\sum_{j=N_1}^{N_2} [y(t+j) - w(t+j)]^2 + \sum_{j=1}^{N_U} \rho[\Delta u(t+j-1)]^2 \right) \tag{15}$$

where

$w(t+j)$	is a sequence of future setpoints
N_1	is the minimum prediction horizon
N_2	is the maximum prediction horizon
N_U	is the control horizon
ρ	is a control weighting factor

The minimization of J in Eq. (15) requires j-step ahead predictions of $y(t+j)$. Most often, a CARIMA model of the form below is used to describe the process:

$$A(q^{-1})y(t) = B(q^{-1})u(t-d) + C(q^{-1})e(t)/\Delta \tag{16}$$

Consider the Diophantine equation:

$$C(q^{-1}) = A\Delta F_j(q^{-1}) + q^{-j}G_j(q^{-1}) \tag{17}$$

The j-step ahead prediction for $y(t+j)$ can be written as:

$$\hat{y}(t+j) = [G_j y(t) + BF_j \Delta u(t+j-d)]/C(q^{-1}) \tag{18}$$

In the r.h.s. of Eq. (18) terms can be respectively regrouped in components that are known at time t and those that are unknown at time t. Regrouping all j-step predictors for the $(N_2 - N_1 + 1)$ sampling instants of the prediction horizon, we can write,

$$\hat{\underline{y}} = R\tilde{\underline{u}}(t) + \underline{f} \tag{19}$$

where $\underline{f}^T = [f_{N_1}(t) \quad f_2(t) \ldots f_{N_2}(t)]$, contains the known components, and $\tilde{\underline{u}}(t) = [\Delta u(t) \ \Delta u(t+1) \ldots \Delta u(t+N_U-1)]$. R is a $(N_2 - N_1 + 1) \times N_U$ matrix, whose elements are the impulse response parameters or Markov parameters r_i of the plant,

$$R = \begin{bmatrix} r_{N_1-1} & 0 & \cdots & \cdots & 0 \\ r_{N_1} & r_{N_1-1} & 0 & \cdots & 0 \\ r_{N_1+1} & r_{N_1} & r_{N_1-1} & 0 & \vdots \\ \vdots & \vdots & \vdots & \ddots & \vdots \\ r_{N_1+N_U-2} & r_{N_1+N_U-3} & \cdots & \cdots & r_{N_1-1} \\ \vdots & \vdots & & & \vdots \\ r_{N_2-1} & r_{N_2-2} & \cdots & \cdots & r_{N_2-N_U} \end{bmatrix}$$

It is then trivial to show that the control law is:

$$\tilde{\underline{u}}(t) = [R^T R + \rho I]^{-1} R^T (\underline{w} - \underline{f}) \tag{20}$$

However, this is a receding control law and only the first term in $\tilde{\underline{u}}(t)$ is applied, i.e. if $\underline{r}^T$ is the first row of $[R^T R + \rho I]^{-1} R^T$,

$$\Delta u(t) = \underline{r}^T (\underline{w} - \underline{f}) \tag{21}$$

A computationally attractive parameter setting is $N_U = 1$, in which case R reduces to a vector, i.e. $\underline{R}^T = [r_{N_1-1} \ldots r_{N_2-1}]$, and $\tilde{\underline{u}}$ to a scalar. Thus no matrix inversion is required for computing Δu, as

$$\Delta u = [\underline{R}^T \underline{R} + \rho]^{-1} \underline{R}^T (\underline{w} - \underline{f})$$

In [21], an efficient recursive scheme for solving the $(N_2 - N_1 + 1)$ Diophantine equations required by the predictors is proposed. It significantly reduces the computational burden of the algorithm.

The adaptive version of GPC uses an explicit identification scheme, *i.e.* the parameters in Eq. (16) are identified and used to design the controller. When the process noise is colored, i.e. $C \neq 1$, C could be identified using either recursive extended least-squares or approximate maximum- likelihood. Noting that Eq. (16) can be written as,

$$A(q^{-1})\Delta y_f(t) = B(q^{-1})\Delta u_f(t - d) + e(t) \qquad (22)$$

with $y_f = y/C(q^{-1})$ and $u_f = u/C(q^{-1})$, it is proposed in [21] to use ordinary recursive least-squares to estimate A and B in Eq. (22). However in general C is unknown and it is replaced by a fixed observer polynomial C_e, *i.e.* $y_f = y/C_e(q^{-1})$ and $u_f = u/C_e(q^{-1})$. Note that in theory C_e should also replace C in the Diophantine equation Eq. (17). In [22], Mohtadi shows that in the presence of unmodelled dynamics, it may be advantageous to use different polynomials for control and for estimation. To handle unknown dead time, an extended B-polynomial is used in Eq. (22).

3. Laguerre-Based Adaptive Control

a. **Modelling with Laguerre functions.** One way to design a robust adaptive control requiring minimal a priori information and capable of handling time-delay plants (common in process control) is to abandon the usual transfer function models and instead develop an unstructured adaptive control scheme using an orthonormal series representation. This is the approach recently developed in our laboratory [23], [24]. The set of Laguerre functions is particularly appealing because it is simple to represent and is similar to transient signals. It also closely resembles Padé approximants. The continuous Laguerre functions, a complete orthonormal set in $L_2[0, \infty)$, can be represented by the simple and convenient ladder network shown in Figure 3 and can be described in the frequency domain by:

$$F_i(s) = \sqrt{2p}\frac{(s - p)^{i-1}}{(s + p)^i} \qquad , \quad i = 1, .., N \qquad (23)$$

where i is the order of the function $(i = 1, ..N)$, $p > 0$ is the time-scale, and $\mathcal{L}_i(x)$ are the Laguerre polynomials. Based on the continuous network compensation method, the Laguerre ladder network of Fig. 3 can be expressed in a stable, observable and controllable state-space form [23] as,

$$\underline{l}(t + 1) = A\underline{l}(t) + \underline{b}u(t) \qquad (24)$$

$$y(t) = \underline{c}^T\underline{l}(t) \qquad (25)$$

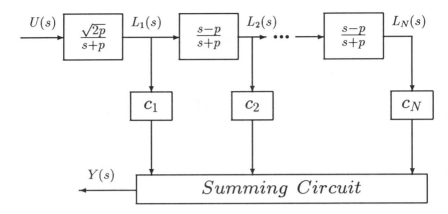

Fig. 3. Laguerre ladder network.

with $\underline{l}^T(t) = [\ l_1(t)\quad l_2(t)\quad \ldots\quad l_N(t)\ ]^T$, and $\underline{c}_0^T = [c_1\quad c_2 \ldots c_N]$. The l_i's are the outputs from each block in Fig. 3, and $u(t)$, $y(t)$ are the plant input and output respectively. A is a lower triangular $N \times N$ matrix where the same elements are found respectively across the diagonal or every subdiagonal , $\underline{b}$ is the input vector, and $\underline{c}$ is the Laguerre spectrum vector. The vector $\underline{c}$ gives the projection of the plant output onto the linear space whose basis is the orthonormal set of Laguerre functions (for details, see [22]). Some of the advantages of using the above series representation are that, (a) because of its resemblance to the Padé approximants time-delays can be very well represented as part of the plant dynamics, (b) theoretically the model order N does not affect the coefficients c_i, and (c) extensions to multivariable schemes do not require the use of interactor matrices [25].

Suppose that the output of the plant $y(t)$ can be described by,

$$y(t) = \sum_{i=1}^{N} c_i l_i(t) + w(t) = \underline{c}_0^T \underline{l}(t) + w(t) \tag{26}$$

where $w(t)$ is the process noise. If the input $u(t)$ is white noise then the parameter-vector $\underline{c}$ obtained by using the least-squares estimation technique is proven to be unbiased [26] even if the output of the plant is corrupted by coloured noise but as long as it is uncorrelated with the the outputs l_i's. If u is generated by feedback, then an external white noise process $v(t)$ independent of $w(t)$ must be injected in the system to guarantee unbiased estimates. Note that for the adaptive controller considered here, no external signal is injected in the system.

b. Predictive Control. Although a standard state feedback control technique could be used, we settled for a predictive control law. The major advantage is simplicity of use, intuitive appeal and easy handling of varying time-delay and non-minimum phase behaviour. Because the nominal model will generally be non-minimum phase, zero cancellation must be avoided, and thus predictive control is attractive. For the h-steps ahead output function we can write,

$$y(t + h) = y(t) + \underline{c}^T[\underline{l}(t + h) - \underline{l}(t)] \tag{27}$$

Under the assumption that

$$u(t) = u(t + 1) = \cdots = u(t + h - 1) \tag{28}$$

the h-step ahead predictor is:

$$y(t + h) = y(t) + \underline{k}^T\underline{l}(t) + \beta u(t) \tag{29}$$

where,

$$\underline{k}^T = \underline{c}^T(A^h - I) \tag{30}$$

$$\beta = \underline{c}^T(A^{h-1} + \cdots + I)\underline{b} \tag{31}$$

Note that β is the sum of the first h Markov parameters of the system. Thus, for a minimum-phase plant with delay d, $h > d$ and $\beta \neq 0$ are equivalent. However, for non-minimum phase systems, if one does not look sufficiently beyond the non-minimum phase behaviour, it is possible that $\beta = 0$. In practice, one has to choose h such that β is of the same sign as the process static gain, and of sufficiently large amplitude. The criterion to be satisfied is then

$$\beta\mathrm{sign}(\underline{c}^T(I - A)^{-1}\underline{b}) \geq \epsilon|\underline{c}^T(I - A)^{-1}\underline{b}| \tag{32}$$

with $\epsilon \approx 0.5$. Note that $(I - A)^{-1}\underline{b}$ can be precomputed as it depends only on the Laguerre filters. We can define a first-order reference trajectory:

$$y_r(t + h) = \alpha^h y(t) + (1 - \alpha^h)y_{sp} \tag{33}$$

Setting $y(t+h) = y_r(t+h)$, and equating the right-hand parts of the equations (29) and (33) we solve for the required control input $u(t)$:

$$u(t) = (y_r - y(t) - \underline{k}^T\underline{l}(t))/\beta \tag{34}$$

or, in velocity form:

$$\Delta u(t) = (y_r - y(t) - \underline{d}^T\Delta\underline{l}(t))\beta^{-1} \tag{35}$$

where $S = (A^{h-1} + \cdots + I)$ and $\underline{d}^T = \underline{c}^T S A$.

c. Robustness of the Design. It takes an infinite Laguerre series to exactly represent a system in L_2. In practice, we use a truncated series of relatively low dimension N. Thus, there will be some unmodelled dynamics. From the Riesz-Fisher theorem, we know that the sequence $\{c_i\}$ is a Cauchy sequence that converges to zero, [27]. The completeness property of the Laguerre set implies that for any arbitrary accuracy $\epsilon > 0$, there exists a finite N that describes a signal in L_2 within that accuracy. In practice, computers have finite word length, and typical analog-to-digital converters have 12 to 16 bit resolution. So for all practical purposes, the real system will be exactly represented by a large but finite Laguerre series of dimension M.

Let the actual plant be

$$P = P_N + H = \sum_{i=1}^{N} c_i L_i(q^{-1}) + \sum_{i=N+1}^{M} c_i L_i(q^{-1}) \tag{36}$$

where P_N is the nominal plant and H is the unmodelled dynamics and $M >> N$. Note that this implies that the unmodelled dynamics lies in L_2, i.e. H has an impulse response bounded by a decaying exponential. This is a standard assumption. With $L = H/P_N$, one can write the plant in the standard multiplicative uncertainty form

$$P = (1 + L)P_N \tag{37}$$

From equation (34), we can write the nominal closed-loop transfer function as

$$G_{CN} = \frac{G_0}{1 + G_0} \tag{38}$$

where

$$G_0(q^{-1}) = \frac{\sum_{i=1}^{N} c_i L_i(q^{-1})}{\beta + \sum_{i=1}^{N} k_i L_i(q^{-1})} \tag{39}$$

We know [15] that a sufficient condition for stability of the perturbed closed-loop system is,

$$|L(q^{-1})| \left| \frac{G_0(q^{-1})}{1 + G_0(q^{-1})} \right| < 1 \tag{40}$$

on the unit circle. With the above definitions, this sufficient condition becomes

$$\left| \sum_{i=N+1}^{M} c_i L_i(q^{-1}) \right| \left| \frac{1}{\beta + \sum_{i=1}^{N} (c_i + k_i) L_i(q^{-1})} \right| < 1 \tag{41}$$

on the unit circle. This means that the effect of the unmodelled dynamics must be negligible in the bandwidth of the nominal closed-loop plant. The second term in the inequality is affected both by N and h. Decreasing h reduces β and thus increases the sensivity to unmodelled dynamics. Because $\{c_i\}$ converges to zero, the first term of the inequality can be made arbitrarily small by increasing N. The convergence rate of the $\{c_i\}$ sequence is also affected by p. Thus p also influences the robustness of the design. Note that $c_i + k_i$ converges to zero as d goes to infinity. Hence, if for a given N the inequality

$$| \sum_{i=N+1}^{M} c_i L_i(q^{-1})| \left| \frac{1}{P_N(1)} \right| < 1 \tag{42}$$

cannot be satisfied on the unit circle, then no finite h will satisfy condition (41), meaning that N has to be increased.

d. Adaptive Control. An explicit adaptive control scheme based on the above formulation uses the recursive least-squares (RLS) method to identify the parameter vector c. The control law (34) is then computed at every sampling instant. Theorems proving the global convergence and stability as well as a preliminary robustness analysis of this scheme are presented in [24]. The choice of the parameter p in the Laguerre functions is not crucial, however it influences the accuracy of the approximation of the plant dynamics as a truncated series. For a given plant, there exists an optimal p that minimizes the number of filters required to achieve a given accuracy. Noting that

$$\exp(-s\tau) = \lim_{N \to \infty} \left(\frac{(1 - s\tau/2N)^N}{(1 + s\tau/2N)^N} \right) \tag{43}$$

we see that the chain of all-pass filters in the Laguerre network provides good representation of the time delay, in particular when $p = 2N/\tau$. A dead time τ introduces a phase lag $\omega\tau$. For accurate representation of a dead time τ sampled at interval T, one must describe a phase lag upto $\pi\tau/T$. Knowing that each pair of all-pass filters can contribute a maximum phase lag of 2π, this gives an estimate of how many filters are needed. Finally, because the plant dynamics are also represented by the Laguerre nerwork, p should be chosen around the crossover frequency of the plant. For closed-loop stability one only needs a good approximation of the delay over the bandwidth of the closed-loop plant. Thus in practice one could do with only a couple of all-pass filters for a delayed well damped system. Because pure oscillators are not in L_2, the presence of poorly damped modes will require more filters. Note that the actual plant order has little bearing on the number of filters N. These guidelines are used to choose N and p, however both can be changed on-line. The horizon h is automatically adjusted on-line to satisfy the criterion (32) with $\hat{c}$. Note that for a stable plant, increasing h will increase the robustness as it will decrease the closed-loop bandwidth.

C. APPLICATIONS

1. Kamyr Digester Chip Level

A Kamyr digester is a vertical plug flow continuous reactor in which wood chips are cooked in the presence of a sodium hydroxide solution, see Fig. 4. The chips and the cooking liquor are fed to the top, and the pulp is extracted from the bottom. The retention time is a function of the plug velocity and of the column height. The velocity is largely determined by the production rate, thus at constant production rate, a constant chip level is equivalent to a constant residence time. A constant residence time is, with a constant temperature important to maintain the desired degree of delignification. Thus, chip level control is important. Although, it would probably be better controlled from the top, chip level control is often achieved by manipulating the flow of pulp out of the bottom of the digester (known as blow flow). Unmeasured changes in chip size and density, wood species, cooking conditions affect the movement and elasticity of the chip column, and hence the process dynamics. this is the main justification for applying adaptive control to this loop. In [28], a self-tuning controller based on the Clarke-Gawthrop algorithm was successfully implemented on an industrial digester. More recently, an adaptive GPC scheme was commissioned on another industrial Kamyr digester, see [29] for more details.

Preliminary identification experiments indicated that the process could reasonably be represented by,

$$(1 - a_1 q^{-1})\Delta y(t) = b_0 \Delta u(t-2) + (1 - c_1 q^{-1} - c_2 q^{-2})e(t) \qquad (44)$$

However, because from experience we know that this process is subject to dead- time changes, the B-polynomial was extended:

$$(1 - a_1 q^{-1})\Delta y(t) = (b_0 + b_1 q^{-1} + b_2 q^{-2})\Delta u(t-2) + (1 - c_1 q^{-1} - c_2 q^{-2})e(t) \quad (45)$$

In order to reduce the number of estimated parameters, the parameters of $A(q^{-1})$ and $C(q^{-1})$ were fixed, thus enabling the adaptive controller to track only dead-time changes which are thought to be the main problem with this process. Following the suggestion in [21], the process input and output were both filtered through $1/C_e$. The parameters are then estimated via RLS using the following model:

$$(1 - 0.98 q^{-1})\Delta y_f(t) = (b_0 + b_1 q^{-1} + b_2 q^{-2})\Delta u_f(t-2) + e(t) \qquad (46)$$

with $C_e(q^{-1}) = (1 - 1.48 q^{-1} + 0.49 q^{-2})$. The filter Δ/C_e has a band-pass characteristic, thus filtering out both low frequency disturbances and high-frequency noise and unmodelled dynamics. For this application, it was found

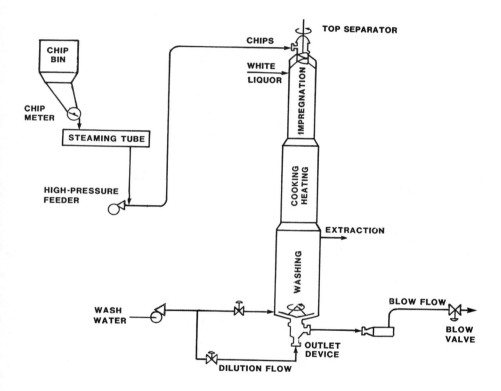

Fig. 4. Simplified Diagram of a Kamyr Digester.

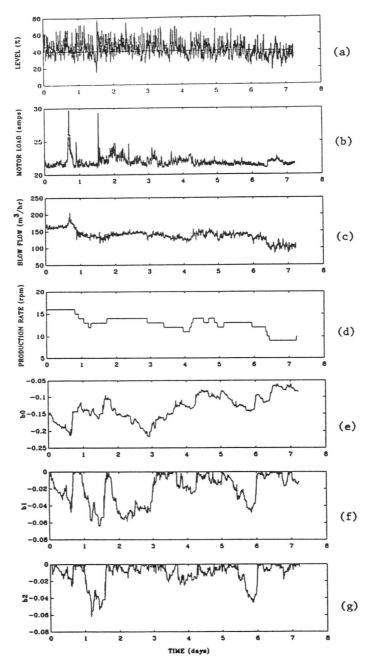

Fig. 5. Adaptive control of Kamyr digester chip level using GPC [29].

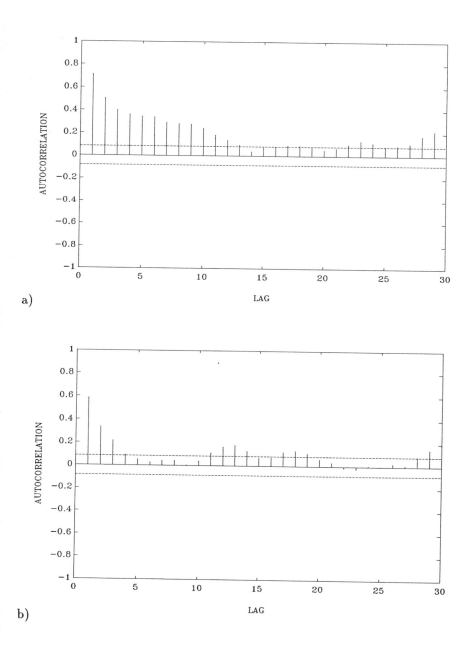

Fig. 6. Chip level autocorrelation under a)conventional control, b)GPC [29].

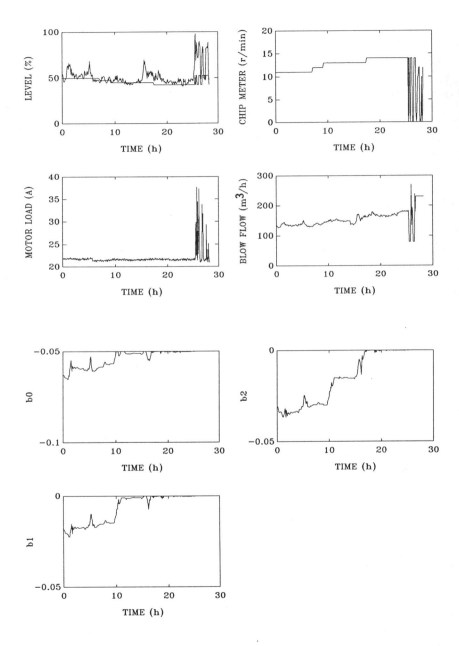

Fig. 7. Advance notice of hangup via parameter estimates [29].

that the exponential forgetting and resetting algorithm (EFRA) of [30] version of the estimator did much to improve robustness. After a simulation study, and some preliminary industrial experience, the following parameters were chosen for GPC, $N_1 = 1$, $N_2 = 15$, $N_U = 1$ and $\rho = 0$. Figure 5 shows a one-week period of GPC control under a highly variable production rate. Figure 6 shows the autocorrelation functions of the chip level under the previously used control scheme and GPC. The improvement is obvious. An aditional benefit of using adaptive control on this loop is that it may provide advance warning of a serious operational problem, chip column hangup. Hangups occur when a protion of the chip column separates and plugs the digester. If not detected at an early stage, recovery measures are costly and very disruptive. However, from first principles we can predict that a hangup will result in a zero process gain. Because, the adaptive controller estimates this gain, it has the potential of providing such a warning. Figure 7 shows a period of operation leading to a hangup and to shutdown at $t = 25$ hours. It is seen that the b-parameters start drifting toward zero at $T = 10$, and all three are zero (except b_0 which is purposely constrained at -0.05) by $t = 15$, $i.e.$ 10 hours before the problem became noticeable to the operator, by which time it was too late to take corrective measures other than shutdown.

2. Bleach Plant pH Control

The main objectives of bleaching are, to remove the colouring compounds still present in the fibers, to increase the brightness of the pulp, and to produce a white pulp of satisfactory physical and chemical properties for the manufacture of printing or tissue papers. Modern kraft pulp bleaching is achieved in a multistage plant, using expensive chemicals such as chlorine, chlorine dioxide, caustic and oxygen. A typical North American bleach plant consists of chlorination, alkaline extraction, chlorine dioxide bleaching, alkaline extraction and chlorine dioxide bleaching in that sequence. When unbleached woodpulp arrives at the bleach plant, it still contains a significant amount of lignin and other chromophores. To remove as much of the residual lignin as feasible without damaging the pulp, it is first chlorinated. The objective of the caustic extraction stage that follows is to remove the alkali-soluble portion of the lignin from the wood pulp. Three steps are generally required following the washing of the chlorinated pulp: (a) mixing of caustic solution (sodium hydroxide) with the pulp , (b) heating to the desired temperature, and (c) retention to complete the reaction. The reaction time varies from 50 to 100 minutes depending on the grade. One of the most important variables affecting the pulp quality is the pH at the end of that stage, i.e. at the exit of the first caustic extraction tower. The pH of the feed stream is around 2, while the target pH after the tower is usually between 10 and 10.5. Low pH degrades the pulp quality and

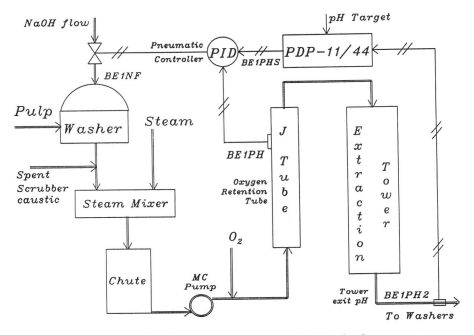

Fig. 8. Alkaline extraction stage of a bleach plant.

increases chemical consumption in further bleaching stages. High pH proves to
be of no substantial benefit while the additional caustic consumption is costly.
Thus, it is desirable to minimize caustic consumption while maintaining pulp
quality. Lab tests indicate that the titration curve of the reaction displays
the characteristics of a strong acid, strong base reaction. The control problem
is complicated by the fact that the buffering effect provided by the aqueous
system may vary and by the time-varying nature of the dynamics. In particular, because of the propensity of the flowdown tower for channelling, the dead
time in the process can be highly variable. Fig. 8 shows the process and instrumentation diagram of the first alkaline extraction stage of a mill's bleaching
process. The industrial control scheme uses 2 PI controllers in cascade. This
cascade configuration is justified by the long retention time in the tower. The
tuning of the internal pneumatic PI controller is relatively easy since it does
not involve any long dead time. The tuning of the outer digital PI presents
serious problems due to the long retention time involved in the first caustic
tower (40 to 70 minutes) and the inherent strong non-linearity of the loop. Its
tuning constants are set on a gain scheduling basis depending on the value of
the exit pH. The nominal dead time is about 40 minutes but is known to vary
from 25 min. to 40 min. and the settling time 50 to 70 minutes. Because of
this, the outer loop is in manual mode more often than not. Fig. 9 shows the

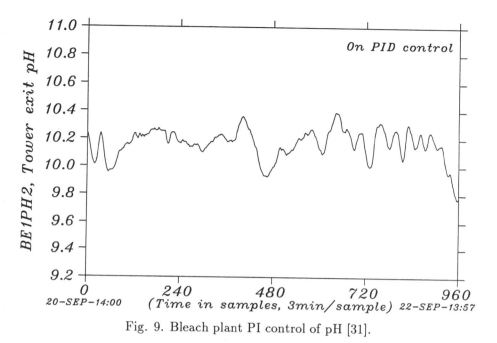

Fig. 9. Bleach plant PI control of pH [31].

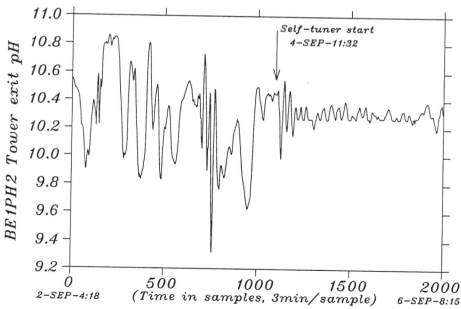

Fig. 10. Bleach plant Laguerre-based control of pH [31].

tower exit pH under PI computer control, when the latter is performing at its best. Note the existence of an offset, as the setpoint is 10.2.

For the implementation of the adaptive controller, the outer PI was removed and the new self-tuner was applied in its place, see [31] for more details. The inner PI loop was left intact since it is an integral part of the control valve that manipulates the caustic flow and it never presented any problems. A process step response from the system's database was used to compute the initial parameter estimate vector $\hat{c}(0)$. The sampling interval for the PID scheme was 8 min., and was kept the same for the adaptive controller. The time constant of the plant dictates a choice of p between 0.05 and 0.5. The choice of the number of filters depends primarily upon the dead time of the process, which can be as long as 50-60 min. Given the sampling interval of 8 min., it means that we must describe phase shifts up to 1350 deg. To accurately describe such a phase shift upto the Nyquist frequency, at least 8 all-pass filters are required, i.e. at least 9 Laguerre functions. To be on the safe side, we chose $N = 15$. Now, from equation (43) to represent dead time of 60 min, the Laguerre gain distribution can be centered around the 7th all-pass filter, by choosing $p = 0.25$. The following parameters were thus chosen: sampling time $T = 8$ min , number of filters $N = 15$, Laguerre pole $p = 0.25$. Because this choice should provide good approximation of the delay over the whole freqency range, the control scheme should be very robust. For control, the prediction horizon is $h = 9$, and the driver block pole $\alpha = 0.5$.

Figure 10 compares, on the same graph, the performance of the new self-tuning scheme to the previously used one. The target exit pH (BE1PH2) was 10.3. The graph shows a total of 2000 points, each point representing 3 min, from September 2, 04:18 a.m, to September 6, 08:15 a.m. The new self-tuner in its final form was applied on September 4, 11:32 a.m. (1106th point on the graph). The initial transient period of the Laguerre self-tuner, clearly shown on the graph, indicates the adaptation period to the dynamics of the plant. The period that follows is indicative of the good regulation performance of the self-tuner. The short time period between the 684th and 792th point on the graph (i.e. September 3, 14:30 - 16:18) is an earlier experimental test of the new self-tuning algorithm based on zero initial parameter estimates.

III. CONTROL OF WEB PROCESSES

A. THE PAPER MACHINE

The paper machine is by far the most automated of all pulp and paper processes. Sophisticated sensors are available to measure various properties of the sheet, ranging from simple basis weight and moisture to physical properties such as strength. Although most functions on paper machines are still controlled using very simple control algorithms, many of those control problems are not trivial. Indeed, control of web processes is a class of process control problem on its own.

The paper machine is a very complex system. Fig. 11 outlines some features of a particular type of paper machine. The purpose of the headbox is to distribute the proper amount of fibre suspension on the paper machine wire at the right speed and in a uniform way. Most modern headboxes are pressurized, meaning that the air pad between the surface of the stock and the top of the headbox is kept under pressure. The total head, *i.e.* air pad pressure plus level, at the slice determines the velocity of the outlet jet. An extensive range of online sensors installed at the dry end of the paper machine is available to measure critical properties such as basis weight, moisture, thickness, etc. The paper machine is by far the best instrumented process in the industry.

B. PAPER MACHINE MODELS

Because of the complexity of the various underlying processes, it has so far proved impossible to derive first-principle dynamic models for paper machine. Instead, based on [32], we shall consider transfer function descriptions of both the process and the stochastic disturbances that affect it. Those models will be linear whenever possible and valid for small deviations around fixed operating conditions.

1. Dry Weight

The dry-weight model describes the temporal (MD) and spatial (CD) variations of the fiber distribution on the sheet. The MD actuator is the stock flow, the CD actuators are the slice screw actuators bending the slice lip, thus affecting the distribution of fibres across the wire. The dynamics of those actuators is generally neglected in view of the other dynamics present. In a linear model, the effects of MD and CD actuators is considered to be additive. A typical model is, [32]:

$$y_i(t) = \frac{bq^{-1}}{1 - aq^{-1}} v_1(t - k) + \underline{K}_i^T \underline{u}(t - d) + w_i(t) \qquad (47)$$

where

$y_i(t)$ is the dry weight at CD position i and time t
v_1 is the stock flow
u contains the CD actuator positions
$\underline{K}_i$ response at CD position i to CD actuator moves
k delay from stock valve to reel
d delay from CD actuators to reel, $d < k$
w_i is the disturbance at CD position i

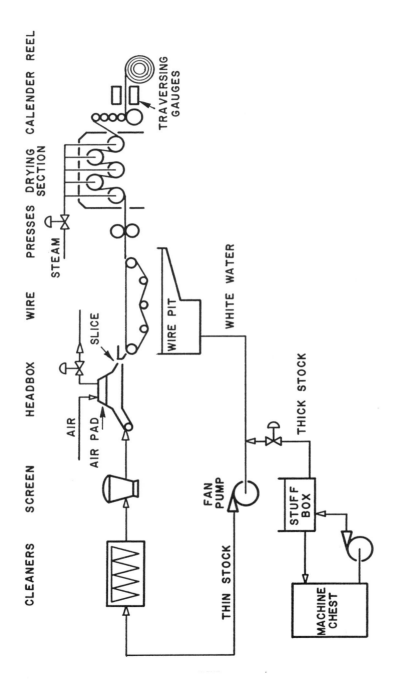

Fig. 11. Simplified Diagram of a Paper Machine.

A model describing the sheet in its entire width at once can be written as:

$$\underline{y}(t) = G v_1(t - k) + K \underline{u}(t - d) + \underline{w}(t) \tag{48}$$

$\underline{y}(t)$ contains the dry weight at all CD positions and time t
K is the interaction matrix whose rows are $\underline{K}_i$
G is a vector whose elements are $bq^{-1}/(1 - aq^{-1})$
$\underline{w}$ is the disturbance vector

The interaction matrix K reflects the spatial response of the paper machine. The underlying phenomena are quite complex. A first factor is the flexibility of the slice lip. Attempts to model the lip as a loaded beam have given some insight into the effect of various slice lip design factors on the spatial spread of actuator moves. As a general rule, the more rigid the slice lip is, the larger the influence zone of a given actuator is. The second factor is the possibility of crossflow across the wire. Indeed it has been observed that at points of low slice lip opening, the flow out of the headbox is laterally divergent. However, the flow observed are so too complex to be modeled. As a general rule, twin-wire former are less susceptible to crossflow than fourdrinier machines. Also, lighter grades such as newsprint exhibit less crossflow than heavier grades such as linerboard. The common approach considers the spread across the web in response to an actuator move is finite, and limited to a few CD positions on each side of the actuator (typically 3 for newsprint, 6 for sack paper and 10 for linerboard), identical for all actuators, and symmetric. The response can therefore be represented by a small number of coefficients and the K matrix is a band diagonal matrix such as:

$$
\begin{bmatrix}
a & b & c & 0 & \cdots & 0 & 0 & 0 & 0 \\
b & a & b & c & 0 & \cdots & 0 & 0 & 0 \\
c & b & a & b & c & 0 & \cdots & 0 & 0 \\
0 & c & b & a & b & c & 0 & \cdots & 0 \\
\vdots & & & \ddots & \ddots & & & & \vdots \\
0 & 0 & \cdots & 0 & b & a & b & c & 0 \\
0 & 0 & \cdots & 0 & c & b & a & b & c \\
0 & 0 & 0 & \cdots & 0 & c & b & a & b \\
0 & 0 & 0 & 0 & \cdots & 0 & c & b & a
\end{bmatrix}
$$

The description of the disturbance vector $\underline{w}(t)$ requires some insight into the source of those disturbances, [33]. By definition, the MD disturbance occurs simultaneously across the web. MD disturbances generally originate upstream of the headbox and are due to poor mixing in tanks, poor consistency regulation and pressure pulsations in the approach system. Some periodic MD variations may also be caused by faults in the wire. The CD component of the disturbance is primarily due to hydrodynamics effects within the headbox and its distribution system causing localised flow and consistency variations across the slice. These variations tend to be locally cross-correlated. Although various models are possible, the following ARIMA process has been proposed [32]:

$$D(q^{-1})\Delta \underline{w}(t) = \underline{e}(t) \tag{49}$$

where $\underline{e}(t)$ is a multivariate white noise sequence with covariance:

$$R = \begin{bmatrix} 1 & r & r^2 & \cdots & \cdots & \cdots & r^n \\ r & 1 & r & \cdots & \cdots & \cdots & r^{n-1} \\ r^2 & r & 1 & r & \cdots & \cdots & r^{n-2} \\ \vdots & \vdots & \vdots & \ddots & \ddots & & \vdots \\ r^{n-2} & \cdots & \cdots & \cdots & 1 & r & r^2 \\ r^{n-1} & \cdots & \cdots & \cdots & r & 1 & r \\ r^n & \cdots & \cdots & \cdots & r^2 & r & 1 \end{bmatrix} \times \sigma^2 \tag{50}$$

where r is the correlation coefficient between two adjacent points. The term Δ in the left-hand side of Eq. (49) implies that the MD or temporal disturbance is nonstationary. This is a common assumption in process control to force integral action in the controller. The polynomial D is generally of low order, typically 1 or 2.

2. Moisture

The moisture model is slightly more complex for two reasons. Although the MD actuator for moisture is the steam pressure, the stock flow is also affecting the moisture, as the drainage rate and the drying rate are decreasing functions of the dry weight. Also, because of the nonlinearity of the drying process during falling rate drying, a realistic model should be nonlinear. Indeed, it has been shown in [34] that the CD variability increases with the average moisture. By the same token, the process gain decreases as the average moisture decreases. A further feature of the moisture model is that it must account for the interaction from the stock flow. Indeed, basis-weight variations affect teh moisture content. Because of this, the model model is more complicated than the basis-weight model.

C. PROFILE ESTIMATION

While the gauges traverse the web, the sheet moves in the machine direction at very high speed. For instance, on a 4m wide newsprint sheet moving at

900 m/min, during a 20s scanning time, 300 m of paper has passed under the gauge. Thus, the path of the gauge on the sheet is at a very shallow angle (0.76 deg) relative to the machine direction. Therefore, the raw profile measurement contains a vary significant machine direction component. The problem is then: is it possible to retrieve the true profile from this raw signal?

On a machine with a steady profile and nearly white machine direction variations, the retrieval of the profile is a rather trivial task and the standard method which follows works well. Let $x_i(t)$ denote the raw measurement obtained at the measurement box i during the scan s. The standard method is simply an exponential filter

$$y_i(s) = (1 - a)y_i(s - 1) + ax_i(s) \tag{51}$$

where $y_i(s)$ is the estimated profile for the measurement box i and the scan s, and $0 < a < 1$ is the exponential filter pole. The exponential filter pole a is generally set such that $a \leq 0.3$, which means that after a step change in the profile it takes about 8 scans to get within 5% of the new profile. The simple filter described above is crude and rather slow to converge to the true profile. Moreover, it is not optimal. Because of the inherent process nonlinearity, for moisture profiles it will give biased estimates. Efforts to overcome this problem have resulted in a significant amount of patented work, all of which belongs to the vendors of paper machine control system.

More recently in our laboratory, Natarajan et al. [35] proposed a scheme consisting of a least-squares parameter identifier for estimating CD profile deviations and a Kalman filter for estimating MD profiles. With a sampling interval equal to the travel time between measurement boxes, the Lindeborg model for moisture deviations can be written as, [34]:

$$y_i(t) = p_i + (1 + Bp_i)u(t) + v(t) \tag{52}$$

$$u(t) = \bar{u} + \xi(t) \tag{53}$$

where

$y_i(t)$ is the measured profile deviation from the reference level at CD position i and time instant t

p_i is the percentage deviation from the reference level at the CD position i.

B is a constant

$u(t)$ is the percentage MD variation at time t

$v(t)$ is sensor noise assumed to be Gaussian white noise

$\bar{u}$ is the mean moisture content in the MD

ξ is a zero mean stochastic process described by
$\xi(t + 1) = a\xi(t) + w(t)$ where a is known and w is a Gaussian, zero-mean white noise with variance q

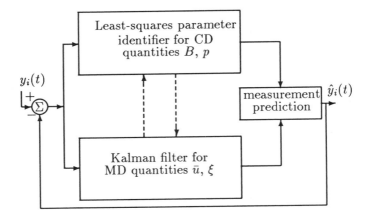

Fig. 12. Structure of the profile estimation algorithm [35].

This model can be expressed in a state-space form given by:

$$\underline{x}(t+1) \;=\; A\underline{x}(t) + \underline{W}(t) \tag{54}$$
$$y_i(t) \;=\; p_i + \underline{C}_i^T \underline{x}(t) + v(t) \tag{55}$$

where:

$$\underline{x}(t) = \left[\begin{array}{c} \bar{u} \\ \xi(t) \end{array} \right]; \qquad A = \left[\begin{array}{cc} 1 & 0 \\ 0 & a \end{array} \right];$$

$$\underline{W}(t) = \left[\begin{array}{c} 0 \\ w(t) \end{array} \right];$$

$$\underline{C}_i^T = \left[\begin{array}{cc} (1 + Bp_i) & (1 + Bp_i) \end{array} \right] \tag{56}$$

If p_i and B are known, by using Equations (54) and (55) the estimation of $\bar{u}$ and $\xi(t)$ can be approached as a Kalman filtering problem. Conversely, if $\bar{u}$ and $\xi(t)$ are known, then the estimation of p_i and B can be attempted as a least squares parameter identification problem. The proposed algorithm is a bootstrap algorithm combining these two ideas. Using the present estimates of p_i and B in a Kalman filter, we predict $\bar{u}$ and $\xi(t+1)$ (at next instant), then using this prediction and the measurement $y_{i\pm1}(t+1)$, we update $p_{i\pm1}$, B and so on. The overall structure of the algorithm is shown in Fig. 12. For every measurement box, the least-squares algorithm provides an estimate of the CD profile deviation and the Kalman filter provides an estimate of the MD variation. Simulation and industrial test results show the algorithm to be more accurate and faster to converge to the true profile than other techniques,

[35]. It can be tuned to provide optimal (in the least-squares sense) estimates. Any shortening of the paper machine, for example by successful use of impulse drying, will shorten dead time, making MD control at the current rate of once a scan unattractive. Because it gives several estimates of the MD variations per scan, this filter will allow MD control at a much increased rate. Obviously, the use of full CD non-scanning gauges would render such a filtering algorithm redundant.

C. CONTROL

1. Combined MD CD Control

It is possible to rewrite the model Eq. (48 in a state-space form. Because the sensor is scanning, only one component $y_i(t)$ of $\underline{y}(t)$ is available at any one time. Thus the measurement vector changes at every sampling time with a period equal to twice the scanning time. Thus, if we design an LQG control scheme for this system, the steady-state Kalman and controller gains will each converge to a sequence of $2 \times n$ gains where n is the number of samples per scan, [32]. These will repeat with a period equal to twice the scanning time. Because in practice the numbers of CD actuators and of samples per scan are very large and of the resulting high dimensionality, this method is never used. Instead, all practical schemes rely on the assumption that MD variations are well separated from CD ones on the time scale. More precisely, it is assumed that CD variations are very slow compared to MD variations. This is generally the case on a well designed paper machine. Then, the MD and CD control problems can be treated independently.

2. MD Control

Machine direction control considers the control of the scan averages of basis-weight and moisture. Control of dry weight manipulates the thick stock flow setpoint to obtain the desired fiber flow to the headbox. Usually, the thick stock consistency is controlled and feedforward to the thick stock flow setpoint is provided as well. Moisture is generally controlled from the steam pressure in the last dryer section. As seen earlier, the dynamics of the basis-weight and moisture control loops is heavily dominated by dead time. Also, on a newsprint machine, the setpoints for moisture and in particular basis-weight are rarely changed. Thus, for these loops one should really solve the regulation problem, which requires characterization of the process disturbance as a stochastic process.

However, in practice most systems installed in mills are based on deterministic control design and often tuned for the servo problem, *i.e.* setpoint tracking. The Dahlin algorithm is used extensively on paper machines. It is a simple and straightforward way to perform dead-time compensation. Depending on the process disturbance characteristics, it may in some instances be tuned to near minimum-variance performance, even though it does not use a model of the process noise. To solve the desired control problem for the basis-weight and moisture control loops requires minimum-variance control. Indeed, these crucial qualities of the final product must meet stringent specifications, hence it is important to obtain good control. In addition, any reduction in the moisture variations allows an increase in the moisture content, thus saving steam or permitting a production increase.

The first application of minimum-variance control to a paper machine was reported in 1967 by Åström [3]. In practice, true minimum variance is rarely desirable because of the generally excessive control energy it requires. This is particularly true on a paper machine where actuators are valves that should not be subjected to excessive wear. Another problem with minimum-variance control is that it is impossible to tune manually. It relies on model of the plant, which because of furnish or grade changes, or other variations of the equipment or the environment, may need to be updated from time to time.

One way to automate this task is through adaptive control. Self-tuning control for the paper machine was proposed very early. Indeed two of the first industrial tests of self-tuning controllers were performed on a paper machine [4], [5]. In both cases MD control of basis weight and moisture was considered. For moisture control, even the gains of the feedforward loop from the couch vacuum for decoupling control were adapted, using a controller of the form:

$$\Delta u(t) = b_1 \Delta u(t-1) + b_2 \Delta u(t-2) + a_0 y(t) + a_1 y(t-1) + c_0 \Delta v(t) + c_1 \Delta v(t-1)$$
$$(57)$$

where Δu is the incremental control signal, y is the error in moisture content, and Δv is the incremental couch vacuum.

Attempts by a vendor to develop a system based on the above work did not lead to a commercial product for reasons not disclosed [36]. In Canada, a pulp and paper company attempted to apply the above methodology to a linerboard machine but failed to develop a scheme reliable enough for continous operation. The intended scheme used feedforward from several variables, resulting in the estimation of a large number of parameters and identifiability problems, [37]. This emphasizes the fact that adaptive control is not yet an *off-the-shelf* technology. Nevertheless, another pulp and paper company in Canada claims to have successfully developed and operated a self-tuning regulator for MD control of a fine paper machine for several years [38].

3. CD Control

The most impressive progress in paper machine control over the past 10 years or so has been in the area of CD control. There are basically two approaches for CD control design. The first uses an interaction matrix, the second a non-causal spatial impulse response [39].

The first approach is based on the model previously derived. Then, given an initial deviation from target profile $\nabla \underline{y}(t-1)$, where each element corresponds to a particular measurement box, and $\Delta \underline{u}(t)$ is a vector containing the actuators moves, the resulting deviation from target profile $\nabla \underline{y}(t)$ is given by

$$\nabla \underline{y}(t) = K \Delta \underline{u}(t-d) + \nabla \underline{y}(t-1) \tag{58}$$

where K is the previously defined band diagonal matrix. Ignoring noise, Eq. (58) can written in a predictor form:

$$\nabla \underline{y}(t+d) = K \Delta \underline{u}(t) + K \sum_{j=1}^{d-1} \Delta \underline{u}(t-j) + \nabla \underline{y}(t) \tag{59}$$

If the interaction matrix K is square, and if at time $t + d$, zero-deviation from the target profile is desired, then from Eq. (59) we obtain the following deadbeat controller:

$$\Delta \underline{u}(t) = -K^{-1} \nabla \underline{y}(t) - K \sum_{j=1}^{d-1} \Delta \underline{u}(t-j) \tag{60}$$

This equation is the equivalent to a deadbeat controller. Because in practice, CD is performed at a very slow rate, $d = 1$, Eq. (60 becomes:

$$\Delta \underline{u}(t) = -K^{-1} \nabla \underline{y}(t) \tag{61}$$

As seen on Figures 13 and 14, even for a simple K matrix with a fairly narrow interaction band, the K^{-1} matrix is a full and rather complex matrix. This also raises the question of the conditioning of K and its invertibility. This is crucial, as our knowledge of K is generally imperfect, and its identification is not straightforward. In practice, as the number of measurement boxes is generally larger than the number of actuators, the matrix K is non-square and thus non-invertible. Moreover, large control actions are also undesirable as they put the integrity of the slice lip at risk. Thus a better problem is to find $\Delta \underline{u}$ that minimizes J where

$$J = \nabla \underline{y}^T Q \nabla \underline{y} + \Delta \underline{u}^T P \Delta \underline{u} \tag{62}$$

and P and Q are symmetric, positive definite matrices. Differentiating with respect to $\Delta \underline{u}$ and equating to zero gives the control law:

$$\Delta \underline{u}(t) = -(K^T Q K + P)^{-1} K^T Q \nabla \underline{y}(t) \tag{63}$$

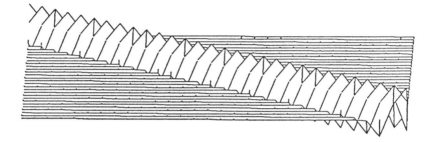

Fig. 13. 3-D plot of the band diagonal interaction matrix.

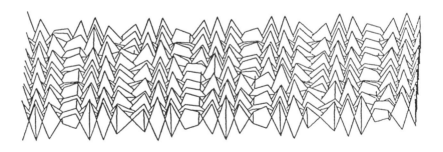

Fig. 14. 3-D plot of the inverse of the band diagonal interaction matrix.

Note that, when $P = 0$ and $Q = I$, one obtains

$$\Delta \underline{u} = -(K^T K)^{-1} K^T \nabla \underline{y}(t)$$

Obviously, to implement this control law, one needs to know the matrix K. One approach is to use a least-squares estimator to update it on-line, *i.e* having a self-tuning scheme able to track changes in the profile response, see [41]. This is one of the most successful commercial applications of adaptive control. The fact that it is static and only three parameters are estimated may have a lot to do with that success.

The other approach to modelling is to view the spread across the web as a spatial impulse response, which differs from the familiar temporal impulse response by the fact that on the negative x-axis it is non-zero, and thus is non-causal. Assuming symmetry of the response, such a system can be approximated by:

$$\sum_{i=-n}^{n} a_i \Delta y_i = \sum_{j=-m}^{m} b_i \Delta u_j \tag{64}$$

where Δy_i is the profile deviation at CD location i, and Δu_j is the movement of actuator j. One can then design a linear quadratic controller that minimizes the following performance index:

$$J = \sum_{i=0}^{n} [(\Delta y(ih))^2 + \rho \left. \frac{d^2 \Delta u(x)}{dx^2} \right|_{x=ih}] \tag{65}$$

where h is the actuator spacing. The second term in this performance index, the second derivative of the slice lip deflection curve is proportional to the bending moment, and is there to protect the slice lip from permanent damage. For computation, this term is approximated by finite differences, see [39].

V. CHIP REFINER CONTROL

A. The Problem

A wood-chip refiner consists of two rotating grooved plates, with pressure exerted on one of them by a hydraulic cylinder, see Fig. 15. Wood chips and dilution water are fed near the axis and forced to move outward between the plates by centrifugal and frictional forces. Steam produced by evaporation and chips broken down into fibres by mechanical action create a few hundred micrometer thick pad between the plates. Steam and pulp are discharged at the periphery. The specific energy, or energy per mass unit of wood fibres, is a major factor controlling the pulp quality, thus both wood feedrate and motor load need to be controlled. To control the motor load, the plate gap is adjusted

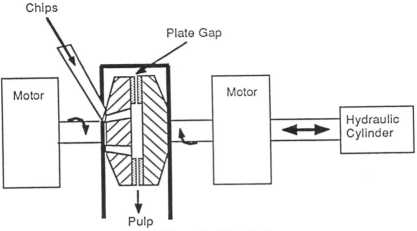

Fig. 15. Chip Refiner.

by manipulating the hydraulic pressure. Although plate gap sensors are now available, the common practice is to simply measure the shaft displacement to indicate the plate movement. However, because of plate wear and thermal expansion, this does not provide an absolute measurement of the gap. The hydraulic pressure is generally manipulated by sending a pulse to a microdial commanding a servo-valve directing oil to the cylinder.

The process dynamics are both non-stationary and non-linear. The non-stationarity is due to the variability in the feed characteristics, such as wood species, chip size and density as well as to plate wear. As the plates wear, the noise level increases and the gain of the process drops dramatically. Wear generally occurs gradually over a period of several hundred hours, but can also occur quickly in case of plate clash, i.e. metal to metal contact between the two plates. It is highly undesirable as it disrupts production and damages the plates. At any given time, the static relationship between motor load and plate gap looks as shown on Fig. 16. The refiner must be operated in the region where closing the gap increases the load. As the gap decreases, a maximum load is reached. If the gap is further reduced, then the load sharply drops. At this point, the pad collapses and can no longer sustain the pressure. This also corresponds to the point where the refiner starts cutting fibers. Then, for the pad to rebuild and the gain to become negative again the gap has to be open past the point where the collapse occured. This corresponds to the hysteresis pattern of Figure 16. Note that as explained above, this curve is non-stationary and thus the collapse point is not predictable. The chip refiner dynamics are essentially due to the hydraulic system and can be represented

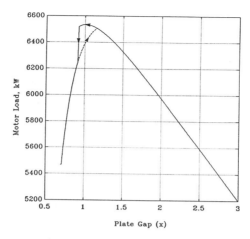

Fig. 16. Chip refiner motor load versus plate gap [42].

by a discrete linear system with an output nonlinearity.

$$\Delta y(t) = a\Delta y(t-1) + b(x)\Delta u(t-d) \tag{66}$$

$$b(x) = k(1-a)f(x) \tag{67}$$

where Δu the duration of the pulse sent to hydraulic cylinder, x the plate gap, a the process pole, and d the dead time. Typical values of the dead time and the time constant are respectively 4s and 7s. With exact knowledge of the gain k, the plate gap x and and the nonlinearity $f(x)$, the control of the above system is straightforward. Unfortunately, as explained before, not only is the exact nonlinearity unknown, but it is also time-varying. Adaptive control provides a way of controlling such a system. The simplest approach is to consider b in Eq. (66) as a time-varying parameter and to estimate it on-line. This is the approach taken in [41] where an indirect adaptive controller, using recursive least-squares with variable forgetting factor is used to estimate b. When $\hat{b} < 0.05$, a Dahlin algorithm is used to control the load, otherwise the plates are open at maximal speed until $\hat{b} < 0$. Although during a 6-month trial that scheme proved capable of tracking both the slow gain drift and the change in sign, continuous and unsupervised operation requires further refinements. We shall now discuss a more recent approach that makes use of qualitative knowledge about the process non-linearity to give a nonlinear adaptive control scheme with interesting features, [42].

C. Nonlinear Adaptive Control

In order to simplify the discussions, we shall consider the delay-free case.

$$\Delta y(t) = a\Delta y(t-1) + b\Delta u(t-1) + \sigma e(t) \tag{68}$$

for which the minimum-variance controller has the form

$$\Delta u(t) = [-(1+a)y(t) + ay(t-1)]/b \tag{69}$$

For the design of a nonlinear adaptive controller, the choice of the performance index is crucial as it must reflect several important aspects of the control problem. The previous control laws developed for this problem are all linear in Δu, and because of that, require addition of logic for handling of pad collapses. Although the exact nonlinearity is not known, the a-priori knowledge includes the general shape of it. It is thus natural to use a performance index that yields a nonlinear control law. An important feature of the control criterion is that control in the positive gain zone must be severely penalized if not prohibited. When the gain estimate is negative and of relatively large amplitude, closing the gap should obviously be allowed, however as the estimate approaches zero, negative control actions should be increasingly penalized. Furthermore, when the estimate is close to zero and the uncertainty is large, probing should not tend to further close the plates. Finally, around steady-state behaviour, true minimum-variance control is not desirable, as the resulting control moves will probably be of large amplitude, thus increasing the risk of entering the undesirable zone. Given all these considerations, a criterion of the form below is chosen

$$
\begin{aligned}
J_T &= E\{ \sum_{k=t+1}^{T} [(y(k) - y_r)^2 + q\Delta u^2(k-1) \\
&\quad + h(b(k))g(\Delta u(k-1))] \}
\end{aligned}
\tag{70}
$$

where y_r is the setpoint, $q \geq 0$ allows weighting of the control effort, using Δu to force integral action. The function $h(b)$ should be of small amplitude when b is negative, and should increase as b approaches zero and then becomes positive. As for $g(\Delta u)$, it should be small for positive Δu and increase rapidly as Δu approaches zero and becomes negative. Various choices are possible but the functions $h(b) = \exp(m\ b)$ with $m > 0$ and $g(\Delta u) = \exp(-\Delta u)$ satisfy the above requirements while being simple.

It is instructive to look at some of the properties of the one-step ahead performance index and expected control law. Taking the expected value of J_1 one can write

$$
\begin{aligned}
J_1 &= [(1+a)y(t) - ay(t-1) + \hat{b}\Delta u(t) - y_r]^2 \\
&\quad + \sigma^2 + [P(t) + q]\Delta u^2(t) \\
&\quad + \exp[m\hat{b}(t+1) + m^2 P(t+1)/2]\exp[-\Delta u(t)]
\end{aligned}
\tag{71}
$$

With (from now the time argument is dropped when it is equal to t)

$$\hat{b}(t+1) = \hat{b} + P\Delta u[\sigma^2 + P\Delta u^2]^{-1}\varepsilon(t+1) \tag{72}$$

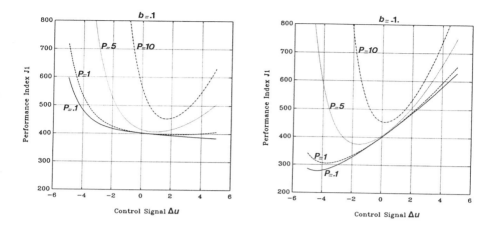

Fig. 17. Typical curves for $J_1(\Delta u)$ when $y < y_r$ [42].

$$P(t+1) = P\sigma^2[\sigma^2 + P\Delta u^2]^{-1} + \varrho^2 \tag{73}$$

and knowing that $\varepsilon(t+1) = N(0, \sigma^2 + P\Delta u^2)$, the performance index becomes

$$
\begin{aligned}
J_1 ={} & [(1+a)y(t) - ay(t-1) + \hat{b}\Delta u(t) - y_r]^2 \\
& + \sigma^2 + [P(t) + q]\Delta u^2(t) \\
& + \exp[m^2\varrho^2/2]\exp[m\hat{b} + m^2 P/2]\exp[-\Delta u] \tag{74}
\end{aligned}
$$

Typical curves of this performance index for various values of b, P and for both a positive and a negative control error are shown on Fig. 17 . As seen, the influence of $h(b)g(\Delta u)$ is negligible when the estimator is confident that the refiner is operating well inside the desirable zone. Then, as that confidence decreases and the gain estimate increases, one can see the movement of the optimal control signal toward positive values. It is also interesting to note that there is no dual effect due to $h(b)$. However, due to $g(\Delta u)$, there is no turn-off phenomenon, as Δu does not migrate toward zero with increasing P. Of course, dual effect will be obtained with increasing horizon. If dual action is desired even with the myopic control law, then $h(b)$ has to be modified. For example the following modification penalizes inaction in presence of large uncertainty, as seen from Fig. 18 which depicts J_{d1} as defined below, compare with Fig. 17.

$$
\begin{aligned}
J_{d1} ={} & [(1+a)y(t) - ay(t-1) + \hat{b}\Delta u(t) - y_r]^2 \\
& + \sigma^2 + [P(t) + q]\Delta u^2(t) \\
& + \exp[m^2\varrho^2/2]\exp[m\hat{b} + m^2\sigma^2 P/2\gamma^2]\exp[-\Delta u] \tag{75}
\end{aligned}
$$

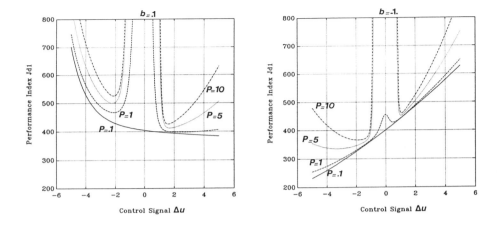

Fig. 18. Typical curves for $J_{d1}(\Delta u)$ when $y < y_r$ [42].

with $\gamma^2(t) = \sigma^2 + P(t)\Delta u^2(t)$. From both Figs. 17 and 18, it is seen that the resulting control law will not attempt to control the refiner with a collapsed pad, and that indeed, it will probably refuse to close the plates when such a move is likely to induce a collapse. Furthermore, no heuristic logic should be needed, making the resulting controller more attractive. Both performance indices give the same nonlinear control law when $P = 0$. Note that, even when b is known exactly, the control law cannot be computed analytically, but as for J_S, a Newton optimization scheme can be used. For J_1, as there is only one minimum the choice of Δu_{nom} is not crucial. For J_{d1}, as there are two minima, it is important to start on the side of the global one. An empirical method, consisting in evaluating J_{d1} at six different locations and starting at that giving the lesser value for J_{d1} was used satisfactorily in the simulations.

In simulations, the refiner non-linearity was represented by the following nonlinearity with $g_1 = 7600$, $g_2 = 800$, $\alpha = 1$ in normal state, and $\alpha = 0.77$ in collapsed state when $x < 0.9$.

$$f(x) = g_1 - g_2 \frac{(\alpha x)^4 + 1/3}{(\alpha x)^3} \tag{76}$$

Other settings are process pole $a = -0.75$, sampling interval $T = 2s$, $\sigma = 3$, $\rho = 0.15$, $m = 4$ and $q = 1$. The initial covariance P is set to 10. Fig. 19 shows that the myopic control law based on J_{d1} keeps the load as close as possible to the unreachable setpoint while not closing the gap below 0.9, the collapse point. When the certainty equivalence principle is used in conjunction with J_1, a collapse occurs, stressing the importance role of the term $\exp(P(t))$ in the

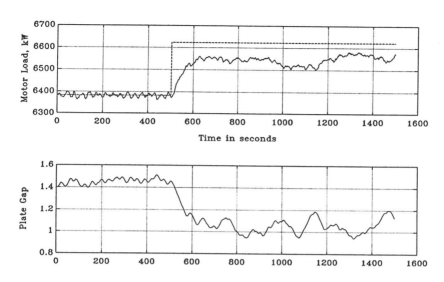

Fig. 19. Unreachable setpoint, adaptive control using J_{d1} [42].

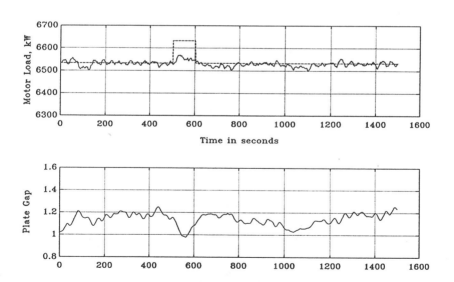

Fig. 20. Control near collapse, adaptive control using J_1 [42].

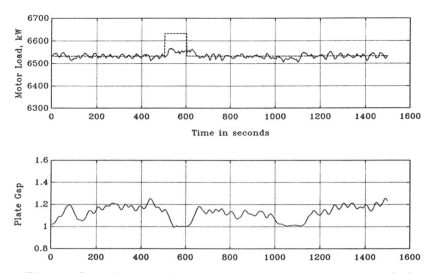

Fig. 21. Control near collapse, control using J_1 when b known [42].

performance index. When the refiner is operating near maximum load, small setpoint changes or just the regular process noise might produce collapses. Fig. 20 shows the behavior of a scheme based on J_1. For $500 < t < 600$, the setpoint is raised by 100 kW to an unreachable value and around $t = 1050$, the process noise is such that if the regulator closes the plates to compensate, it will produce a pad collapse. This situation is quite common in practice, as the refiner is very often operated in that region, in order to maximize production. The performance is remarkably close to that of the ideal nonlinear controller with known parameter b shown on Fig. 21.

V. WOOD SPECIES VARIATIONS

A. The Problem

A major challenge in pulping is to produce a uniform product from a highly variable raw material. Because wood is a natural material, both its chemical composition and physical properties are highly variable. For instance, hardwoods have less lignin, and thus delignify faster than softwoods during kraft pulping. Similarly, hardwoods have shorter fibers than softwoods and behave differently in mechanical pulping and in papermaking. Variations within individual species also exist, not only between trees, but also within the same tree, due to factors such as tree age and even location of the fiber within the tree. For instance, juvenile wood, *i.e.* within the center of the tree has lower density, shorter fiber length and more lignin than mature wood. Another important parameter is the springwood to summerwood ratio, as springwood is thin-walled and lighter. Thus, both the latitude and the altitude of the site where the tree grew influence the behavior of the fibers during the pulping and papermaking operations.

There currently exists no on-line sensor capable of wood species identification. Ignoring the in-species for the time being, we want to explore the following problem. Can we design a control scheme that automatically compensate for species variations, and by the same token provide an indication of the species mix in the feed? As we now proceed to show, a technique known as multimodel adaptive control may prove useful for solving this problem.

B. Multimodel Adaptive Control

The principle of multimodel adaptive control (MMAC) was first enunciated by Athans et al. in [43], and is illustrated in Figure 22. The unknown plant is assumed to lie within a finite set of models. The plant is characterized by a weighted sum of all those models. The weights vary from zero to one, their sum is one, and they reflect the probability of a particular model to adequately represent the plant. The adapter adjusts the weights on-line to minimize a performance index, generally the squared modelling error summed over a finite window. The main challenge is to develop a simple search algorithm that guarantees convergence to the global minimum, as the performance index is likely to be a multimodal function.

A multimodel adaptive controller presents various advantages. The bank of models might correspond to a model parameterized as a function of an unknown physical parameter assuming different probable values. This is particularly appealing in the case of a multivariable system, as it will likely result in fewer parameters to be estimated. Simulations show such multimodel systems have the ability to track very fast changes in dynamics. Another advantage of this approach is its inherent robustness. Indeed, even with a fixed weight distribution, such a control system will be rather insensitive to process dynamics changes.

MMAC may be useful when the raw material is known to consist of an unknown mix of a finite number of known species. If adequate models can be developed for individual species, then a bank of models can be built and the weights adjusted on-line. This not only provides automatic species compensation to the controller but, perhaps more importantly, automatic wood species identification to the operator. We illustrate this through a simple example involving chemical cooking of an unknown mixture of two wood species.

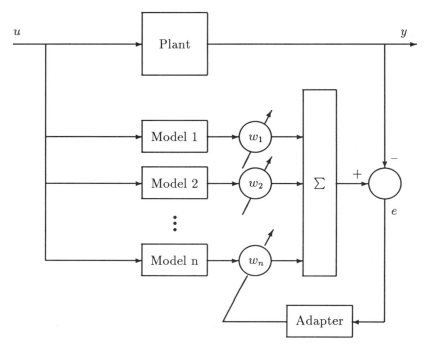

Fig. 22. Multimodel adaptive control.

In kraft pulping, reasonably accurate empirical predictive models exist. The most popular one is the Hatton's equation [44] relating the final Kappa number κ to the H-factor and the initial effective alkali EA for an individual species:

$$\kappa = \alpha - \beta(\log H)(EA)^n \tag{77}$$

The parameters α, β and n have been determined experimentally for a variety of species, [44]. Assume that the pulping takes place in a hypothetical reactor that can be described by first-order dynamics with pole a plus dead time d. Now, assume the feed stream is composed of a mix of two species and let w_i be the weight fraction of species i. The Kappa number out of the reactor is then described by:

$$\kappa_t = a\kappa_{t-1} + \sum_{i=1}^{2} w_i b_i u_{t-d} + c \sum_{i=1}^{2} v_{i,t-d} \tag{78}$$

where

$$b_i = -(1 - a)[\beta_i(EA)^{n_i} \log H]_{t-d}$$

$$c = 1 - a$$

$$v = \alpha_i$$

$$u = \log H$$

and α_i, β_i and n_i are the parameter values for species i in Eq. (77).

From this model, we can easily design a κ-number multispecies controller based on a Dahlin controller and feedforward:

$$u_t = \frac{1 - aq^{-1}}{\sum_{i=1}^2 b_i w_i} \frac{1 - p}{1 - pq^{-1} - (1 - p)q^{-d}}(\kappa^* - \kappa_t) - \frac{c}{\sum_{i=1}^2 b_i w_i} \sum_{i=1}^2 w_i v_{i,t} \qquad (79)$$

where κ^* is the setpoint and p is the desired closed-loop pole. It is easily shown that the multispecies controller can be expressed as a linear combination of the single species controllers using the same control error signal, i.e:

$$u_t = \sum_{i=1}^2 \lambda_i u_{i,t} \qquad (80)$$

where

$$\lambda_i = \frac{b_i w_i}{\sum_{i=1}^2 b_i w_i} \qquad (81)$$

The multispecies controller above assumes knowledge of the feed stream composition, i.e. of the w_i. If the feed stream composition is unknown, then we must find the right values for λ_i or w_i. In this example, we choose the second option, i.e. direct identification of the w_i. For this, we choose the simple projection algorithm:

$$\hat{w}_{1,t} = \hat{w}_{1,t-1} + \frac{\gamma \kappa_{i,t-1}}{1 + \phi_{t-1}^T \phi_{t-1}}[\kappa_t - \phi_{t-1}\hat{\theta}_{t-1}] \qquad (82)$$

where γ is a constant and

$$\phi_t = [\kappa_{1,t}\kappa_{2,t}]^T$$

$$\hat{\theta}_t = [\hat{w}_{1,t}\hat{w}_{2,t}]^T$$

$$\hat{w}_{2,t} = 1 - \hat{w}_{1,t}$$

Figure 23 shows the behaviour of such a scheme when controlling the chemical pulping of a variable mixture of hemlock and cedar. Noise has been superimposed to make this simulation a bit more realistic. It is seen that the Kappa number is well controlled despite wide variations in the species mix, and that the composition of the feed is accurately estimated. Although this example concerns chemical pulping, it is straightforward to apply it to a species-dependent papermaking control problem, such as refining. Note, however that we have simulated an ideal case, where we have exact models for individual species. In practice, the performance of such a scheme is going to be limited by the presence of in-species variations. Nevertheless, this approach merits to be explored further.

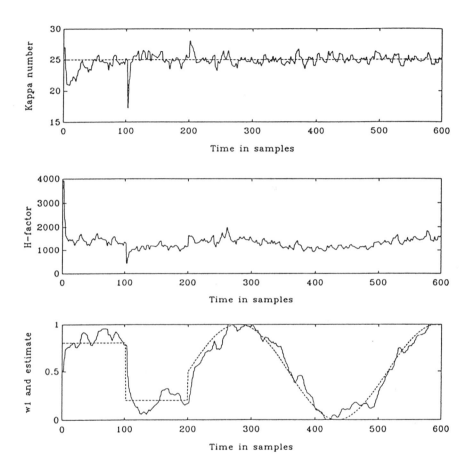

Fig. 23. Adaptive species compensation in kraft pulping of a two-species mix.

VI. CONCLUSIONS

This paper has attempted to describe some control problems of importance to the pulp and paper industry and some of the control techniques for solving them. We have seen that the problem of variable dead time has and still is attracting a lot of attention. On the other hand, the area of web process control, although very important has probably not generated the amount of research it deserves. This is an area of great technical and economical interest not only to the paper industry, but also to the sheet metal and plastic film industries. Because of its specificity, it would deserve a systematic study. The use of nonlinear active adaptive control has been demonstrated on the chip refiner motor load control problem. The nonlinear characteristic of this process has some similarities with the characteristic of bioreactors subject to washout. Thus, the method presented here may have some application on bioreactors. Finally, we have scratched the surface of the wood species compensation problem and have shown a potential application of multimodel adaptive control. The problems and the techniques discussed here by no means represent an exhaustive list of challenging control problems and interesting control techniques encountered in the pulp and paper industry. However, it is the hope of the author that this paper has aroused the curiosity of some process control researchers sufficiently to make them look at the challenges and opportunities the pulp and paper industry has to offer.

REFERENCES

1. K.J. Åström, "Control Problems in Papermaking", Proc. of IBM Scientific Computing Symp. on Control Theory and Applic., Yorktown Heights, N.Y., 135-167 (1964).

2. F. Bolam (Editor), "Papermaking Systems and their Control", The BP & BMA (1970).

3. K.J. Åström, "Computer Control of a Paper Machine — An Application of Linear Stochastic Control Theory" IBM J. Res. Dev., 11, 389-405 (1967).

4. V. Borisson, and B. Wittenmark, "An Industrial Application of a Self-Tuning Regulator", 4th IFAC Symp. on Digital Comp. Appl. to Proc. Control, Zurich (1974).

5. T. Cegrell, and T. Hedqvist, "Successful Adaptive Control of Paper Machines", Automatica, 11, 53-60 (1975).

6. D. Brewster, and A. Bjerring, "Computer Control in Pulp and Paper –
   - 1961–1969". Proc. IEEE, 58, 49–60 (1970).

7. D. Church, "Current and Projected Pulp and Paper Industry Problems
   in Process Control and Process Modelling". AICHE Symp. Series, 72,
   19-39 (1976).

8. M. Perron, and A. Ramaz, "A Survey of Control Strategies in Chemical
   Pulping Plants", Automatica, 13, 383-388 (1977).

9. R.B. Michaelson, "A Summary of Thermomechanical Pulping Plant Advanced Control Appllications". ISA PUPID/PMCD Symp., Portland
   Oregon (1978).

10. J.W. Gee, and R.E. Chamberlain, "Digital Computer Applications in the
    Pulp and Paper Industry", Dig. Comp. Appl. to Process Control, The
    Hague (1977).

11. G.A. Dumont,"Application of Advanced Control Methods in the Pulp
    and Paper Industry", Automatica, 2, 143-153 (1986).

12. G.A. Dumont, "System Identification and Adaptive Control in Paper-
    making", Fundamental Research Symposium, Cambridge, London, Eng-
    land, September 18-22, 1151-1181 (1989).

13. G.A. Dumont, "Analysis of the Design and Sensitivity of the Dahlin
    controller", 69th CPPA Annual meeting, Montréal, B37-B50 (1983).

14. K.J. Åström, and B. Wittenmark, "On Self-Tuning Regulators", Automa-
    tica, 9, 185-199 (1973).

15. K.J. Åström, and B. Wittenmark, "Adaptive Control", Addison-Wesley,
    Reading, Massachusetts, 1989.

16. E.F. Vogel, and T.F. Edgar, "An Adaptive Dead-Time Compensator for
    Process Control", ISA/80, Houston (1980).

17. G.A. Dumont, "Adaptive Dead-Time Compensation", 6th IFAC Symp.
    Ident. Syst. Param. Estim., Arlington, Va., 397-402 (1982).

18. A. Elnaggar, G.A. Dumont, and A.L. Elshafei, "Application of the Vari-
    able Regression Estimation Technique to Chemical Processes Having Un-
    known Delay", AICHE Annual Meeting, San-Francisco (1989).

19. A. Elnaggar, G.A. Dumont, and A.L. Elshafei, "Recursive Estimation
    for Systems of Unknown Delay", 28th IEEE CDC, Tampa, Flda (1989).

20. D.W. Clarke, C. Mohtadi, and P.S. Tuffs, "Generalized Predictive Control – Part I. The Basic Algorithm", Automatica, 23, 137-148 (1987).

21. D.W. Clarke, C. Mohtadi, and P.S. Tuffs, "Generalized Predictive Control – Part II. Extensions and Interpretations", Automatica, 23, 149-160 (1987).

22. C. Mohtadi, "On the Role of Prefiltering in Parameter Estimation and Control", Intern. Workshop on Adaptive Control Strategies for Industr. Use, Kananaskis, Alta, Canada (1988).

23. G.A. Dumont, and C.C. Zervos, "Adaptive Control Based on Orthonormal Series Representation", 2nd IFAC Workshop on Adaptive Systems in Signal Processing and Control, Lund, Sweden, 371-376 (1986).

24. C.C. Zervos, and G.A. Dumont, "Deterministic Adaptive Control Based on Laguerre Series Representation", Int. J. Control, 48, 2333-2359 (1988).

25. C.C. Zervos, and G.A. Dumont, "Multivariable self-tuning control based on Laguerre series representation", Intern. Workshop on Adaptive Control Strategies for Industr. Use, Kananaskis, Alta, Canada (1988).

26. C.C. Zervos, P.R. Bélanger, and G.A. Dumont, " Controller tuning using orthonormal series identification", Automatica, 24, 165-175 (1988).

27. A.H. Zemanian, "Generalized Integral Transformations", Dover Publications, Inc., New York, 1987.

28. P.R. Bélanger, L. Rochon, G.A. Dumont, and S. Gendron, "Self-Tuning Control of Chip Level in a Kamyr Digester", AICHE Journal, 32, 65-74 (1986).

29. B.J. Allison, G.A. Dumont, L. Novak, and W. Cheetham, "Adaptive-Predictive Control of Kamyr Digester Chip Level Using Strain Gauge Level Measurement", AICHE Annual Meeting, San Francisco (1989).

30. M.E. Salgado, G.C. Goodwin, and R.H. Middleton, "Modified Least-Squares Algorithm Incorporating Exponential Resetting and Forgetting", Int. J. Control, 47, 477-491 (1988).

31. C.C. Zervos, G.A. Dumont, and G. Pageau, "Laguerre-Based Adaptive Control of pH in an Industrial Bleach Plant Extraction Stage", Automatica, in press, (1989).

32. L.G. Bergh, and J.F. MacGregor, "Spatial Control of Sheet and Film Forming Processes", Canadian Journ. Chem. Eng., 148-155 (1987).

33. J. Mardon, et al.," Analysis of Paper Machine Stability and Performance by Means of Basis Weight Investigation", Canadian Pulp and Paper Assoc. Monograph, Montréal, May 1973.

34. C. Lindeborg,"A Process Model of Moisture Variations", Pulp Paper Canada, 87(4), T142-147 (1986).

35. K. Natarajan, G.A. Dumont, and M.S. Davies,"An Algorithm for Estimating Cross and Machine Direction Moisture Profiles on Paper Machines", 1988 IFAC Symp. on Identification and Syst. Param. Estimation, Beijing, China, 1503–1508 (1988).

36. W.H. Kelly,"Self-Tuning Control Strategies: Modern Solutions to Papermaking Control problems", Can. Industr. Comp. Soc. Conf., Hamilton (1982).

37. R.F. Sikora, and W.L. Bialkowski, "A Self-Tuning Strategy for Moisture Control in Papermaking", ACC, San Diego, 54-61 (1984).

38. A.F. Gilbert, and M.V. Venhola,"Application of Self-Tuning regulators at Great Lakes Forest Products", Can. Industr. Comp. Soc. Conf., Ottawa, 48.1-48.5 (1986).

39. R.G. Wilhelm, and M. Fjeld, "Control Algorithms for Cross Directional Control", 5th Int. IFAC/IMEKO Conf. on Instrum. and Autom. in the Paper, Rubber, Plastics and Polymerization Industries, Antwerp, Belgium, 163-174 (1983).

40. S.C. Chen, R.M. Snyder, and R.G. Wilhelm,"Adaptive Profile Control for Sheetmaking processes", 6th Int. IFAC/IMEKO Conf. on Instrum. and Autom. in the Paper, Rubber, Plastics and Polymerization Industries, Akron, OH, 77-83 (1986).

41. G.A. Dumont, "Selt-Tuning Control of a Chip Refiner Motor Load", Automatica , 18, 307-314 (1982).

42. G.A. Dumont, and K.J. Åström,"On Wood Chip Refiner Control", IEEE Control Systems Magazine, 8(2), 38-43 (1988).

43. M. Athans, D. Castañon, K-P. Dunn, C.S. Greene, W.H. Lee, N.R. Sandell, and A.S. Willsky, "The Stochastic Control of the F-8C Aircraft Using Multiple Model Adaptive Control (MMAC)–Part I: Equilibirum Flight", IEEE Trans. Aut. Contr., AC-22, 768-780 (1977).

44. J.V. Hatton, "Development of Yield Prediction Equations in Kraft Pulping", Tappi, 56(7), 97-100 (1973).

ADVANCES IN THE THEORY AND PRACTICE OF PRODUCTION SCHEDULING

K. PRESTON WHITE, JR.

Department of Systems Engineering
University of Virginia
Charlottesville, Virginia 22903

I. INTRODUCTION

Production scheduling is the allocation of resources over time for the manufacture of goods. Scheduling problems arise when a common set of resources--labor, material, and equipment--must be shared to make a variety of different products during the same period of time. The objective of scheduling is to find an efficient and effective way to assign and sequence the use of these shared resources, such that production constraints are satisfied and production costs are minimized.

Interest in new approaches to production scheduling has been stimulated by a variety of practical concerns, most especially the increasingly competitive world markets for manufactured goods. Better production schedules provide a competitive advantage through gains in resource productivity and related efficiencies in operations management. Competition also has motivated the introduction of sophisticated and capital-intensive new manufacturing systems made possible by the declining cost and increasing power of industrial computers and robots. Most notable among the new manufacturing technologies are automated systems for materials

115

handling, storage, and retrieval (AS/RS), flexible manufacturing (FMS), computer-aided engineering (CAE), and computer-integrated manufacturing (CIM) [1],[2]. These new systems have created a range of new operational problems, further quickening the pace of scheduling research.

Interest in new approaches to production scheduling also has been stimulated by theoretical considerations. The development of complexity theory and maturation of artificial intelligence (AI) have begun to redirect the body of scheduling research. Sequencing and scheduling theory long has been preoccupied with the design of constructive solutions and optimization algorithms for highly simplified problems. Theoretical advances now appear to have legitimized research on innovative heuristic search procedures which are applied to more realistic scheduling problems. These. problems and procedures appear to be more robust than optimization-based machine scheduling and, for this reason, hold greater promise for commercial adaptation. Taken as a whole, current market, technological, and theoretical developments have made solutions to both long-standing and newly-emerging scheduling problems the subject of intense applied and theoretical research.

Recent advances in the theory and practice of production scheduling transcend traditional disciplinary boundaries. Different research communities have begun to address different aspects of the problem, bringing to bear a variety of different research traditions, problem perspectives, and analytical techniques. As a consequence, the scheduling literature has escaped its traditional locus in operations research, management science, and industrial engineering. Production research has been reported in proceedings and journals principally concerned with control theory, artificial intelligence, system simulation, man-machine interaction, large-scale systems, and other branches of engineering and computer science. The sheer diversity and momentum of activity has made developments in production scheduling increasingly difficult to track and assimilate.

This chapter updates a survey [3] which grew out of the desire to assess the breadth and practical implications of contemporary production scheduling research and practice. We seek to provide a structured overview, a high-level bibliography, and a limited critique of seven production scheduling paradigms which we believe are historically significant or particularly promising. The first purpose of this chapter is to provide a means by which researchers working within traditional disciplinary bounds can explore in outline some of the more novel formulations and technical approaches to scheduling problems. A second purpose of this chapter is to stimulate researchers new to production research, by providing access to a broad spectrum of contemporary production scheduling literature.

The purpose and scope of this survey have required compromise and compression in its presentation. We have attempted to be comprehensive, although the pace of current activities prohibits guarantees against even significant omissions. Wherever possible, we rely on reference to original research papers and reports and to prior surveys and current texts to provide accurate and detailed exposition of the rich substructures within the various paradigms surveyed. Our treatment of specific techniques is necessarily shallow. We attempt neither a critique of specific technical methods, nor a definitive comparison of the various areas of research endeavor. Instead, we attempt to convey the sense and direction of research conducted within these areas and to offer our summary observations concerning the overall strengths and weaknesses of the various research paradigms as applied to practical scheduling problems.

In the following section, we develop a precise problem definition, describe the important characteristics of the production scheduling environment, and elicit the corresponding research issues. A summary and brief assessment of each of seven alternative approaches to the problem are presented in the third section. These include approaches used in current industrial practice and approaches based on traditional and novel theoretical paradigms. The final section offers some concluding thoughts and remarks. The references at the end of the chapter provide a starting point for additional reading and are categorized as follows--general and problem definition [1]-[20], current industrial practice [21]-[33], machine sequencing and scheduling theory [34]-[46], resource-constrained project scheduling [47]-[61], dynamic systems and control theory [62]-[88], discrete-event simulation [89]-[101], queueing theory and stochastic optimization [102]-[136], and artificial intelligence [137]-[157].

II. PRODUCTION SCHEDULING PROBLEMS

Scheduling problems typically involve a set of jobs to be completed, where each job comprises a set of operations to be performed. Operations require machines, labor, and/or other material resources and must be performed according to some feasible technological sequence defined by the jobs. Schedules are influenced by such diverse factors as job priorities, due-date requirements, release dates, cost restrictions, production levels, lot-size restrictions, machine capabilities, operation precedences, resource and labor requirements, and resource and labor availabilities. Performance criteria typically involve tradeoffs between holding inventory for partial and completed jobs, frequent production changeovers, satisfaction of production-level requirements, satisfaction of due dates, and utilization of labor and capital resources.

Developing a production schedule involves selecting a sequence of operations (or process routing) which will result in the completion of a job, designating the resources needed to execute each operation in the routing, determining the run quantity (or lot or batch size) for each job, and assigning the times at which each operation in the routing will start and finish execution. Routings and resource assignments typically are the product of *process planning. Master production scheduling* concerns the determination of run quantities and the approximate release times and due dates for each lot of each job type. *Machine scheduling* refers to the detailed activity of timetabling the operations for individual job lots within the framework provided by the master schedule [11].

A distinction typically is drawn between *open shop* and *closed shop* scheduling, based on the source of production orders [12]. In an open shop, all jobs are made to customer order and inventories of finished jobs are not maintained. Run quantities are determined by individual customer orders. In its simplest form, open shop scheduling reduces to a *sequencing problem*, in which the processing order of operations belonging to open jobs must be determined for each of the machines. In a closed shop, customer orders are filled from inventories and jobs are made to replenish inventories. Closed shop scheduling requires the solution of a *lot-sizing problem*, in which the strategy for replenishing inventories is determined based on anticipated product demand. The solution to this problem determines the master production schedule. Sequencing decisions must then be made for the individual operations of each job lot, just as for open shops.

An important distinction also can be made between static and dynamic scheduling disciplines. Static approaches seek to develop a complete production schedule for a production facility processing a finite number of jobs, prior to schedule implementation. Dynamic approaches typically seek to control the flow of jobs and operations in real time on the factory floor, evolving or modifying the production schedule as time progresses.

To illustrate the art of production scheduling, consider a specific example of the static, deterministic, single-stage, job-shop or machine sequencing problem, which is the meat of traditional deterministic scheduling theory. Table I provides job processing times and machine routing data for three jobs which are to be processed on three machines [19]. Figure 1 is a Gantt chart which shows one choice for scheduling the jobs. Under the usual assumptions that all jobs are ready at the start and that all operations require the exclusive use of a machine and cannot be split, this schedule represents the sequence of operations that minimizes the total time required to complete the processing of all jobs (the makespan). Note that the required ordering of operations within each job (the technological sequence) is preserved and that the ordering of operations on the machines has been selected so as to achieve the desired objective.

Table I. Data for a three-job, three-machine sequencing problem.

| Job | Operations (Processing Time/Machine Routing) | | |
	1	2	3
1	$1/M_1$	$8/M_2$	$4/M_3$
2	$6/M_2$	$5/M_3$	$3/M_3$
3	$4/M_1$	$7/M_3$	$9/M_2$

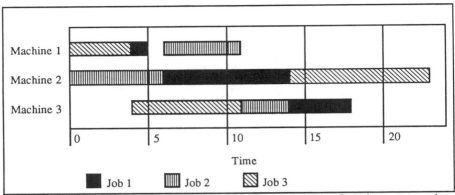

Figure 1. Minimum makespan production schedule for the sequencing problem.

Although the difficulty of determining the optimal machining sequence for problems of this type is well established (the problem is known to be NP-hard, i.e., the time required to solve the problem to optimality increases exponentially with increasing problem size [16]), the formal machine scheduling problem itself is an almost trivial abstraction of the multitude of complex, dynamic, probabilistic, and multi-objective scheduling problems which must be resolved in actual production environments. A given scheduling approach might be classified according to how it represents and deals with these complexities. Since to a greater or lesser degree every production environment is unique, the appropriateness of a given scheduling approach might be assessed by how well its assumptions correlate with the important features of a particular target production environment.

In this survey, we have selected five attributes of typical production scheduling environments which we have found especially useful in distinguishing among the paradigms considered. These attributes are:

Boundary states. A production schedule must be implemented at some given point in time. Except from cold start-up, it is rare (if not undesirable) that all job inventory levels are zero, all machines are idle and ready to process any job, and all required labor and material resources are fully available. It is commonly observed that the scheduling problem encountered in production is really one of *rescheduling*, i.e., one of modifying an existing production schedule to account for new jobs, new job priorities, or unscheduled disruptions in production. To provide for dynamic rescheduling, a general scheduling approach must be capable of accommodating all relevant initial states for the production system.

Similarly, a production schedule generally spans some finite time horizon. For an closed shop or cyclical production environment (where the same or similar sets of jobs are repeated in a regular cycle), this time horizon typically is keyed to the calendar, with weekly or monthly production levels adjusted to account for available capacity. For an open shop, where the number and composition of future orders are unknown, this time horizon might correspond to the latest due-date for any current job or to the makespan for all current jobs. Except for long-term shutdown, it is unlikely that either type of production system (or any combination of these) actually will be empty and idle at the end of the scheduling horizon. The ability to accommodate or even specify final system states can be a desirable attribute of a scheduling approach.

Batch sizes and setup costs. Discrete parts are grouped into one or more batches for processing. Larger batches reduce the number of setups and changeovers required during the scheduling horizon, while smaller batches decrease in-process inventories and increase the number of scheduling options. Minimum batch sizes typically are established by management decisions based on the products, technologies, and resources involved (down to single part batches for flexible manufacturing systems). Maximum batch sizes typically are governed by performance needs and the amount of work to be done. The most general and flexible scheduler would need to determine batch sizes as a product of scheduling constraints and objectives, would accommodate both fixed and variable batch sizes, would represent setups and changeovers explicitly, and would account for sequence-dependent setup times and costs.

Process routings. In the classical machine-scheduling problem, every job is processed on every machine exactly once, in a strict technological (within job) sequence. In actual manufacturing environments, process routings can be far more complex and even dynamic. Rarely is every job processed on every machine. Frequently, the technological sequence is a semi-order with some limited choices among which operation of a job can be processed next. The process routing can change dynamically given the state of the plant (e.g., the availability of resources) or the condition of the job (e.g., the need for rework).

Typically, there also exists a choice of machines on which certain operations can be performed. These machines may be identical with identical processing times for a given operation, as in a common machine bank. Alternatively, machines may be nonidentical with widely varying processing times and even processing characteristics and resource requirements. Examples include choices between newer and older equipment, between automated and manual equipment, and between special-purpose and general-purpose equipment (e.g., robots). The process routing and even the technological sequence of operations may change depending on the choice of machines. Ideally, a general scheduler would need to be highly flexible with respect to the types of process routings it could capture.

Random events, disturbances, and status information. In the classical machine-scheduling formulation, process times, release times, and due dates are deterministic and known *a priori.* Jobs, machines, labor, and material resources are available at all times. This can be far from the case in actual production environments. A general scheduler would need to be capable of representing the stochastic scheduling environment. This includes random events and disturbances such as the timing of hot jobs, machine failures, operator unavailabilities, and material stockouts, as well as variable job process times, release times, and due dates. A general scheduler also would need to compensate for the lack of accurate and timely status information, including hard and soft feedback that may be incomplete, ambiguous, biased, outdated, or inaccurate.

Performance criteria and multiple objectives. The classical machine-scheduling formulation usually specifies a single optimality criterion, such as minimum makespan or minimum tardiness. Such criteria tend implicitly to maximize machine utilization over the (unspecified) scheduling time horizon. Actual production environments clearly embrace multiple, conflicting, and sometimes noncommensurate constraints and performance objectives. While management typically seeks to minimize costs and maximize the utilization of expensive machines and resources, scheduling also frequently includes objectives directed toward minimizing operating stresses or satisfying hidden agenda. Examples include improving schedule stability, reducing confusion, and placating a demanding customer. Related objectives actually can imply deliberate underutilization of machines to reduce queues and inventories or to insure reliability. A completely general scheduler would need to capture and balance a great variety of performance criteria.

It is unlikely that a single scheduling technique usefully can represent all of these complex problem attributes in their full richness. Attempts to develop general-purpose scheduling tools probably are destined to fall from their own weight, inhibited by massive data requirements, cumbersome output, slow execution, and even the mistrust of the human schedulers that might be compelled to use these. Useful tools are most likely to capture only the most important features of a given environment, as viewed by the human scheduler, ignoring the least important entirely and dealing with those of intermediate importance in an aggregate way. For the most part, each scheduling paradigm considered in the following section addresses only a subset of these attributes. Selecting a specific paradigm to use in the development of a scheduling aid most likely depends upon how the production scheduling problem of interest can best be structured.

III. SCHEDULING APPROACHES

A. INDUSTRIAL PRACTICE

Production scheduling in discrete-parts manufacturing is generally acknowledged to be skilled craft practiced by experienced human schedulers. Despite the comparatively recent emergence of a number of industrial database and software tools for automated scheduling, it is perhaps still accurate to say that actual production schedules are largely generated by hand, using paper, pencil, and graphical aids (such as the familiar Gantt chart). Indeed, in a field study McKay *et al.* [30] found that many practicing schedulers cope with the dynamics and complexity of the scheduling environment by applying simple dispatching rules for very short time horizons, avoiding long-term detailed machine scheduling altogether. Clearly, knowledge and intuition gained through first-hand experience are the principal tools employed by the scheduler in generating and maintaining satisfactory production schedules.

To assist the human scheduler and to improve the quality, consistency, and acceptability of production schedules, major manufacturers have developed or purchased database systems which track raw-materials and work-in-process (WIP) inventories. Many of these database systems also incorporate software tools which, to a greater or lesser degree, automate some aspect of schedule generation. These commercial tools are generally classified by the scheduling technique or the underlying scheduling philosophy employed. Among the most current of the scheduling philosophies and associated software packages are Materials Requirements Planning (MRP) and its successor Manufacturing Resource Planning (MRP II), Just-in-Time (JIT) or Zero-Inventory (ZI) production and its implementation using *Kanban*, and Optimized Production Timetables (OPT) [28].

MRP systems are perhaps the most widely installed in industry today [21],[27],[31]. For a fixed planning horizon, MRP systems determine (1) the quantities of each item that will be used in the production of a prescribed volume of end-products and (2) the times at which each of these items must be purchased or manufactured in order to meet prescribed due-dates for the end-products. MRP works approximately as follows. First, for each end-product, the quantities of all components and subassemblies which are used in the manufacture of that product are determined. Using prescribed processing times and working backwards from the date for final assembly, MRP next determines the latest time at which these components and subassemblies should be made or ordered. Finally, MRP performs a more detailed capacity requirements analysis, determines an operation sequence, and sizes production lots.

MRP systems are highly detailed and an excellent means for determining and tracking materials requirements. As a means for production scheduling, however, MRP systems leave a good deal to be desired. The principal shortcoming of MRP is that it determines capacity *requirements* based on prescribed product volumes, release (starting) dates, and due dates. Actual *installed* or *available* production capacity is ignored, with the result that MRP schedules can prescribe machine loadings in excess of 100% utilization. Production volumes and due dates must be adjusted manually in order to achieve feasible schedules. This shortcoming is compounded by the fact that MRP is entirely deterministic and cannot anticipate the impact on schedules of variable processing times and random events. In production control, MRP is sometimes referred to as a type of *push system*, since a static production schedule is devised off-line and is used to push jobs through the facility, irrespective of current production status information.

JIT production is a scheduling philosophy that dictates reduced materials inventories and minimum WIP inventories in order to aid process improvement and reduce process variability [24],[27],[29],[31]-[33]. The obvious benefit of JIT schedules is the reduced capital cost associated with holding inventories, both in terms of reduced inventory storage requirements and reduced investment in raw materials and intermediate goods on hand. Its more subtle benefits reside in associated improvements in process flow and floor control, particularly with respect to the early detection of rejects and immediate isolation of the associated culprit operation.

The success of JIT production is empitomized by Japanese automobile manufacturing, a relatively stable production environment characterized by high-volumes and repetitive tasks. The JIT scheduling philosophy also demands total quality control, including highly reliable suppliers, workforce, and repair facilities, since buffer stocks are essentially eliminated. In the present context, however, JIT has the singular disadvantage that it purely descriptive. While JIT is commonly misclassified as a "method" that achieves minimal WIP with a lot size of one, in fact there is no prescriptive theory or technique for deriving JIT schedules or achieving achieving JIT goals.

Instead, the JIT philosophy is implemented in part by dynamic production control systems which do not schedule jobs in advance, but rather authorize production based on current status information. These are known as *pull systems*, since the completion of operations downstream in the production routing triggers the start of appropriate upstream operations, pulling jobs through the facility. The best known pull system is Kanban. Kanban is the Japanese word for the cards, tags, or tickets passed between work centers which authorize either the production of a container quantity of an item, or the movement of a container from the work center where it is made to the work center where it will be used.

OPT is a proprietary software (and hardware) package marketed by Creative Output, Inc., that recently has generated considerable interest as a production planning and scheduling tool for large-scale systems [22],[25],[26],[30]. OPT can be viewed as an alternative to a comprehensive MRP system for production planning, materials planning, and resource scheduling. In concept, OPT works by considering how production resources should be used to meet requirements, sequentially, at fixed intervals of time. OPT first selects a candidate production schedule and then simulates this schedule to determine critical resources or bottleneck machines. Next, operations are scheduled from the critical resources using heuristic procedures to feed the bottleneck during periods when it is starved. Production batches are split as required to insure that bottleneck machines are kept busy. This basic process is repeated until an "optimal" schedule is produced, or until some stopping rule is invoked. Because it uses both pre-planned schedules and current status information to revise these schedules, OPT can been seen to combine elements of both push and pull production control systems.

While there has not been a comprehensive comparison of OPT schedules to those produced using conventional scheduling logic, the underlying philosophy has considerable intuitive appeal. By informal accounts, OPT customers are generally pleased with the quality of schedules produced. The primary disadvantages of OPT appear to derive from its proprietary status. Licenses and maintenance agreements are expensive and custom installation and maintenance must be performed by the vendor. Also, there may be hidden costs associated with excessive work-in-process inventory and with set-up/tear-down on non-bottleneck machines (labor, wear, and tear) that result from batch splitting. OPT also can require significant time to generate schedules and therefore may be inappropriate as a real-time scheduling tool where production is highly variable.

MRP, OPT, and JIT are not distinct scheduling models and were not designed specifically to address the scheduling problem. It is common in comparative analyses of the MRP, JIT, and OPT scheduling philosophies to note that it may not be feasible to develop a schedule for arbitrary combinations of due dates and resource constraints. In such cases, solving the problem requires relaxing one of three conditions: the initial state, the final state, or the state-trajectory constraints. The initial state corresponds to initial inventory levels and schedules of incoming parts and materials. The final state corresponds to production volumes of finished goods with prescribed due dates, as well as desired final material and WIP inventories. The trajectory and control constraints correspond to limitations on maximum WIP inventories and maximum productive capacity. To guarantee a feasible schedule will be found, each of the commercial philosophies employs a different relaxation technique. MRP systems relax trajectory constraints, OPT ignores the final state, and JIT ignores the initial conditions [4].

B. MACHINE SEQUENCING AND SCHEDULING THEORY

The formal job-shop or machine-scheduling problem has been studied extensively over the past several decades [5]-[17],[19],[34]-[46]. The problem may be stated as follows: N jobs are to be processed on M machines. Each job consists of a set of M operations, one operation uniquely associated with each of the M machines. The processing time for an operation can not be split. Technological constraints demand that the operations within each job must be processed in a unique order. The scheduling problem involves determining the sequence and timing of each operation on each machine, such that some given performance criteria is maximized or minimized. Typical performance criteria include minimizing the makespan (i.e., minimizing the time required to complete all of the jobs) and minimizing maximum tardiness (i.e., minimizing the largest difference between completion times and due dates).

The machine-scheduling problem is a highly simplified formalism for the production scheduling problem encountered in practice. The problem is of particular interest, however, precisely because it captures the fundamental computational complexity of the central problem of sequencing jobs on machines, divorced of any side-issues. This general problem has been shown to be NP-hard for instances larger than two jobs and two machines [16]. The time required to compute an optimal schedule increases exponentially with the size of the problem. The standard 10-job, 10-machine benchmark problem first posed by Muth and Thompson [14] in 1963 has only recently been solved to optimality, requiring the generation of 22,000 nodes in five hours on a Prime 2655 computer [34],[38]. Problems of larger than modest size can not in general be solved to optimality, even with computing power that far exceeds the capacities of modern supercomputers.

From a practical vantage, work on the machine-scheduling problem has demonstrated the need for heuristic (non-optimizing) approaches to commercial scheduling problems [11],[34],[36],[39],[41]-[46]. A large number of single-stage heuristics have been advanced and tested in this context. In general, single-stage heuristics select the next operation to be processed based upon some easily computed parameter of the jobs, operations, or machines. These parameters can include processing times, due dates, operation counts, costs, set-up times, arrival times, and machine loadings. Examples are SPT (shortest processing time first), LPT (longest processing time first), FIFO (first-in, first-out), LPR (longest processing time remaining), EDD (earliest due date), and pure random (Monte Carlo) selection. More complicated heuristics are generally built-up from simpler rules. Panwalkar and Iskander [44], for example, cite some 113 scheduling heuristics that have been proposed or actually applied.

Work on the machine-scheduling problem has also provided a wealth of information on solution strategies and approximation algorithms which exploit single-stage heuristics and which may ultimately form the basis for commercially viable tools. For example, partial enumeration techniques which combine heuristics and neighborhood search strategies have been shown to work reasonably well under various conditions [11],[17],[34],[35],[41]. These strategies involve the use of a heuristic to find a good seed or starting schedule, modify the seed, and then evaluate the resulting schedule. A cycle of adjustment and evaluation is repeated until no further progress relative to the performance measure is achieved. Much of the current interest in AI approaches to production scheduling also is closely associated with the development and testing of novel heuristics for combinatorial sequencing problems.

Although these contributions are significant, the machine-scheduling problem itself is perhaps too restrictive a formulation to provide results that are anything more than suggestive for actual production scheduling. McKay *et al.* [30] list the following attributes of a production facility that would allow the direct application of deterministic machine scheduling theory: a stable and well-understood simple manufacturing process; simple manufacturing goals which are not affected by hidden agendas; short cycle times so that work can start and finish without interruption; predictable and reliable set-up and processing times; known delivery quantities, delivery times, and delivery qualities; long times between failures relative to cycle times, and short repair times; and accurate and complete information on processing requirements and the status of the jobs in the computer. While we note that many of these attributes in themselves are highly desirable for improved production control (and, indeed, in many ways reflect the goals of the JIT production philosophy), McKay *et al.* also note that none of the job-shops they studied could be characterized by even a small proportion of these attributes.

We conclude, as do many others, that there is a great need for better scheduling algorithms and heuristics, more realistic models of the scheduling setting, and better understanding of the dynamics inherent in the scheduling environment. Developing an appropriate model for direct use in solving production scheduling problems will require a significant relaxation of the basic assumptions of the classical formulation of the machine-scheduling model, particularly with regard to plant stability and perfect information. Mathematical formulations which exploit the special structure of the scheduling problem offer a readily accessible starting point for research into new areas concerning machine scheduling and algorithm development [16],[18],[19].

C. RESOURCE-CONSTRAINED PROJECT SCHEDULING

The machine scheduling problem is a special case of resource-constrained project scheduling. The conceptual similarity of the machine- and project-scheduling problems has stimulated interest in the reciprocal applicability of solution approaches [47],[48]. These similarities are emphasized when both problems are modeled as networks. When viewed this way, traditional machine sequencing rules are tested on project networks and certain forms of the machine scheduling problems are treated as project scheduling problems.

As with the machine scheduling problem, determining the best resource-constrained project schedule requires the solution of a combinatorial optimization problem. Any approach which seeks an exact solution to a resource-constrained project scheduling problem is really just a different way to formulate and solve this optimization problem. Among the approaches most commonly studied are branch-and bound methods, branch-and-dominate methods, cutting planes, dynamic programming, zero-one integer programming, Lagrange multiplier methods, heuristic search, and simulation optimization [49]-[59].

The resource-constrained project scheduling paradigm is fundamentally more pragmatic than machine scheduling and takes into account many more of the complexities of the scheduling environment. More complex technological sequences and multiple resource requirements are considered and a range of ingenious heuristics have been developed [60]. In addition, there is a large body of commercial software available for project scheduling. The heuristic algorithms embodied within this software might well be adapted to production scheduling.

D. CONTROL THEORY

Gershwin *et al.* [74] present an excellent interpretation of recent progress in manufacturing systems from the perspective of control theory. A control-systems framework for manufacturing systems issues is provided, which includes production scheduling issues. Current practice and research applied to the general manufacturing environment also are summarized.

Control theory seeks scheduling methods which either explicitly reflect the uncertain nature of the available information or give some guarantee as to the insensitivity of the schedule to future information. This modeling paradigm attempts to limit the effect of machine failures, operator absences, material unavailability, surges in demand, or other disruptions upon the scheduling process [4],[62],[63],[73],[77],[78],[87]. Schedules are sought which are robust to disruptions, robust in the absence or inaccuracy of status information, and flexible to change. Control theory views scheduling as a dynamic activity, where the scheduling problem is really one of understanding how to reschedule.

Although control theory has only recently been applied to discrete production scheduling, the underlying problem and fundamental issues are perhaps most naturally described as problems in control. Consider the standard control paradigm which is depicted in Figure 2 as the control system model for the scheduling problem. The system under control operates on a sequence of inputs $u(t)$, yielding a sequence of outputs $y(t)$. The outputs at any time are a function of the state of the system $x(t)$. The system state at time t is defined as the set of variables, such that knowledge of the state at some initial time, together with knowledge of the inputs to the system at this and all future times, is sufficient to determine (given an adequate model) all of the future states of the system. A general statement of the control problem is to determine a means to guide the system state (and thus the system output) through time according to some trajectory which satisfies the constraints imposed by the system model and which simultaneously satisfies some set of performance criteria.

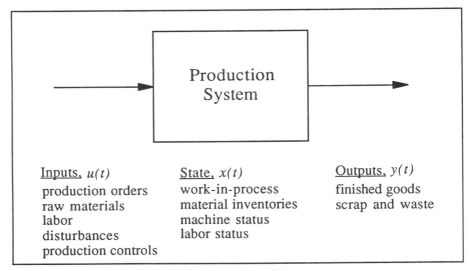

Figure 2. Production scheduling as a dynamic system.

For a manufacturing facility, inputs to the system include production orders, raw material and labor, disturbance inputs, and controlled inputs. Production orders specify the quantities of various jobs to be processed and the dates at which these quantities are due for delivery. Raw material and labor are used in production and without these production is impaired. Disturbance inputs include machine failures or labor outages, which alter the productive capacity of the plant over some period of time, but over which the scheduler has little influence. Controlled inputs include scheduling, maintenance, and overtime decisions, which alter the productive capacity of the plant, but which the scheduler can regulate within certain bounds. The state of a manufacturing facility defines the levels of inventory for all completed and partially completed jobs, the status of all machines (whether idle, active on job, in setup for job, or in repair), availability of labor, and the levels of inventory for all materials. Outputs from the manufacturing facility can be any appropriate combination of the state, for example, the inventory levels of all jobs ready for shipment on a specified due date.

The scheduling problem is then to determine the control, i.e., the size of production jobs, the sequencing of the operations of each job, the scheduling of overtime and maintenance, and the timing of materials orders, which yields a desired state trajectory. The control must also satisfy constraints imposed by limited manufacturing resources and must simultaneously satisfy performance measures. These performance measures may include minimum tardiness of finished jobs, minimum and maximum in-process inventory levels, and minimum production and holding costs. The control must be robust in the face of disturbances and noisy feedback.

Merits of the control paradigm are numerous. First, this modeling paradigm recognizes the need to integrate scheduling activity with planning activity. Second, the paradigm accepts the dynamic environment of the scheduling problem as a given and attempts to find schedules which are robust, flexible, and adaptable to this dynamic environment. Third, the paradigm recognizes that status information may be missing or imperfect and compensates for this fact. The control theory modeling paradigm also provides a wealth of knowledge in defining the scheduling problem and the corresponding scheduling objectives.

While the control paradigm appears to be especially well suited to defining the fundamental problem qualitatively, control theory has yet to develop a set of techniques entirely adequate to the formulation, analysis, or design of scheduling models. The mathematics and techniques of control theory apply to continuous- and discrete-time systems (Markov processes), but are not well-adapted to discrete-event systems (semi-Markov processes). Those aspects of the manufacturing problem which can be usefully approximated by continuous- or discrete-time systems (such as highly aggregated production flows) are the areas most likely to benefit from traditional quantitative applications of control theory, at least in the short run.

The recent emergence of discrete-event dynamic systems (DEDS) as a principal thrust of control research offers considerable promise for future developments, however [78],[87]. A discrete-event system is one in which the state evolves in a piecewise-linear fashion over time. The processes which define the different types of state transition possible are called *events*. In contrast to discrete-time systems, the distinguishing feature of a discrete-event system is that the timing of future events is not prescribed exogenously. This information instead is a part of the current definition of system state and, hence, itself evolves over time. For this reason, the trajectory of a DEDS cannot be determined by the solution of differential or difference equations alone, but also requires some additional logical or procedural information.

Ho [78] suggests the following desiderata for a DEDS model which can embrace: the discontinuous nature of discrete events, the continuous nature of most performance measures, the importance of probabilistic formulation, the need for hierarchical analysis, the presence of dynamics, and the feasibility of computational burden. He suggests that while there has been no lack of attempts a constructing general models for DEDS, there is general agreement that no consensus has developed as to which one of the models (if any) has the potential eventually to serve as the analog of the differential equation for continuous dynamic systems. Among the current candidates are *minimax algebraic models* [66],[70]-[72]; *automata* (including Petri nets and extended state machines) [64],[69],[75],[83]-[86],[88]; *Markov decision processes* (MDP's) [104],[106]-[109],[123],[124]; *queueing networks* [103], [105],[111],[114],[115], [117]-[119],[121], [125]-[127], [129], [131]-[135]; and *generalized semi-Markov processes* (GSMP) (including discrete-event simulation [89]-[101] and related techniques, such as perturbation analysis (PA) and simulation optimization [65],[67],[68],[76],[79]-[81]).

Minimax or path algebra is an algebra of real numbers in which the usual binary operations of addition and multiplication of two numbers (or matrices) are replaced by arithmetic addition and by the max (or min) operator, which selects of the greater (or lesser) of two numbers. This algebra has been shown applicable to many different path-finding problems in networks [66]. In particular, Cohen *et al.* [70] draw an analogy between linear dynamic systems and DEDS and demonstrate that the behavior of closed, periodic, discrete-event systems can be totally characterized by solving an eigenvalue and eigenvector equation in the minimax algebra. They use this approach to analyze production in a periodic flow shop. McPherson [82] derives the same and additional results using conventional algebra and a develops the mathematical properties and a scheduling philosophy for periodic flow lines.

In automation and language theory, a DEDS is modeled by an finite-state machine (an automaton, or, equivalently, a finite directed graph) which generates a string of symbols drawn from a finite, non-empty alphabet. The alphabet corresponds to the set of all possible types of events for the DEDS and each symbol corresponds to an instance of a specific type of event. The symbol strings generated by the automaton therefore correspond to the state trajectory of the DEDS. This model has been used to define and analyze various DEDS properties, including controllability, observability, and predictability. DEDS can also be represented by timed Petri nets, originally developed for the performance evaluation of distributed computer systems. Applied to production scheduling, the Petri net *transitions* correspond to operation processing activities and the net *tokens* correspond to production resources.

MDP's are a well-known class dynamic systems that evolve over time according to the joint effects of probabilistic laws of motion, typically modeled by a finite-state Markov chain, and the sequence of control actions taken, as the result of decisions which affect the future states. The MDP optimal control problem is that of finding a control law (called a policy) which specifies the sequence of decisions which minimizes the expected total cost of the state trajectory, as determined by some combination of the cost of the control actions and the expected cost of the corresponding observed states. The MDP model has been applied to various production scheduling problems, including machine maintenance and replacement scheduling and inventory control. Current research interests include the development of efficient computational procedures for MDP's, the derivation of optimal policies for MDP's with partially observed states, and that adaptation of the discrete-time MDP formalism to DEDS.

GSMP is a second class of stochastic processes that has been found useful for describing fairly general queues, including queueing networks. Recent interest among the control community has been directed toward GSMP as a model which captures the essential dynamical structure of DEDS. This model has been used in attempts to develop a qualitative theory and numerical algorithms.

PA seeks to determine the sensitivity of the performance of a DEDS to changes in its parameters. A performance measure is defined for the system, typically based on the duration of each of the events in the state trajectory. PA assumes that a single finite trajectory of the system is available for analysis, obtained either through observation of the actual DEDS or through a single run of a discrete-event simulation of the system. Using this single trajectory, PA determines how the trajectory (and thus the performance of the system) would change in response to a change in the parameters of the DEDS. PA is sometimes viewed as an efficient technique for analyzing simulation models, since it requires just a single simulation run with fixed model parameters, rather than multiple simulation runs, each with different parameter values. Recent studies of PA have shown that it may be better than repeated simulation in many situations.

E. DISCRETE-EVENT SIMULATION

Most manufacturing systems are too complex to allow realistic models to be evaluated analytically. Simulation models provide the desired realism and can be evaluated numerically over the time period of interest [89]-[101]. Data are gathered as the model is running in order to estimate the characteristics and throughput of the simulation model.

Discrete-event simulation typically is not used as a means for generating production schedules *de novo*. Instead, simulation is applied as a vehicle for testing schedules or comparing the performance of alternative schedules, which have been developed using some other logic and based on some simpler model of the production facility. Simulation similarly is applied as a vehicle for testing and comparing the performance of alternative scheduling heuristics and dispatching rules.

In addition to providing a realistic model for the evaluation of schedules and heuristics, simulation also provides a limited capability for testing modifications to an existing schedule. Since the simulation process provides dynamic output, the user can identify long queues within the model and attempt to eliminate the bottlenecks which may exist for priority jobs on the floor [91]. This supports the concept of a flexible, interactive simulation tool which could be used as a real-time decision aid for scheduling a manufacturing facility. Such a tool would display the the status of the simulated facility and permit the scheduler to use his knowledge and practical experience in order to stop, interact with the model, and try alternative scheduling tactics [89],[94],[95],[98].

An advantage of the simulation approach to scheduling is that it can model the effects of such factors as policy changes, which cannot be accounted for in an analytic model [99]. Another advantage of discrete-event simulation is that it can provide the user with the opportunity of performing exploratory tests upon the schedules being produced [90]. The experimental nature of simulation is also its principal disadvantage, however. Simulation studies are difficult to generalize beyond the specific experimental setting employed and, as such, contribute little to the theory of production scheduling. While new tools and techniques for simulation analysis (such as PA, GSMP, and simulation optimization) hold promise for reducing the cost of such analyses, using simulation to produce schedules is likely to remain expensive, both in the computer time used to generate schedules and in the human modeling effort required to design and run the simulation model. The accuracy of a simulation is limited by the judgment and skill of a programmer and even highly accurate modeling does not guarantee that optimal or even good schedules will be found experimentally.

F. STOCHASTIC OPTIMIZATION

Stochastic optimization addresses a number of types of manufacturing systems problems by applying the research practices of queueing theory [103], [105], [111], [114], [115], [117]-[119], [121], [125]-[127], [129], [131]-[135], reliability theory [104],[107]-[109],[120],[123], lot-sizing techniques [102],[109],[112],[116],[120],[122],[128],[130], and inventory theory [109],[113],[120],[123],[136]. These practices emphasize the probabilistic nature of production environments. A common shortcoming of stochastic optimization models, however, is that these typically require highly simplified models in order to pursue analytical solutions.

Queueing theory pictures each machine center as a set of servers, with each machine a server and each job a customer. The wide variety of jobs and the complex routing of operations is represented by assuming that processing times and arrival of operations at a center are governed by probability distributions. The application of queueing theory to manufacturing was considerably enhanced by the development of network-of-queues theory [111],[114]. However, only the more recent development of efficient computational algorithms and good approximation methods has enabled the implementation of reasonable "first-cut" evaluative models of fairly complex manufacturing systems [74],[119],[129].

An important reason for the increasing popularity of queueing models in manufacturing is their proven usefulness in the area of computer/communication systems modeling. Queueing models tend to give reasonable estimates of system performance, require relatively little input data, and do not use much computer time. For these reasons, queueing models can be used interactively to arrive quickly at preliminary decisions. More detailed models can then refine these decisions. There also is currently much interested in the area of queueing control.

The disadvantages of queueing network models are that these must represent certain aspects of the system in an aggregate way and fail to represent certain other features at all (such as limited buffer space). Also, the output measures produced are average values, based on a steady-state operation of the system. Thus, queueing network models are not useful for modeling transient effects which result from infrequent but severe disruptions such as machine failures. Also, there are many queueing situations in which potentially useful queueing models are mathematically intractable or in which mathematical models are difficult or impossible to derive.

Reliability theory is concerned with maintaining a stable schedule in the face of machine failures [123],[124]. Reliability is the probability that a machine performs adequately over an interval. Knowing the reliability of machines may help to limit the disruptions caused by machine failures. Determining the reliability of machines requires knowledge of the time-varying failure rate, however. The assumption of constant failure rate, which appears to be used frequently in practice, may lead to crude results [109].

Lot-sizing techniques and inventory theory attempt to determine lot sizes which will maximize production throughput and assure that due dates are met and production and inventory costs are minimized. Traditional lot- sizing models trade-off the cost of setting up a machine against the cost of holding inventory, on an individual machine basis. Lot size decisions are frequently required for many different products and demand quantities, which creates the need for a practical, efficient decision rule [128],[136]. Finding the optimal lot size which minimizes the total of production, setup, holding, shortage, and scrap costs, however, is not currently feasible. Most techniques either cannot guarantee the generation of a feasible solution or are computationally prohibitive [120].

Inventory theory models and analyzes the effects of uncertainty to derive optimal stocking policies. Inventory stocking policies assume that each item stocked has an exogenous demand, modeled by some stochastic process, and attempt to find the best stocking policy for each item. There are many inventory situations that possess complications which must still be taken into account, e.g., interaction between products. Several complex models have been formulated in an attempt to fit such situations, but even these efforts leave a large gap between practice and theory.

G. ARTIFICIAL INTELLIGENCE

Artificial intelligence (AI) approaches typically depict the scheduling problem as the determination and satisfaction of the large number and variety of hard and soft constraints which are found in the scheduling domain [139]. AI is used to extend knowledge representation techniques to capture these constraints, to integrate constraints into a search process, to relax constraints when a conflict occurs, and to diagnose poor solutions to the scheduling problem.

AI search methods hold promise for scheduling applications. These methods employ heuristic rules to guide the search and may offer efficient search procedures for finding good solutions to computationally complex problems [140]. Ow and Morton [148], for example, describe how a beam-search strategy can be used as an efficient search method and discuss the importance of heuristics in obtaining maximum advantage from the search technique and limited knowledge about the problem. Similarly, current research on genetic algorithms, simulated annealing, and learning systems may hold potential for improving the speed and accuracy of various production scheduling approaches.

Production scheduling also may benefit from other research areas within AI, perhaps including rule-based and knowledge-based (expert) systems [138],[139],[141]-[143],[145]-[147],[149]-[157]. The scheduling problem appears appropriate for expert systems work because it is heuristic in nature, that is, it requires the use of rules of thumb to achieve acceptable solutions . Expert systems, however, appear to be best suited to developing a diagnosis within a slowly evolving knowledge domain. This may result in poor adaptation of the scheduling problem to the expert system domain. Because each production production facility is different, the expert systems approach may not be sufficiently robust to handle new production and scheduling situations [145].

Other difficulties exist as well. First, expert systems are expensive and time-consuming to develop. The costs of developing expert schedulers tailored to specific production environments may well be prohibitive. Second, expert systems for reasonably-sized problems may result in very slow computational speeds [144]. The costs of using expert schedulers, therefore, may also be prohibitive. Finally, expert systems strive to automate decisions that are made by genuine human experts. To reach this goal, it is necessary to capture the high level of expertise of individuals currently involved in the solution of scheduling problems [155]. Unfortunately, the quality of human performance in many scheduling tasks is suspect and expert human schedulers frequently may not always exist in the production scheduling environment.

One may conclude that AI and conventional operations research techniques need to be appropriately combined in order to alleviate some of the difficulties discussed above [142]. Bruno et al. [138] provide an example of one such combination, which uses expert systems techniques for knowledge representation and heuristic problem solving, an activity-scanning scheduler adapted from discrete-event simulation, and a closed queueing-network algorithm for schedule analysis and performance evaluation. The authors report that this system is currently in use in a plant that produces several different types of air compressors.

Shaw [150]-[152] and Shaw and Whinston [153] have developed an interesting AI approach to scheduling which is characterized by knowledge-based organization, symbolic representation, and state-space inferencing for schedule generation and which has the capability for dynamic scheduling and plan revision. This approach has been implemented on a LISP machine in a prototype system for FMS scheduling. The approach appears to provide a foundation for integrating intelligent planning, scheduling, and machine learning in FMS.

Other specific AI projects reported in the literature are ISA, DEVISER, ISIS, ISIS-2, OPIS, GENSCHED, FIXER, and PLANNET. ISA (Intelligent Scheduling Assistant) was developed for master planning and scheduling at Digital Equipment Corporation [145]. The ISA system loads orders for the assembly of computers, sequentially, one job at a time. Approximately 300 rules are employed to build and modify the evolving schedule, relaxing scheduling constants as required.

DEVISER is a general purpose planner/scheduler that generates a scheduling network, specifying nominal starting times for operations [157]. Discrete events are defined and schedules are tailored around those events which are fixed in time. A "start-time window" for each operation is updated dynamically, as the schedule evolves, in order to maintain consistency with the windows and processing times of adjacent operations. The system is goal directed, adjusting windows to achieve production milestones within imposed time constraints, while respecting fixed events. DEVISER was developed specifically to plan and schedule those actions on-board an unmanned spacecraft (such as the Voyager) which are required to return pictures of objects in deep space.

The best known AI production scheduler is ISIS (Intelligent Scheduling and Information System), a prototype system which uses "constraint-directed reasoning" to construct shop schedules [141]. The system selects a sequence of operations needed to complete an order, determines start and end times, and assigns resources to each operation. Although very much larger than ISA (the knowledge-base occupies over 10 megabytes of disk space), ISIS also can act as an intelligent assistant, using its expertise to help plant schedulers maintain schedule consistency and identify decisions that result in unsatisfied constraints [156]. Developed for scheduling production at a Westinghouse facility manufacturing turbine blades, the prototype system was not implemented, in part because of difficulties in integrating the scheduling system with existing databases and information systems. Extensions of the underlying constraint-directed scheduling concept subsequently have been embodied in ISIS-2 and OPIS (Opportunistic Intelligent Scheduler) [154].

GENSCHED (General Scheduling Knowledge-Based System) is a multi-domain, intelligent scheduling system which uses a mixture of optimization techniques, hierarchical planning, and heuristic search to provide planning and scheduling capabilities for a wide range of problems [149]. Efforts to date have concentrated on developing the conceptual model for applications in a manufacturing environment, building a petroleum routing scheduler, and producing a hierarchical planner and scheduler. Reduction of the GENSCHED conceptual model to actual practice, however, does not as yet appear to have been demonstrated.

AI concepts and techniques also were used in the prototype Fault Identification and Expediting Repair (FIXER) system [142]. FIXER was developed for aircraft repair-job scheduling. The scheduling of aircraft repair jobs can be characterized as stochastic, dynamic, job-shop scheduling, with varying objectives and constraints. The current version of the FIXER prototype uses heuristics to avoid the combinatorial problem. The heuristics are the loading rules found in the operations research scheduling literature.

PLANNET (Planning and Scheduling Expert System) is a knowledge-based system for scheduling cargo operations for NASA's space shuttle program. A set of rules are applied to an initial schedule in order to improve upon it. Current rules of the system deal with allocating overtime and extra resources when there exists a possibility that a mission completion time will be missed. Manual or semi-manual scheduling are also possible and the user has immediate feedback as to how the changes affect the schedule [149].

IV. SUMMARY AND FUTURE DIRECTIONS

No single modeling paradigm currently appears to offer the basis for unified theory of production scheduling, or to provide an appropriate calculus for generating schedules, or even to support a complete representation of the attributes of the complex production scheduling environment. Control theory, simulation, and AI knowledge-representation schemes appear to capable of capturing a wide range of problem attributes, but fail as yet to provide insights on workable solution strategies. Machine sequencing, resource-constrained project scheduling, and AI search techniques offer insight into possible solution approaches, but do not nearly address the richness of the complex scheduling environment. At least in the near term, we suspect that a synthesis of paradigms will be required.

One such synthesis might rely on the control paradigm to define the nature and objectives of scheduling problems, serving as a problem framework. Within this framework, combined solution heuristics from machine scheduling, resource-constrained project scheduling, and AI search could be used to generate candidate schedules. The performance of candidate schedules could be verified using simulation models, where the simulations could be manipulated interactively by an expert scheduler. Such a synthesis would seemingly wed the rich modeling aspects of control and simulation and with the solution-oriented techniques of heuristic algorithms. This and other syntheses appear to merit further exploration.

ACKNOWLEDGEMENT

The author is very much indebted to Fred Rodammer, his former student co-author on an earlier version of this chapter.

REFERENCES

1. K. P. White, Jr., "Computer-Aided Design and Manufacturing," in "Encyclopedia of Science and Technology," 7th ed., McGraw- Hill, New York, 1992.

2. K. P. White, Jr., and C. M. Mitchell, "Manufacturing Systems Engineering: A Second Industrial Revolution?," IEEE Transactions on Systems, Man, and Cybernetics 19, 161-163 (1989).

3. F. A. Rodammer and K. P. White, Jr., "A Recent Survey of Production Scheduling," IEEE Transactions on Systems, Man, and Cybernetics 18, 841-851 (1989).

4. C. Abraham, B. Dietrich, S. Graves and W. Maxwell, "A Research Agenda for Models to Plan and Schedule Manufacturing Systems," working paper, IBM Watson Research Center,(1985).

5. K. Baker, "Introduction to Sequencing and Scheduling," Wiley, New York, 1974.

6. R. Bellman, A. O. Esogbue and I. Nabeshima, "Mathematical Aspects of Scheduling and Applications," Pergamon, New York, 1982.

7. E. G. Coffman, Jr., ed., "Computer and Job Shop Scheduling," Wiley, New York, 1976.

8. R. W. Conway, W. L. Maxwell and L. W. Miller, "Theory of Scheduling," Addison-Wesley, Reading, MA, 1967.

9. J. E. Day and M. P. Hottenstein, "Review of Sequencing Research," Naval Research Logistics Quarterly 17 (1970).

10. M. A. H. Dempster, J. K. Lenstra and A. H. G. Rinnooy Kan, eds., "Deterministic and Stochastic Scheduling," Reidel, Dordrecht, 1982.

11. S. French, "Sequencing and Scheduling: An Introduction to the Mathematics of the Job-Shop," Wiley, New York, 1982.

12. S. C. Graves, "A Review of Production Scheduling," Operations Research 29, 646-675 (1981).

13. A. Kusiak and W. E. Wilhelm, eds., "Analysis, Modelling, and Design of Modern Production Systems," Annals of Operations Research 17, 1989.

14. J. F. Muth and G. L. Thompson, eds., "Industrial Scheduling," Prentice-Hall, Englewood Cliffs, NJ, 1963.

15. S. S. Panwalkar, R. A. Dudek and M. L. Smith, "Sequencing Research and the Industrial Scheduling Problem" in "Lecture Notes in Economics and Mathematical Systems: Symposium on the Theory of Scheduling and Its Applications," Springer-Verlag, New York, 1973.

16. A. H. G. Rinnooy Kan, "Machine Scheduling Problems: Classification, Complexity and Computations," Martinus Nijhoff, The Hague, 1976.

17. F. A. Rodammer, "A Production Scheduling Procedure For Realistic Manufacturing Systems," Ph.D. Dissertation, University of Virginia, Department of Systems Engineering, Charlottesville, VA (1987).

18. F. A. Rodammer and K. P. White, Jr., K. P., "Design Requirements for a Real-Time Production Scheduling Decision Aid", in "Real-Time Optimization in Automated Manufacturing Facilities," (R. H. F. Jackson and W. T. Jones, eds.) National Bureau of Standards Special Publication 724, Gaithersburg, MD (1986).

19. R. V. Rogers, "Generalizations Of The Machine Scheduling Problem," Ph.D. Dissertation, University of Virginia, Department of Systems Engineering, Charlottesville, VA (1987).

20. G. Buxey, "Production Scheduling: Practice and Theory," European Journal of Operations Research 39, 17-31 (1989).

21. K. A. Fox, "MRP-II Providing a Natural Hub For Computer-Integrated Manufacturing System," Industrial Engineering, 44-50 (1984).

22. E. M. Goldratt, "Computerized Shop Floor Scheduling," International Journal of Production Research 26, 443-455 (1988).

23. S. C. Graves, "Safety Stocks in Manufacturing Systems," Journal of Manufacturing and Operations Management 1, 67-101 (1988).

24. W. R. Hall, "Zero Inventories," Dow Jones-Irwin, Homewood, IL, 1983.

25. F. R. Jacobs, "OPT Uncovered: Many Production Planning and Scheduling Concepts Can Be Applied With or Without the Software," Industrial Engineering, 32-41 (1984).

26. F. R. Jacobs, "The OPT Scheduling System: A Review of a New Production Scheduling System," Production and Inventory Management 24, 47-51 (1983).

27. J. J. Kanet, "MRP96: Time to Rethink Manufacturing Logistics," Production and Inventory Management Journal 29, 57-61 (1988).

28. R. J. Levulis, "Finite Capacity Scheduling and Simulation Systems," Report MTIAC TA-85-04, Manufacturing Technology Information Analysis Center, Chicago, IL, (1985).

29. O. Kimura and H. Terada, "Design and Analysis of Pull System: A Method of Multi-Stage Production Control," International Journal of Production Research 19, 241-253 (1981).

30. K. N. McKay, F. R. Safayeni, and J. A. Buzacott, "Job-Shop Scheduling Theory: What is Relevant?," Interfaces 18, 84-90 (1988).

31. G. W. Plossl, "Production and Inventory Control: Principles and Techniques," 2nd ed., Prentice-Hall, Englewood Cliffs, NJ, 1985.

32. R. J. Schonberger, "Just-In-Time Production Systems: Replacing Complexity With Simplicity In Manufacturing Management," Industrial Engineering, 52-63 (1984).

33. R. J. Schonberger, "World Class Manufacturing: The Lessons of Simplicity Applied," The Free Press, New York, 1986.

34. J. Adams, E. Balas and D. Zawick, "The Shifting Bottleneck Procedure for Job Shop Scheduling," Management Science 34, 391-401 (1988).

35. J. Barker and G. McMahon, "Scheduling the General Job Shop", Management Science 31, 594-598 (1985).

36. J. H. Blackstone, D. T. Phillips and G. L. Hogg, "A State-of-the-Art Survey of Dispatching Rules for Manufacturing Job Shop Operations," International Journal of Production Research 20, 27- 45 (1982).

37. E. H. Bowman, "The Schedule-Sequencing Problem," Operations Research 7, 621-624 (1959).

38. J. Carlier and E. Pinson, "An Algorithm for Solving the Job-Shop Problem," Management Science 35, 164-176 (1989).

39. W. S. Gere, "Heuristics in Job Shop Scheduling," Management Science 13, 167-190 (1966).

40. B. Giffler and G. L. Thompson, "Algorithms for Solving Production-Scheduling Problems," Operations Research 8, 487-503 (1960).

41. J. R. King and A. S. Spachis, "Heuristics For Flow-Shop Scheduling," International Journal of Production Research 18, 345-357 (1980).

42. A. S. Manne, "On the Job-Shop Scheduling Problem," Operations Research 9, 219-223 (1960).

43. P. Mellor, "A Review of Job Shop Scheduling," Operational Research Quarterly 17, 161-171 (1966).

44. S. S. Panwalkar and W. Iskander, "A Survey of Scheduling Rules," Operations Research 25, 45-61 (1977).

45. Y. B. Park, C. D. Pegden and E. E. Enscore, "A Survey and Evaluation of Static Flowshop Scheduling Heuristics," International Journal of Production Research 22, 127-141 (1984).

46. A. S. Spachis and J. R. King, "Job-Shop Scheduling Heuristics With Local Neighborhood Search," International Journal of Production Research 17, 507-526 (1979).

47. E. Balas, "Project scheduling with Resource Constraints," in "Applications of Mathematical Programming," (E.M.L. Beale, ed.) The English University Press, London, 187-200, 1971.

48. M. Bartusch, R. H. Mohring and F. J. Radermacher, "Scheduling Project Networks with Resource Constraints and Time Windows," Annals of Operations Research 16, 201-240 (1988).

49. G. E. Bennington and L. F. McGinnis, "A Critique of Project Planning With Constrained Resources," in "Lecture Notes in Economics and Mathematical Systems: Symposium on the Theory of Scheduling and Its Applications," Springer-Verlag, New York, 1973.

50. E. W. Davis, "Project Scheduling Under Resource Constraints-- Historical Review and Categorization of Procedures," AIIE Transactions 5, (1973).

51. R. F. Deckro and J. E. Herbert, "Resource Constrained Project Crashing, " OMEGA 17, 69-79 (1989).

52. S. E. Elmaghraby, "Activity Networks: Project Planning and Control by Network Models," Wiley, New York, 1977.

53. D. Golenko-Ginzburg, "Controlled Alternative Activity Networks for Project Management," European Journal of Operations Research 37, 336-346 (1988).

54. J. J. Moder, C. R. Phillips and E. W. Davis, "Project Management with CPM, PERT and Precedence Diagramming," Van Nostrand Reingold, New York, 1983.

55. R. P. Mohanty, "Multiple Projects-Multiple Resources-Constrained Scheduling: Some Studies," International Journal of Production Research 27, 261-280.

56. R. H. Mohring, "Minimizing the Costs of Resource Requirements in Project Networks Subject to Fixed Completion Time," Operations Research 32, 89-120 (1982).

57. F. J. Radermacher, "Scheduling of Project Networks," Annals of Operations Research 4, 227-252 (1986).

58. C. T. Ragsdale, "The Current State of Network Simulation in Project Management Theory and Practice," OMEGA 17, 21-25 (1989).

59. J. Stinson, E. W. Davis and B. M. Khumawala, "Multiple Resource-Constrained Scheduling using Branch and Bound," AIIE Transactions 10, (1978).

60. J. D. Weist, "A Heuristic Model for Scheduling Large Projects with Limited Resources," Management Science 15, (1967).

61. B. M. Woodworth, "Identifying the Critical Sequence in a Resource Constrained Project," International Journal of Project Management 6, 89-96 (1988).

62. R. Akella, Y. Choong and S. B. Gershwin, "Performance of Hierarchical Production Scheduling Policy," IEEE Transactions on Components, Hybrids, and Manufacturing Technology 7, 225-238 (1984).

63. J. C. Ammons, T. Govindaraj, and C. M. Mitchell, "Decision Models for Aiding FMS Scheduling and Control," IEEE Transactions on Systems, Man, and Cybernetics 18, 744-756 (1988)

64. X. R. Cao, "The Predictability of Discrete-Event Systems," IEEE Transactions on Automatic Control 34, 1168-1171 (1989).

65. X. R. Cao and Y. C. Ho, "Sensitivity Analysis and Optimization of Throughput in a Production Line With Blocking," IEEE Transactions on Automatic Control 32, 959-967 (1987).

66. B. A. Carre, "An Algebra for Network Routing Problems," Journal Inst Math Appl 7, 273-294 (1971).

67. C. G. Cassandras and Y. C. Ho, "An Event Domain Formalism for Sample Path Perturbation Analysis of Discrete Event Dynamic Systems," IEEE Transactions on Automatic Control 30, 1217-1985 (1985).

68. C. G. Cassandras and S. G. Strickland, "A General Approach for Sensitivity Analysis of Discrete Event Dynamic Systems," IEEE Transactions on Automatic Control 32, (1988).

69. C. Cieslak, C. Declaux, A. Fawaz and P. Varaiya, "Supervisory Control of Discrete Event Systems with Partial Observations," IEEE Transactions on Automatic Control 33, 249-260 (1988).

70. G. Cohen, D. Dubois, J. P. Quadrat and M. Voit, "A Linear System-Theoretic View of Discrete-Event Processes and Its Use for Performance Evaluation in Manufacturing," IEEE Transactions on Automatic Control 30, 210-220 (1985).

71. R. A. Cuninghame-Green, "Describing Industrial Processes and Approximating Their Steady-State Behavior," <u>Operations Research Quarterly 13</u>, 95-100 (1962).

72. R. A. Cuninghame-Green, "Projections in Minimax Algebra," <u>Mathematical Programming 10</u>, 111-123 (1976).

73. S. B. Gershwin, R. Akella and Y. F. Choong, "Short-Term Production Scheduling of an Automated Manufacturing Facility," <u>IBM Journal of Research and Development 29</u>, 392-400 (1985).

74. S. B. Gershwin, R. R. Hildebrant, R. Suri and S. K. Mitter, "A Control Perspective on Recent Trends in Manufacturing Systems," <u>IEEE Control Systems</u>, 3-15 (1986).

75. H. P. Hillion and J. Proth, "Performance Evaluation of Job-Shop Systems Using Timed Event Graphs," <u>IEEE Transactions on Automatic Control 34</u>, 3-9 (1989).

76. W. B. Gong and Y. C. Ho, "Smoothed (Conditional) Perturbation Analysis of Discrete Event Dynamical Systems," <u>IEEE Transactions on Automatic Control 32</u>, 858-866 (1987).

77. Y. C. Ho, "A New Approach to the Analysis of Discrete Event Dynamic Systems," <u>Automatica 19</u>, 149-167 (1983).

78. Y. C. Ho, "Editorial: Basic Research, Manufacturing Automation, and Putting the Cart Before the Horse," <u>IEEE Transactions on Automatic Control 32</u>, 1042-1043 (1987).

79. Y. C. Ho, "Performance Evaluation and Perturbation Analysis of Discrete Event Dynamic Systems," <u>IEEE Transactions on Automatic Control 33</u>, 563-572 (1988).

80. Y. C. Ho, "Perturbation Analysis Explained," <u>IEEE Transactions on Automatic Control 33</u>, 761-763 (1988).

81. Y. C. Ho and S. Li, "Extensions of Infinitesimal Perturbation Analysis," <u>IEEE Transactions on Automatic Control 33</u>, 427-438 (1988).

82. R. F. McPherson, "Scheduling the Period Flow Line," Ph.D. Dissertation, University of Virginia, Department of Systems Engineering, Charlottesville, VA (1988).

83. J. L. Peterson, "Petri Net Theory and the Modeling of Systems," Prentice-Hall, Englewood Cliffs, NJ, 1981.

84. P. J. Ramadge and W. M. Wonham, "Supervisory Control of a Class of Discrete Event Processes," SIAM Journal of Control and Optimization 25, 206-230 (1987).

85. P. J. G. Ramadage, "Some Tractable Supervisory Control Problems for Discrete-Event Systems Modeled by Buchi Automata," IEEE Transactions on Automatic Control 34, 10-19 (1989).

86. G. Tadmor and O. Maimon, "Control of Large Discrete Event Dynamic Systems: Constructive Algorithms," IEEE Transactions on Automatic Control 34, 1164-1168 (1989).

87. C. C. White, III, "Final Report of an NSF-Sponsored Workshop Entitled Decision and Control of Discrete Event Dynamic Systems," Airlie, VA, 1988. Abridged versions of this report appear in IEEE Control Systems Magazine 9, 78 (1989), and OR/MS Today 16, 20-22 (1989).

88. W. M. Wonham, "On Control of Discrete Event Systems," in "Computational and Combinatorial Methods in System Theory," (C. I. Byrnes and A Lindquist, eds.), Elsevier, New York, 1986.

89. A. D. Amar and J. N. D. Gupta, "Simulated Versus Real Life Data in Testing the Efficiency of Scheduling Algorithms," IIE Transactions 18, 16-25 (1986).

90. C. T. Baker and Dzielinski, "Simulation of a Simplified Job Shop," Management Science 6, 311-323 (1960).

91. M. H.Bulkin, J. L. Colley and H. W. Steinhoff, Jr., "Load Forecasting, Priority Sequencing, and Simulation in a Job Shop Control System," Management Science 13, (1966).

92. B. L. Detrich and B. M. March, "An Application of a Hybrid Approach to Modeling a Flexible Manufacturing System," Annals of Operations Research 3, 393-402 (1985).

93. P. W. Glynn and D. L. Iglehart, "Simulation Methods for Queues: An Overview," Queueing Systems 3, 221-256 (1988).

94. R. D. Hurrion, "An Investigation of Visual Interactive Simulation Methods Using the Job-Shop Scheduling Problem," Journal of the Operational Research Society 29, 1085-1093 (1987).

95. A. M. Law, "Simulation Series; Part I: Introducing Simulation: A Tool for Analyzing Complex Systems," Industrial Engineering, 46-63 (1986).

96. A. M. Law and W. D. Kelton, "Simulation Modeling and Analysis," McGraw-Hill, New York, 1982.

97. R. G. Sargent, "Event Graph Modelling for Simulation with and Application to Flexible Manufacturing Systems," Management Science 34, 1231-1251 (1988).

98. R. E. Shannon, "Artificial Intelligence and Simulation," 1984 Winter Simulation Conference Proceedings, 1984.

99. K. E. Stecke and J. J. Solberg, "Loading and Control Policies for a Flexible Manufacturing System," International Journal of Production Research 19, 481-490 (1981).

100. R. Suri and J. W. Dille, "On-Line Optimization of Flexible Manufacturing Systems Using Perturbation Analysis," Proceedings of the First ORSA/TIMS Special Interest Conference on Flexible Manufacturing Systems, Ann Arbor, Michigan, 379-384 (1984).

101. B. D. Ripley, "Simulation Methodology--An Introduction for Queueing Theorists," Queueing Systems 3, 201-220 (1988).

102. P. Afentakis, "Simultaneous Lot Sizing and Sequencing for Multistage Production Systems," IIE Transactions 17, 327-331 (1985).

103. A. Arbel and A. Seidmann, "Performance Evaluation of Flexible Manufacturing Systems," IEEE Transactions on Systems, Man, and Cybernetics 14, 606-617 (1984).

104. D. P. Bertsekas, "Dynamic Programming and Stochastic Control," Academic Press, New York, 1976.

105. J. A. Buzacott, "Optimal Operating Rules for Automated Manufacturing Systems," IEEE Transactions on Automatic Control 17, 80-86 (1982).

106. C. Derman, "Finite State Markovian Decision Processes," Academic Press, New York, 1970.

107. E. B. Dynkin and A. Yushkevich, "Controlled Markov Processes," Springer-Verlag, New York, 1979.

108. D. Heyman and M. Sobel, "Stochastic Models in Operations Research," McGraw-Hill, 1982.

109. F. S. Hillier and G. J. Lieberman, "Introduction to Operations Research," Holden-Day, Oakland, CA, 1980.

110. Y. C. Ho, X. Cao, and C. Cassandras, "Infinitesimal and Finite Perturbation Analysis for Queueing Networks," Automatica 19, 439-445 (1983).

111. J. R. Jackson, "Jobshop-Like Queueing Systems," Management Science 10, 131-142 (1963).

112. U. S. Karmarkar, S. Kekre, S. Kekre and S. Freeman, "Lot-Sizing and Lead-Time Performance in a Manufacturing Cell," Interfaces 15, 1-9 (1985).

113. K. Kivenko, "Managing Work-In-Process Inventory," Marcel Dekker, New York, 1981.

114. L. Kleinrock, "Queueing Systems," Wiley, New York, 1976.

115. E. Koeningsberg, "Twenty-Five Years of Cyclic Queues and Closed Queueing Networks," Journal of the Operational Research Society 33, 605-619 (1982).

116. J. B. Lasserre, C. Bes and F. Roubellat, "The Stochastic Discrete Dynamic Lot Size Problem: An Open-Loop Solution," Operations Research 33, 684-689 (1985).

117. A. J. Lamoine, "Networks of Queues--A Survey of Equilibrium Analysis," Management Science 24, 464-481 (1972).

118. W. Lin and P. R. Kumar, "Optimal Control of a Queueing Systems with Two Heterogeneous Servers," IEEE Transactions on Automatic Control 29, 696-703 (1984).

119. W. G. Marchal, "The Performance of Approximate Queueing Formulas: Some Numerical Remarks," <u>Proceedings of the First ORSA/TIMS Special Interest Conference on Flexible Manufacturing Systems</u>, Ann Arbor, Michigan, 123-128 (1984).

120. J. O. McClain and L. J. Thomas, "Operations Management: Production of Goods and Services,"Prentice-Hall, Englewood Cliffs, NJ, 1985.

121. F. H. Moss and A. Segall, "An Optimal Control Approach to Dynamic Routing in Networks," <u>IEEE Transactions on Automatic Control 27</u>, 329-339 (1982).

122. A. S. Raturi and A. V. Hill, "An Experimental Analysis of Capacity Sensitive Lotsizing Methodologies," working paper, University of Minnesota, April, 1986.

123. S. M. Ross, "Applied Probability Models with Optimization Applications," Holden-Day, San Francisco, 1970.

124. S. M. Ross, "Introduction to Stochastic Dynamic Programming," Academic Press, New York, 1983.

125. K. W. Ross and D. D. Yao, "Optimal Dynamic Scheduling in Jackson Networks," <u>IEEE Transactions on Automatic Control 34</u>, 47-53 (1989).

126. A. Seidmann and P. J. Schweitzer, "Part Selection Policy for a Flexible Manufacturing Cell Feeding Several Production Lines," <u>IIE Transactions 16</u>, 355-362 (1984).

127. A. Seidmann and A. Tenenbaum, "Optimal Queueing Systems Controls with Finite Buffers and with Multiple Component Cost Functions," <u>IEEE Transactions on Systems, Man, and Cybernetics 19</u>, 356-364 (1989).

128. M. Sepehri, E. A. Silver and C. New, "A Heuristic for Multiple Lot Sizing For an Order Under Variable Yield," <u>IIE Transactions 18</u>, 63-69 (1986).

129. K. E. Stecke and J. J. Solberg, "The Optimality of Unbalanced Workloads and Machine Group Sizes in Closed Queueing Networks of Multiserver Queues," <u>Operations Research 33</u>, 882-910 (1985).

130. E. Steinberg and H. A. Napier, "Optimal Multi-Level Lot Sizing for Requirements Planning Systems", Management Science 26, 1258-1263 (1980).

131. S. Stidham, "Optimal Control of Admission to a Queueing System," IEEE Transactions on Automatic Control 30, 705-713 (1985).

132. D. D. Yao and J. A. Buzcott, "Modeling a Class of State-Dependent Routing in Flexible Manufacturing Systems," Annals of Operations Research 3, 153-168 (1985).

133. D. D. Yao and J. A. Buzcott, "Modeling the Performance of Flexible Manufacturing Systems," International Journal of Production Research 23, 1141-1151 (1985).

134. D. D. Yao and Z. Schechner, "Decentralized Control of Service Rates in a Closed Jackson Queue," IEEE Transactions on Automatic Control 34, 236-245 (1989).

135. D. D. Yao and J. G. Shanthikumar, "The Optimal Input Rate to a System of Manufacturing Cells," INFOR 25, 57-65 (1987).

136. P. R. Winters, "Constrained Inventory Rules for Production Smoothing," Management Science 8, 470-481 (1962).

137. P. Alpar and K. N. Srikanth, "A Comparison of Analytic and Knowledge-Based Approaches to Closed-Shop Scheduling," Annals of Operations Research 17, 347-362 (1989).

138. G. Bruno, A. Elia and P. Laface, "A Rule-Based System to Schedule Production," IEEE Computer 19, 32-40 (1986).

139. V. Chiodini, "A Knowledge-Based System for Dynamic Manufacturing Replanning," Proceedings of the Symposium on Real-Time Optimization in Automated Manufacturing Facilities, National Bureau of Standards Publication 724, Gaithersburg, MD, 357-372 (1986).

140. L. Davis, "Job Shop Scheduling with Genetic Algorithms," Proceedings of an International Conference on Genetic Algorithms and Their Applications, Carnegie-Mellon University, Pittsburgh, PA, 1985.

141. M. S. Fox, B. P. Allen, S. F. Smith and G. A. Strohm, "ISIS: A Constraint-Directed Reasoning Approach to Job Shop Scheduling: System Summary," Carnegie-Mellon University, The Robotics Institute, CMU-RI-TR-83-8, 1983.

142. T. J. Grant, "Lessons for O.R. From A.I.: A Scheduling Case Study," Journal of the Operational Research Society 37, 41-57 (1986).

143. R. A. Herrod and B. Papas, "Artificial Intelligence Moves Into Industrial and Process Control," Industrial and Process Control Magazine, 45-60 (1985).

144. R. H. F. Jackson and A. T. Jones, "An Architecture for Decision Making in the Factory of the Future," Interfaces 17, 15- 28 (1987).

145. J. J. Kanet and H. H. Adelsberger, "Expert Systems in Production Scheduling," European Journal of Operational Research 29, 51-57 (1987).

146. A. Kusiak, "Designing Expert Systems for Scheduling of Automated Manufacturing," Industrial Engineering, July, 42-46 (1987).

147. A. Lamatsch, M. Morlock K. Neumann, and T. Rabach, "SCHEDULE--An Expert-Like System for Machine Scheduling," Annals of Operations Research 16, 425-438 (1988).

148. P. S. Ow and T. E. Morton, "An Investigation of Beam Search for Scheduling," working paper, Carnegie-Mellon University, Pittsburgh, PA, 1985.

149. A. C. Semeco, S. P. Roth, B. D. Williams and J. Gilmore, "The GENSCHED Project: General Scheduling Knowledge Base System," Final Report, Georgia Institute of Technology, 1986.

150. M. J. Shaw, "Knowledge-Based Scheduling in Flexible Manufacturing Systems," Texas Instruments Technical Journal, Winter, 54-61 (1987).

151. M. J. Shaw, "A Pattern-Directed Approach to FMS Scheduling," Annals of Operations Research 15, 353-376 (1988).

152. M. J. Shaw, "FMS Scheduling as Cooperative Problem Solving," Annals of Operations Research 17, 323-346 (1989).

153. M. J. Shaw and A. B. Whinston, "An Artificial Intelligence Approach to the Scheduling of Flexible Manufacturing Systems," IIE Transactions 21, 170-183.

154. S. F. Smith, M. S. Fox and P. S. Ow, "Constructing and Maintaining Production Plans: Investigations into the Development of Knowledge-Based Factory Scheduling Systems, AI Magazine 7, Fall, 45-61 (1986).

155. A. Thesen and L. Lei, "An Expert System For Scheduling Robots In A Flexible Electroplating System With Dynamically Changing Workloads," Technical Report, University of Wisconsin-Madison, 1986.

156. D. A. Waterman, "A Guide to Expert Systems", Addison-Wesley Publishing Company, Reading, Massachusetts, 1986.

157. S. A. Vere, "Planning in Time: Windows and Durations for Activities and Goals," IEEE Transactions on Pattern Analysis and Machine Intelligence 5, 246-267 (1983).

DETECTION AND DIAGNOSIS OF PLANT FAILURES: THE ORTHOGONAL PARITY EQUATION APPROACH

Janos J. Gertler

with

Xiaowen Fang and Qiang Luo

George Mason University
School of Information Technology and Engineering
Fairfax, VA 22030

1. INTRODUCTION

The detection and isolation (diagnosis) of failures, whether they occur in the basic technological equipment or in the measurement and control system, has always been one of the major tasks of computers supervising complex production plants, such as chemical plants, oil refineries or power stations. With more stringent requirements on performance and economy and, especially, with the proliferation of the microprocessor, similar systems have appeared more recently on airplanes, automobiles and even on household appliances.

In practice, the most frequent approach to failure detection and diagnosis implies the comparison of plant measurements to set limits. While this approach is, no doubt, very simple, it is overly cautious since it does not take into account how those limits vary according to the state of the system. A substantial improvement in failure sensitivity can be achieved by taking the mathematical plant model into account. Such model based methods utilize the notion of analytical redundancy, meaning that the actual plant measurements are checked for consistency with the mathematical model. The resulting residuals are further analyzed, usually involving statistical testing, to arrive at a diagnostic decision.

A significant class of model based failure detection and diagnosis methods uses Kalman filters [1,2,3] or observers [4] to generate residuals. This approach leads to relatively complex algorithms and may not support failure isolation in a very natural way. More recently, techniques relying on parity equations for residual generation have gained popularity [5,6,7]. Parity equations are rearranged and usually transformed variants of the input–output model equations. They are relatively easy to generate and use and, as it will be shown below, can be well utilized in failure isolation.

In this paper, a general model–based failure detection and diagnosis methodology will be described. The approach uses parity equations that return residuals orthogonal to certain failures in the system. Isolability considerations lead to special structural requirements on the diagnostic models formed from these equations. Transformation techniques will be introduced and discussed in some detail that generate parity equations in the desired structure. The residuals are subjected to individual statistical testing. The computation of test thresholds will be treated and failure sensitivity measures introduced, together with residual filtering schemes aimed at improving sensitivity performance. The concepts and results first developed for additive type failures will later be extended to multiplicative failures. This will also lead to the definition of measures of robustness in the face of modeling errors. A design procedure* will be described that generates diagnostic models satisfying the structural isolability requirement and optimal for some selected sensitivity or robustness measure. Finally, the design will be illustrated on two examples: the models of an automotive engine and of a distillation column.* A more detailed description of the assumptions and objectives of this effort follows the mathematical definition of the problem in Section 2.

Several aspects of the work described here have been reported in previous conference and journal publications [7,8,9]. Related results in parity equation generation by Chow and Willsky [6] have been utilized in our design procedure. Isolability considerations, somewhat similar to ours, have been reported by Ben–Haim [10] and Massoumnia et al [11]. Different techniques to accommodate for modeling errors in the design have been suggested by Leininger [12], Horak [13], Lou et al [14] and Emami–Naeini et al [3]. A detailed treatment of the relevant literature can be found in [15].

---

*These parts have been reprinted from J. Gertler and Q. Luo: Robust isolable models for failure diagnosis, *AICHE Journal*, *31*, 1856–1868 (1989), with the kind permission of the American Institute of Chemical Engineers.

2. PROBLEM STATEMENT

2.1. System Description

We are concerned with the detection and isolation (diagnosis) of failures in multiple–input multiple–output linear dynamic systems. Static systems are included as a special case. Non–linear systems are to be represented by a linearized model.

The system is described by a discretized input–output model. With $u(t)=[u_1(t),..., u_k(t)]^T$ and $y(t)=[y_1(t),...,y_m(t)]^T$ denoting the vector of input and output variables, respectively, the ideal input–output relationship is described as

$$G^*(z)\ u(t) = H^*(z)\ y(t) \tag{2.1}$$

Here $G^*(z)$ and $H^*(z)$ are polynomial matrices of degree n in the inverse shift operator z^{-1} (where n is the system order):

$$G^*(z) = G_o^* + G_1^* z^{-1} + ... + G_n^* z^{-n}$$

$$H^*(z) = H_o^* + H_1^* z^{-1} + ... + H_n^* z^{-n} \tag{2.2}$$

The $H^*(z)$ matrix is also diagonal. This by no means restricts the system, it simply reflects the fact that any output is expressed as a sole function of the inputs. Such input–output equations may be obtained, for example, by solving the usual state–equations for the outputs.

For convenience, the input–output equation (2.1) will be written as

$$F^*(z)\ q(t) = 0 \tag{2.3}$$

where

$$F^*(z) = [G^*(z)\ |\ -H^*(z)] \tag{2.4}$$

and

$$q(t) = \begin{bmatrix} u(t) \\ y(t) \end{bmatrix} \tag{2.5}$$

While Eq. (2.3) contains the true inputs and outputs $q(t)$, computationally the equation can only be applied to the variable values $\tilde{q}(t)$ available on–line. Also, the known system matrix $\hat{F}^*(z)$ is, in general, different from the true model $F^*(z)$. Thus the computational equivalent of Eq. (2.3) is

$$\hat{F}^*(z) \; \tilde{q}(t) = e^*(t) \neq 0 \tag{2.6}$$

where $e^*(t)$ is the vector of (primary) residuals. Equation (2.6) will be referred to as the computational form of the (primary) set of parity equations. The quantities $e^*(t)$ these parity equations return will be termed (primary) residuals; as it will be shown below, they are the combined result of failures, noise and modeling errors.

Three kinds of input variables will be considered:

1. Measured inputs $u_M(t)$ (Fig. 1.a.). These are input variables that are not controlled, at least locally, but measured by some sensor. The measurements are subject to sensor bias Δu_M and sensor noise δu_M, with the available (measured) value $\tilde{u}_M$ being

$$\tilde{u}_M(t) = u_M(t) + \Delta u_M(t) + \delta u_M(t) \tag{2.7}$$

2. Controlled inputs $u_C(t)$ (Fig. 1.b.). These are input variables controlled by some actuators. For these, the command value $\tilde{u}_C$ is available on–line, while the true control inputs u_C are contaminated with actuator bias Δu_C and actuator noise δu_C. Thus

$$\tilde{u}_C(t) = u_C(t) - \Delta u(t) - \delta u_C(t) \tag{2.8}$$

3. Disturbance inputs $u_D(t)$ (Fig. 1.c.). These represent the additive plant failures (leaks, loads, etc.) Δu_D and any plant noise δu_D, that is,

$$u_D(t) = \Delta u_D(t) + \delta u_D(t) \tag{2.9}$$

The disturbance inputs are nominally zero; their actual value is not known. To facilitate uniform treatment with the other inputs, a "measured" value will be defined as

$$\tilde{u}_D(t) = u_D(t) - \Delta u_D(t) - \delta u_D(t) = 0 \tag{2.10}$$

The outputs $y(t)$ are measured with sensor bias $\Delta y(t)$ and sensor noise $\delta y(t)$, so their available (measured) value $\tilde{y}(t)$ is

$$\tilde{y}(t) = y(t) + \Delta y(t) + \delta y(t) \tag{2.11}$$

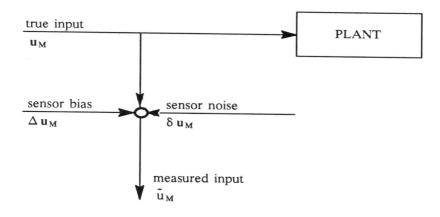

a. Measured input variable

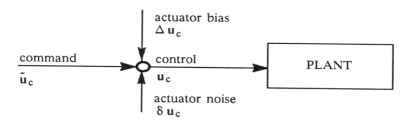

b. Controlled input variable

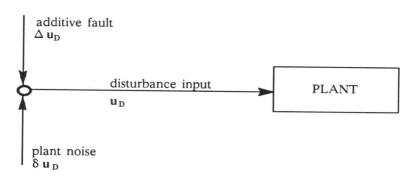

c. Disturbance input variable

Figure 1.

Combining Eq. (2.5) with (2.7) through (2.11) yields

$$
\tilde{q}(t) =
\begin{bmatrix}
\tilde{u}_M \\
\tilde{u}_C \\
\tilde{u}_D \\
\tilde{y}
\end{bmatrix}
=
\begin{bmatrix}
u_M \\
u_C \\
u_D \\
y
\end{bmatrix}
+
\begin{bmatrix}
\Delta u_M \\
-\Delta u_C \\
-\Delta u_D \\
\Delta y
\end{bmatrix}
+
\begin{bmatrix}
\delta u_M \\
-\delta u_C \\
-\delta u_D \\
\delta y
\end{bmatrix}
$$

$$
= q(t) + \Delta q(t) + \delta q(t) \tag{2.12}
$$

Of the components of $\tilde{q}(t)$, in accordance with the above discussion, $\tilde{u}_M(t)$ and $\tilde{y}(t)$ are measured values, $\tilde{u}_C(t)$ is the command value and $\tilde{u}_D(t)=0$.

The available system matrix $\hat{F}^*(z)$ is related to the true plant model $F^*(z)$ as

$$
\hat{F}^*(z) = F^*(z) + \Delta F^*(z) \tag{2.13}
$$

where $\Delta F^*(z)$ is the model discrepancy. This includes

a. parametric plant failures, that is, changes of plant parameters unaccounted for in the model (such as surface contamination, performance reduction, etc),

b. modelling errors, that is, inaccuracies in the original representation of the plant.

Substituting Eq. (2.12) and (2.13) into Eq. (2.6) and taking Eq. (2.3) into account yields

$$
e^*(t) = \hat{F}^*(z) \, \Delta q(t) + \hat{F}^*(z) \, \delta q(t) + \Delta F^*(z) \, q(t) \tag{2.14}
$$

presenting the residuals as a combined effect of additive failures $\Delta q(t)$, noises $\delta q(t)$ and model discrepancies $\Delta F^*(z)$. Equation (2.14) will be referred to as the internal form of the (primary) parity equations.

2.2. Outline and Assumptions

Our objective is to detect and isolate (diagnose) failures on the basis of parity equation residuals. Our approach is based on the following assumptions and choices:

A. It will be assumed that failures are either of step–type, occurring at an arbitrary time and staying, or of drift–type, starting at an arbitrary moment and increasing with time, not necessarily linearly. Though explicit failure models will not be used, the above assumptions will be implicitly utilized, particularly in the process of separating failures from noise effects.

B. We will not be interested in the size of the failure, nor in the exact time of its occurrence. The algorithm we are going to develop is meant for on–line use, providing detection and isolation information in real–time, its reaction delay depending somewhat on the failure size.

C. Only single failures will be considered. The extension of the numerical methodology to multiple failures is straightforward but implies a combinatorial increase in certain aspects of the algorithm. Multiple failures are treated in a natural way, at smaller added computational cost, in an evidential reasoning extension of the basic numerical algorithm; this, however, will not be treated here [16].

All noises are assumed to be random, with zero mean and known distribution. This latter assumption may prove unrealistic in practice, however, the method is not very sensitive to noise parameters. To separate noise effects from those of failures, the residuals will be subjected to statistical testing. The tests will be performed separately for each residual, in parallel, in order to facilitate failure isolation. Failure sensitivity of the tests will be considered; to improve sensitivity without increasing the risk of false alarms, residuals may be filtered before testing.

Diagnostic methods first to be developed for additive failures (measurement and actuator bias and leak/load type plant faults) will later be extended to parametric faults. However, while additive failures have constant coefficients (the model parameters) in the internal parity equations, the coefficients of parametric failures are variables, obviously time–varying. This will result in more extensive on–line computations in the handling of parametric failures. Also, parametric faults are manifested the same way as certain modeling errors therefore the proposed algorithm, in its basic form, will not be able to support their distinction. To facilitate their separation, the long–term behavior of the residuals should be monitored, a possible direction to extend the present methodology. By contributing to the residuals, modeling errors interfere even with the diagnosis of additive failures. Therefore, modeling error robustness will be one of the major considerations in the development of the diagnostic algorithm.

Due to the structure of the $F^*(z)$ matrix, each primary parity equation contains only one output variable and, at least potentially, all the input variables. This structure is not advantageous from the point of view of failure isolation. In the sequel, *"isolable"* model structures will be defined and a procedure introduced to generate sets of transformed parity equations, called *"diagnostic models"*, that satisfy these structural requirements. As it will be shown, there is a combinatorial multitude of such sets. A search algorithm will also be presented that finds the diagnostic model which is the "best" for a selected sensitivity or robustness measure while satisfying the structural constraints for isolability. This model design procedure is computationally expensive and, as such, is performed off–line; it is the resulting diagnostic model that constitutes the heart of the on–line algorithm.

3. ISOLABILITY CONDITIONS

The residuals obtained from a set of parity equations are subjected to statistical testing, in parallel, independent of each other. Such statistical testing implies the comparison of the momentary residual values to thresholds, previously established for each parity equation. The residual values depend on the failure size, the gain with respect to to the particular failure in the parity equation and also on the momentary value of the noise and any modeling error. The thresholds are determined by the noise statistics, the noise propagation within the parity equation and the selected false alarm rate; their computation will be discussed in Section 6.

In the following, two isolability concepts will be introduced. Loosely speaking, a failure is considered isolable if it can be uniquely traced back from the results of the tests on the residuals. Both isolability concepts will be based on *orthogonal* residuals. A residual is orthogonal to a failure if the coefficient of that failure is identically zero in the parity equation. No matter how large, a failure will never trigger the test on a residual that is orthogonal to it.

Clearly, the orthogonality of residuals only depends on the position of zero coefficients, that is, the structure of the parity equations. The isolability conditons will also be formulated in terms of the structure of the diagnostic model. First, however, some structural definitions will be given in the following sub–section.

The isolability concepts will be introduced for generalized parity equations (cf. Eqs. (2.6) and (2.14))

$$e(t) = \hat{F}(z)\, \tilde{q}(t) = \hat{F}(z)\, \Delta q(t) + \hat{F}(z)\, \delta q(t) + \Delta F(z)\, q(t) \qquad (3.1)$$

Such generalized parity equations are obtained from the primary set by transformation, with the purpose of generating isolable structures. The transformation procedure will be discussed in Section 4.

The notion of isolability is rather straightforward for additive failures (measurement and actuator bias and additive plant faults). The following treatment will first be restricted to such failures. An extension to multiplicative (parametric) failures will be presented in Section 5.

3.1. Structural Definitions

A. The structure of the model matrix $F(z)$ is characterized by its Boolean incidence matrix

$$\Phi = Inc[F(z)] \qquad (3.2)$$

where the elements ϕ_{ij} of Φ are related to the elements $f_{ij}(z)$ of $F(z)$ as

$$\phi_{ij} = 0 \text{ if } f_{ij}(z) = 0$$

$$\phi_{ij} = 1 \text{ if } f_{ij}(z) \neq 0 \qquad (3.3)$$

B. A Boolean signature vector $\epsilon(t)$ (representing the outcome of the set of statistical tests) is associated with the residual vector $e(t)$ according to the relationship

$$\epsilon_i(t) = 0 \text{ if } e_i(t) \leq \eta_i$$

$$\epsilon_i(t) = 1 \text{ if } e_i(t) > \eta_i \qquad (3.4)$$

where $\epsilon_i(t)$ and $e_i(t)$ are elements of $\epsilon(t)$ and $e(t)$, respectively, and η_i is the threshold specified for $e_i(t)$.

C. A system (set of parity equations) will be called *row–canonical* of order l if each row of its incidence matrix contains l zeroes, each in a different pattern. Similarly, a system is *column–canonical* of order l if each column of its incidence matrix contains l zeroes, each in a different pattern.

3.2. Deterministic (zero–threshold) isolability

If there is no noise in the system, the test thresholds may be set to zero. Each column $\phi._j$ of the incidence matrix is associated with an element $\Delta q_j(t)$ of the vector of additive faults $\Delta q(t)$. With zero thresholds, any non–zero fault will trigger the test on each residual which is not orthogonal to it. That is, on $\Delta q_j \neq 0$, $\epsilon_i = 1$ iff $\phi_{ij} = 1$. Thus the failure signature is identical with the respective column of the incidence matrix:

$$\epsilon(t|\Delta q_j) = \phi._j \qquad (3.5)$$

From this it follows that

a. A single additive fault is undetectible if its column in the incidence matrix contains all zeroes.

b. Two single additive faults can be isolated from each other if their two respective columns in the incidence matrix are different.

c. A system is zero–threshold isolable with respect to single additive faults if all columns of its incidence matrix are different and non–zero.

Note that the concept of zero–threshold isolability can be extended to multiple additive failures, making use of the fact that such failures result in signature patterns that are the Boolean sums of the respective columns of the incidence matrix.

3.3. Statistical (high–threshold) isolability

In noisy systems, the test thresholds are set to non–zero values. In selecting the threshold values, there is a trade–off between the probability of missed detections and that of false alarms. In general, frequent false alarms are considered most undesirable, therefore the thresholds are set relatively high.

Whatever the thresholds, there is a range for each failure, the values "small" relative to the thresholds, that will practically never trigger any of the tests and there is another range, the "large" values, that will practically always trigger all the tests, except of course the ones on residuals that are orthogonal to the concerned failure. There is, however, an intermediate range, that results in some of the tests firing and some others not. This situation, that will be referred to as *"partial firing"*, may happen with significant frequency if the thresholds are high. The potential danger of partial firing is that the resulting degraded signature may be attributed to the wrong failure. We are looking for a model structure, to be named statistically (or high–threshold) isolable, which guarantees that partial firing does not lead to such mis–isolation of the failure.

Under full firing, Eq. (3.5) stands, that is, the signature $\epsilon(t|\Delta q_j)$ is identical with the j-th column $\phi._j$ of the incidence matrix. In case of partial firing upon Δq_j, a degraded version of $\phi._j$ results, in that some *ones* in $\phi._j$ are replaced by *zeroes*. To attain statistical isolability, no column of the incidence matrix should be a degraded version of any other column. This requires that between each pair of columns there must mutually be at least one position that is occupied by a *one* in one column and by a *zero* in the other.

As can be seen, statistical isolability implies deterministic isolability.

While there are other ways to attain statistical isolability, a convenient solution is obtained by selecting a column–canonical structure. In a column–canonical structure, each column has the same number of zeroes, each in a different pattern. Obviously, no column can be obtained from any other column by degradation.

It should be noted that a statistically isolable structure does not cure false alarms or missed detections. What it does is avoiding mis–isolation of a failure if the intermediate failure size (relative to the thresholds) leads to partial firing of the tests. In case of such a degraded signature, the failure decision is considered undefined. (Further diagnostic information may be extracted from degraded signatures in an evidential reasoning framework; see [16].)

Example

Of the following three incidence matrices, the first characterizes a system
that is not isolable in either sense, the second one that is isolable deterministically
but not statistically while the third a statistically isolable system.

$$
\begin{bmatrix} 1 & 1 & 1 & 0 \\ 1 & 1 & 0 & 1 \end{bmatrix}
\qquad
\begin{bmatrix} 1 & 1 & 1 & 0 \\ 1 & 1 & 0 & 1 \\ 1 & 0 & 1 & 1 \end{bmatrix}
\qquad
\begin{bmatrix} 1 & 1 & 1 & 0 \\ 1 & 1 & 0 & 1 \\ 1 & 0 & 1 & 1 \\ 0 & 1 & 1 & 1 \end{bmatrix}
$$

4. MODEL TRANSFORMATION

In the previous section, the conditions for failure isolability were formulated
in terms of the structure of the set of parity equations. The primary set,
in general, does not satisfy these structural conditions. Therefore, in designing
diagnostic models, new sets are sought by the transformation of the original
model. The target set is defined in terms of the desired structure of the
model, that is, the position of zeroes in the parity equations. The transformation
proper is performed either on the primary set of parity equations,
seeking the transformed residuals as linear combinations of the primary ones
[7,8], or it is done on the underlying state–space model, generating the desired
parity equations directly from the state–equations [6,9]. Both approaches
will be discussed below, strating with the input–output model
approach.

Like in case of the isolability conditions, model transformation will first
be introduced with additive failures in mind, then it will be shown in Section
5. how the results are extended to multiplicative (parametric) failures. For
additive failures, the primary set of parity equations is, at least potentially,
row–canonical of order $m-1$, that is, each equation has $m-1$ zeroes, each in
a different pattern. This is because each equation contains only one output
(out of m) and potentially all inputs (if some inputs are also missing then the
number of zeroes is greater than $m-1$). We will restrict our model transformation
considerations to row–canonical equations. Though the isolability requirements
have been formulated in terms of column structure
(column–canonical structures will be sought for statistical isolability), the design
procedure will be simpler if these structures are formed of row–canonical
equations. The results may easily be extended to particular

non–row–canonical situations, however, the great variety of such structures precludes their general treatment.

Once the desired equation structures have been determined on the basis of the overall structure of the target model, each equation is obtained separately. Therefore, we will discuss first the generation of single equations. The complete model design procedure will be covered in Section 8.

4.1. Transformation of the Primary Set

A transformed residual $e_i(t)$ will be sought as the linear combination of the primary residuals $e^*(t)$:

$$e_i(t) = w(z) \, e^*(t) \tag{4.1}$$

where $w(z)=[w_1(z),...,w_m(z)]$ is a row–vector of transforming polynomials. Since

$$e_i(t)=\hat{f}_i.(z) \, \tilde{q}(t) \tag{4.2}$$

where $\hat{f}_i.(z)$ is the i-th row of the new model matrix $\hat{F}(z)$, and $e^*(t)=\hat{F}^*(z) \, \tilde{q}(t)$ (cf. Eq. (1.6)), Eq. (4.1) implies

$$\hat{f}_i.(z) = w(z) \, \hat{F}^*(z) \tag{4.3}$$

In the sequel, we shall outline a procedure to find $w(z)$; once it is known, the new parity equation is obtained by Eq. (4.3). Certain rank deficiencies of the primary system matrix may render some equation structures unattainable (not possible to generate); such restrictions on the transformation will also be discussed below. Finally, the degree conditions of the transformed parity equations will be investigated.

It is important to note that the transforming polynomials depend only on the primary model matrix and on the structure of the desired equation and do not depend on the actual measurement. Therefore, the transformed equation will be time-invariant and the transformation may be performed off–line.

4.1.1. A transformation procedure. The desired equation is defined in terms of the position of its zeroes; if it belongs to a row-canonical set of order $m-1$, it has $m-1$ zeroes. This provides $m-1$ homogeneous equations for the m elements of $w(z)$:

$$w(z) \; \hat{F}_i^*(z) = 0 \tag{4.4}$$

where $\hat{F}_i^*(z)$ is an $mx(m-1)$ matrix, consisting of those columns of $\hat{F}^*(z)$ that correspond to the zeroes in $\hat{f}_{i.}(z)$. Assume that the target equation is to contain s of the m output variables, then $m-s$ of the $m-1$ columns in $\hat{F}_i^*(z)$ will belong to outputs. These columns contain a single non–zero element. For these, Eq. (4.4) may only be satisfied if the respective elements in $w(z)$ are zero. Removing these from $w(z)$ and also the corresponding columns and rows from $\hat{F}(z)$ leaves

$$w'(z) \; \hat{F}_i^{*\,\prime}(z) = 0 \tag{4.5}$$

Here $\hat{F}_i^{*\,\prime}(z)$ is size $sx(s-1)$ and it contains decimated columns (with $m-s$ elements missing) from the $\hat{G}^*(z)$ part of the $\hat{F}^*(z)$ matrix. Eq. (4.5) is a set of $s-1$ homogeneous equations for s unknowns; one unknown may be chosen freely. Selecting the p–th element of $w'(z)$ for arbitrary value assignment yields

$$w''(z) \; \hat{F}_i^{*\,\prime\prime}(z) = - \, w_p'(z) \; [\hat{F}_i^{*\,\prime}(z)]_p \,. \tag{4.6}$$

where $[...]_p$. indicates the p–th row of the respective matrix. This can be solved as

$$w''(z) = - \, w_p'(z) \; [\hat{F}_i^{*\,\prime}(z)]_p \cdot [\hat{F}_i^{*\,\prime\prime}(z)]^{-1} \tag{4.7}$$

provided that $\hat{F}_i^{*\,\prime\prime}(z)$ is invertible. Choosing $-w_p(z) = \mathrm{Det}[\hat{F}_i^{*\,\prime\prime}(z)]$ then yields

$$w''(z) = [\hat{F}_i^{*\,\prime}(z)]_p \cdot \mathrm{Adj}[\hat{F}_i^{*\,\prime\prime}(z)] \tag{4.8}$$

that is guaranteed to be polynomial. Of course, any polynomial multiple is also a solution.

The above result implies that, for parity equations belonging to a row-canonical set of order $m-1$, the equation structure uniquely determines the equation coefficients, with the understanding that any polynomial multiple of a parity equation is fundamentally the same parity equation.

4.1.2. Unattainable equations.

A desired parity equation structure is clearly unattainable if the transformation procedure aborts. Also, if the resulting equation has excess zeroes (relative to the intended structure), the desired structure may not be attained. Such situations follow from deficiencies of the $\hat{G}^*(z)$ part of the primary system matrix. Though they are basically independent of the particular transformation procedure employed, they may be conveniently demonstrated in connection with the procedure introduced in 4.1.1.

Case a. If one and only one of the $(s-1) \times (s-1)$ submatrices of $\hat{F}_i^{*\,\prime}(z)$ has zero determinant then the selected equation structure is not attainable. The manifestation of this fact depends on the selection of the position p. If this selection makes $\hat{F}_i^{*\,\prime\prime}(z)$ singular then Eq. (4.7) has no solution. Otherwise, at least one of the elements of $w''(z)$ comes out as zero from the solution of Eq. (4.8). Such a zero element, combined in Eq. (4.3) with a column of $\hat{F}^*(z)$ having a single non-zero element, results in an excess *zero* in the parity equation.

Case b. If all the $(s-1) \times (s-1)$ submatrices of $\hat{F}_i^{*\,\prime}(z)$ have less than full rank then there are fewer than $s-1$ independent homogeneous equations so more than one element of $w'(z)$ may be selected freely. In this situation, the full linear dependence between two or more columns of $\hat{F}_i^{*\,\prime}(z)$ indicates that orthogonalization with respect to some of the selected failures implies orthogonalization for other failures as well, ones that have also been assigned for such orthogonalization. Thus the equation is attainable but the solution is not unique.

Case c. Finally, if $[\hat{F}_i^{*\,\prime}(z) \mid \hat{f}_{.j}^{*\,\prime}(z)]$ has less than full rank, where $\hat{f}_{.j}^{*\,\prime}(z)$ is any column of $\hat{F}^*(z)$ outside $\hat{F}_i^*(z)$, with $m-s$ elements omitted to match the structure of $\hat{F}_i^{*\,\prime}(z)$, then the desired parity equation is not attainable. In this case, a decimated column outside $\hat{F}_i^{*\,\prime}(z)$ linearly depends on the columns inside, so orthogonalization with respect to the selected failures implies same for a failure that has not been intended for orthogonalization.

The above situations are created by some local deficiencies of the $\hat{G}^*(z)$ matrix, such as partial dependencies among rows or columns and zero elements. Global deficiencies, namely full row or column dependence, may have a more general effect on the attainability of equations.

Case A. If there are (polynomial) linear dependence relationships among groups of full rows of $\hat{G}^*(z)$, this leads to the situation described in Case a. above, whenever such a group becomes part of the decimated $\hat{F}^{*\,\prime}(z)$ matrix. That is, many of the equation structures will prove unattainable. With r linear relationships, $s=m-r$ is the maximum row–count in $\hat{F}^{*\,\prime}(z)$ that still permits any solution at all. So the maximum number of input variables to which any equation may be made orthogonal under these circumstances is $m-r-1$.

Case B. If there are (polynomial) linear dependence relationships among full columns of the $\hat{G}^*(z)$ matrix, then orthogonalization with respect to the corresponding input variables is fundamentally restricted. If the relationship covers a group of $r \leq m$ columns then elimination of $r-1$ of the concerned inputs implies the elimination of the r-th as well. Depending on the position of the input variables in the target equation structure, this leads to either Case b. or Case c. above.

4.1.3. **Degree conditions.** Equations (4.8) and (4.3) suggest that the parity equations obtained by transformation may be of very high degree. This would, of course, be rather undesirable, implying high computational load in the on–line application of the model and extreme numerical sensitivity. It can be shown, however, that the degree of the transformed parity equations is, in fact, quite limited and is related to the order n_s of the minimum–order underlying state–space representation of the system. For systems with the number of outputs $m \leq n_s$, the following general degree limit holds:

$$\text{Deg}[\hat{f}_{i\cdot}(z)] \leq n_s \qquad \text{all } i \tag{4.9}$$

Further, if $\hat{G}_o^* = 0$ in the primary model (that is, if $\mathbf{D}=0$ in the underlying state–space model) then

$$\text{Deg}[\hat{f}_{i\cdot}(z)] \leq n_s - s + 1 \qquad \text{all } i \tag{4.10}$$

where s is the number of outputs present in the equation. If $m > n_s$ then the above results hold for any subsystem with n_s linearly independent outputs.

The proof of Eqs. (4.9) and (4.10) is quite technical and, therefore, is given in the Appendix.

The transformation procedure introduced in subsection 4.1.1. does not produce the parity equations in degree reduced form. To take advantage of the possible reductions, common factors have to be identified and canceled out. This algorithmic complication may be avoided, at the expense of some other complications, if the parity equations are generated directly from the underlying state–space model. Such a procedure, first suggested by Chow and Willsky, will be described in a later section.

Note that the order of the minimum underlying state–space model may be higher than the apparent order of the primary inpur–output model of the system. The discrepancy may be quite significant if the input–output model has been obtained by identification. Therefore, model order reduction may be necessary before creating the parity equations.

4.2. The Effect of Modeling Errors on Orthogonalization

Modeling errors may cause imperfections in the orthogonalization. As it will be shown below, the effect of modeling errors is different on the different types of variables (outputs; measured, controlled and disturbance inputs).

Consider Eq. (2.14). Recall Eqs. (2.12), (2.9) and (2.13) and decompose the $\mathbf{F}^*(z)$ matrix as

$$\mathbf{F}^*(z) = [\mathbf{G}_M^*(z) \mid \mathbf{G}_C^*(z) \mid \mathbf{G}_D^*(z) \mid -\mathbf{H}^*(z)] \tag{4.11}$$

where the $\mathbf{G}_M^*(z)$, $\mathbf{G}_C^*(z)$ and $\mathbf{G}_D^*(z)$ matrices are the coefficients of the $\mathbf{u}_M(t)$, $\mathbf{u}_C(t)$ and $\mathbf{u}_D(t)$ input vectors, respectively. Now Eq. (2.14) may be re-written as

$$e^*(t) = \hat{G}_M^*(z)[\Delta u_M(t) + \delta u_M(t)] + \Delta G_M^*(t) u_M(t)$$

$$- \hat{G}_C^*(z)[\Delta u_C(t) + \delta u_C(t)] + \Delta G_C^*(t) u_C(t)$$

$$- G_D^*(z)[\Delta u_D(t) + \delta u_D(t)]$$

$$- \hat{H}^*(z)[\Delta y(t) + \delta y(t)] - \Delta H^*(z) y(t) \qquad (4.12)$$

Notice that while all the other faults (and noises) are multiplied with the estimate of their respective model matrix, the coefficient of the additive plant fault $\Delta u_D(t)$ (and plant noise $\delta u_D(t)$) is the true $G_D^*(z)$. Recall also that $H^*(z)$ is diagonal.

Now let us consider the different fault types separately.

a. Output variables. Since $H^*(z)$ is diagonal, a transformed residual is completely unaffected by any output variable to which it was made orthogonal. This implies that no sensor failure (and noise) on those variables has any effect on the transformed residual, nor does the true value of the concerned outputs.

b. Measured and controlled inputs. For these variables, sensor and actuator biases (and noise) influence the primary residuals via the estimate of the model matrix and the same estimates are used in the tranformation procedure. Therefore orthogonalization with respect to such failures (and noise) is perfect, even if there are modeling errors. However, orthogonalization does not imply complete elimination of the true values of the concerned input variables, in the presence of a non–zero modeling error matrix, unless the latter is proportional to the estimated model matrix.

c. Disturbance inputs. For these variables, the transformation is performed with the estimated model parameters while the primary residuals depend on the faults (and noise) via the true system parameters. Therefore orthogonalization with respect to such failures (and noise) is not perfect if modeling errors are present.

4.3. Orthogonal Residuals from State Equations

In this sub–section, an alternative method for generating orthogonal residuals will be discussed. It is based on a general procedure by Chow and Willsky [6]. The structural constraints following from the isolability requirements will be superimposed on the original Chow–Willsky technique. This approach is conceptually indirect but computationally simple. The resulting parity equations are the same as the ones obtained by input–output model transformation.

Consider the linear time–invariant plant model in state–space form:

$$x(t) = Ax(t-1) + Bu(t-1) \tag{4.13}$$

$$y(t) = Cx(t) + Du(t) \tag{4.14}$$

where $u(t)$ and $y(t)$ are the input and output vector, as defined earlier, $x(t) = [x_1(t), \ldots, x_n(t)]^T$ is the state vector and A, B, C, D are system matrices of appropriate size. (Strictly speaking, the order of this underlying system is n_s rather than n; this, however, will be neglected for the sake of notational simplicity.)

$y(t+1)$ may be obtained by sustitution as

$$y(t+1) = CAx(t) + CBu(t) + Du(t) \tag{4.15}$$

Similarly $y(t+r)$ for any $r>0$ is obtained as

$$y(t+r) = CA^r x(t) + CA^{r-1} Bu(t) + \ldots + CBu(t+r-1) + Du(t+r) \tag{4.16}$$

Collecting the equations for $r=0 \ldots n$ (and shifting the time variable by n) yields the following scheme:

$$
\begin{bmatrix} y(t-n) \\ y(t-n+1) \\ y(t-n+2) \\ \cdot \\ \cdot \\ \cdot \\ y(t) \end{bmatrix}
=
\begin{bmatrix} C \\ CA \\ CA^2 \\ \cdot \\ \cdot \\ \cdot \\ CA^n \end{bmatrix} x(t-n)
+
\begin{bmatrix} D & & & \\ CB & D & & \\ CAB & CB & & \\ \cdot & \cdot & & \\ \cdot & \cdot & & \\ \cdot & \cdot & & \\ CA^{n-1}B & CA^{n-2}B & \ldots & CB & D \end{bmatrix}
\begin{bmatrix} u(t-n) \\ u(t-n+1) \\ u(t-n+2) \\ \cdot \\ \cdot \\ \cdot \\ u(t) \end{bmatrix}
\tag{4.17}
$$

or in compact form

$$Y(t) = Qx(t-n) + RU(t) \tag{4.18}$$

For a system with m outputs and k inputs, vector $Y(t)$ is $(n+1)xm$ long and vector $U(t)$ is $(n+1)xk$ long. Matrix Q has $(n+1)xm$ rows and n columns while matrix R has $(n+1)xm$ rows and $(n+1)xk$ columns.

Pre–multiplying Equation (4.18) with a row–vector ω^T, $(n+1)xm$ long, yields a scalar equation:

$$\omega^T Y(t) = \omega^T Q x(t-n) + \omega^T R U(t) \tag{4.19}$$

In general, this equation will contain a mix of input, output and state variables. It will qualify as a parity equation only if the state variables disappear. This requires that

$$\omega^T Q = 0 \tag{4.20}$$

That is, the $(n+1)xm$ elements in ω^T have to satisfy a set of n homogeneous equations. If the system is observable, these n equations are independent. (Matrix Q may be recognized as the observability matrix, with one row, CA^n added. If the system is observable, Q has rank n, that is, the n columns are independent.) Therefore, $(n+1)xm-n$ elements may be chosen freely. However, to obtain a non–trivial solution, at least one of the free elements must be non–zero. This leaves $(n+1)x(m-1)$ potential zero elements.

To make a parity equation orthogonal to an output $y_i(t)$, all occurances of y_i have to disappear from the equation, that is, $y_i(t)$, $y_i(t-1),...,y_i(t-n)$. This requires $n+1$ zeroes in ω^T, appropriately positioned. Thus a parity equation may be made orthogonal to up to $m-1$ outputs. In fact, choosing one output at a time to stay, this technique re–generates the input–output equations (2.1). In other words, it produces the "primary" parity equations.

Orthogonality to an input $u_j(t)$ requires zeroes in the vector

$$\zeta^T = \omega^T R \tag{4.21}$$

To eliminate all occurances of u_j takes $n+1$ appropriately positioned zeroes, that is, $n+1$ homogeneous equations on the elements of ω^T. This adds to the original n homogeneous equations and reduces the number of free elements. That is, the total number of variables, inputs and outputs, that a parity equation may be made orthogonal to is $m-1$.

Orthogonalization to inputs implies manipulations with columns of matrix R. The good behavior of these columns in general is not guaranteed. Irregularities in matrix R (total or partial zero column or linear dependence) may lead to the abortion of the procedure or to orthogonality to more variables than intended. In either case, the desired parity equation is not attainable.

Note that the procedure described above makes no distinction between accessible (measured or controlled) inputs and additive plant disturbances. This is desirable since orthogonalization is to apply to the latter as well. For the computation of residuals, however, the disturbance variables have to be dropped from the resulting equations.

5. PARAMETRIC FAILURES

In this section, it will be shown how the notion of isolability and the method of linear model transformation, introduced previously for additive failures, may be applied to parametric (multiplicative) failures.

Recall that multiplicative failures are discrepancies (errors) in some model parameters that have not always been there but appear as a result of some change in the plant. In general, model discrepancies may be characterized (cf. Eq. (2.13)) as

$$\hat{F}(z) = F(z) + \Delta F(z) \tag{5.1}$$

If this equation is used to describe parametric failures, then $\hat{F}(z)$ is the nominal model while $F(z)$ is the actual plant. With no failure, $\hat{F}(z) = F(z)$. If a parametric failure occurs, $\hat{F}(z)$ remains the same and $F(z)$ changes. That is, $F(z) - \hat{F}(z) = -\Delta F(z)$ is the parametric failure.

Recall now Eq. (3.1) showing how parametric failures influence the residuals and note that this equation may be written in several variants depending where the bi-linear failure terms $\Delta F(z) \, \Delta q(t)$ and $\Delta F(z) \, \delta q(t)$ are placed:

$$e(t) = F(z)\Delta q(t) + F(z)\delta q(t) + \Delta F(z)q(t) + \Delta F(z)[\Delta q(t) + \delta q(t)] \tag{5.2-a}$$

$$= \hat{F}(z)\Delta q(t) + \hat{F}(z)\delta q(t) + \Delta F(z)q(t) \tag{5.2-b}$$

$$= F(z)\Delta q(t) + F(z)\delta q(t) + \Delta F(z)\tilde{q}(t) \tag{5.2-c}$$

$$= \hat{F}(z)\Delta q(t) + \hat{F}(z)\delta q(t) + \Delta F(z)\tilde{q}(t) - \Delta F(z)[\Delta q(t) + \delta q(t)] \tag{5.2-d}$$

Of the above forms, (5.2-b) is identical with (3.1). The last term in form (5.2-c) represents the full effect of parametric failures via $\Delta F(z)$ (it might, of course, also represent modeling errors).

5.1. Underlying Parameters

We will assume that parametric failures occur relative to the parameters of some underlying model, such as a first–principle physical model or a state–space model. These underlying parameter failures will then manifest themselves in the modification of several parameters in $F(z)$. This underlying parameter assumption is not only plausible but, since this way each parametric failure appears in several equations, it allows for basically the same orthogonalization procedure as the one used for additive failures.

Consider a set of underlying parameters $d=\{d_1, d_2,...,d_g,...\}$. With this, an element of the $F(z)$ matrix may be expressed as $f_{ij}(z)=f_{ij}(z,d)$ and $\Delta f_{ij}(z)= f_{ij}(z,\Delta d)$. Usually, this relationship is non–linear. The effect of parametric failures on the i-th residual is

$$e_i(t|\Delta d) = \sum_{j=1}^{k+m} \Delta f_{ij}(z,\Delta d)\tilde{q}_j(t) \tag{5.3}$$

If the parity equation coefficients are linear functions of the underlying parameters (or if the non–linear relationships are linearly approximated) then

$$\Delta f_{ij}(z) = \sum_g \psi_{ijg}(z)\Delta d_g \tag{5.4}$$

and

$$e_i(t|\Delta d) = \sum_{j=1}^{k+m} \Delta f_{ij}(z,\Delta d)\tilde{q}_j(t) = \sum_g \tilde{p}_{ig}(t)\Delta d_g \tag{5.5}$$

where

$$\tilde{p}_{ig}(t) = \sum_{j=1}^{k+m} \psi_{ijg}(z)\tilde{q}_j(t) \tag{5.6}$$

Eq. (5.2-c) may be re-written using the latest result for linear (linearized) failure propagation as

$$e(t) = F(z) \, \Delta q(t) + F(z) \, \delta q(t) + \tilde{P}(t) \, \Delta d \qquad (5.7)$$

where $\tilde{P}(t) = [\tilde{p}_{ig}(t)]$.

5.2. Isolability

As seen from Eq. (5.3), the residual $e_i(t)$ is orthogonal to the parametric failure Δd_g if none of the coefficients $f_{ij}(z)$, $j=1\ldots k+m$, depends on d_g. In case of linear dependence (see Eq. (5.6)), $\tilde{p}_{ig}(t)=0$ indicates that $e_i(t)$ is orthogonal to Δd_g.

Regard now Eq. (5.7) and associate an incidence matrix with $\tilde{P}(t)$ as

$$\Pi = \text{Inc}[\tilde{P}(t)] \qquad (5.8)$$

Then all the earlier isolability considerations and results apply to parametric failures, with the Π matrix or the combined $[\Phi,\Pi]$ matrix taking the place of the Φ matrix.

For non-linear parametric failure propagation, the incidence matrix Π may be constructed by the inspection of the $f_{ij}(z,d)$ functions. Alternatively, an approximate linear description may be obtained by computing the partial derivatives of the $f_{ij}(z,d)$ functions. Zero local derivatives, however, should not be accepted unless confirmed by the structure of the function, otherwise they should be replaced with the appropriate finite difference ratios.

5.3. Orthogonalization by Model Transformation

The transformation procedure described in Section 4.1, with some modifications, may be used to generate parity equations orthogonal to selected underlying parameter failures. Consider Eq. (5.7), applied to the primary set of parity equations:

$$e^*(t) = F^*(z) \, \Delta q(t) + F^*(z) \, \delta q(t) + \tilde{P}^*(t) \, \Delta d \qquad (5.9)$$

Now a new equation will be sought as

$$e_i(t) = w(t) \, e^*(t) \qquad (5.10)$$

where $w(t)=[w_1(t), w_2(t), \ldots]$ is chosen in a such a way that

$$w(t) \; \tilde{P}_i^*(t) = 0 \tag{5.11}$$

Here $\tilde{P}_i^*(t)$ consists of those columns of the $\tilde{P}^*(t)$ matrix that belong to the underlying parameters d_g to which the transformed residual $e_i(t)$ is to be orthogonal. Notice that the elements of the transforming vector $w(t)$ are numbers (in contrast to polynomials in z^{-1}) but depend on the measurements and thus are time–varying. That is, the transformation procedure is simpler than in case of orthogonalization to additive failures but the resulting parity equation depends on the measurements, therefore the transformation must be performed on–line.

In general, the transformed parity equations are to be orthogonal to a mix of additive and multiplicative failures. That is, the transforming vector has to satisfy

$$w(z,t) \; [\hat{F}_i^*(z) \mid \tilde{P}_i^*(t)] = 0 \tag{5.12}$$

Now the elements of the transforming vector are time–varying polynomials. Though the transformation must be performed on–line and, in general, all elements of the transforming vector depend on the measurements, it is possible to decompose the transformation procedure in such a way that the size of the on–line problem is reduced to the number of time–varying conditions in Eq. (5.12).

Obviously, model transformation may only be done using known quantities, that is, the estimates $\hat{F}^*(z)$ for the model matrix and the measurement-based values $\tilde{P}^*(t)$ for the parametric failure effects. Eqs. (5.2) reveal that, under these circumstances, complete orthogonalization with respect to a mix of additive and multiplicative failures is not possible. The remaining bi–linear failure term, however, will only cause a non–zero residual if additive and multiplicative failure(s) are indeed present simultaneously.

It should be emphasized that the above transformation procedure is predicated on the linear propagation of underlying parameter failures. Usually, the propagation relationships are non–linear; in this case, orthogonalization is valid only in the vicinity of the nominal model.

6. STATISTICAL TESTING

The computed residuals are subjected to statistical testing, individually, in parallel. The on-line testing operation consists simply in comparing each residual to its respective predetermined threshold.

The test thresholds are computed off-line, on the basis of the (known or assumed) statistical distribution of the noise. This computation will be outlined in sub-section 6.1. To determine the threshold values, a false alarm ratio is selected, indicating the probability of the test firing in failure-free situations. Considering that the detection algorithm runs on-line and that the probability of failures is relatively low, this false alarm ratio has to be chosen very small so that test firings due to false alarms do not exceed those due to actual failures.

The overall probability of missed detections may only be computed on the basis of a failure occurance distribution by failure size, but this is very seldom available. However, it is quite straightforward to obtain missed detection probabilities conditioned on hypothetized failure sizes. This idea will be outlined in sub-section 6.2, in the framework of failure sensitivity analysis. Different measures of failure sensitivity are important characteristics of parity equations that play a significant role in the design of diagnostic models (see Section 8.).

Certain failure sensitivity properties can be improved by the filtering of the residuals. Two filtering approaches will be discussed below: one replaces the residual with its average computed over a moving time-window, the other uses some standard recursive algorithm. In either case, the filter is so designed that the mean of the filtered residual is identical with the residual mean without filtering while its variance is reduced. As dynamic elements, residual filters affect the transient behavior of the diagnostic algorithm.

6.1. Computing the Thresholds

Consider the residuals in the absence of any failure (e.g. in Eq. (5.2-c)) and expand $F(z)$ in accordance with Eq. (2.2) as

$$F(z) = F_o + F_1 z^{-1} + \ldots + F_n z^{-n} \tag{6.1}$$

Then

$$e(t) = F_o \delta q(t) + F_1 \delta q(t-1) + \ldots + F_n \delta q(t-n) \tag{6.2}$$

The zero–shift covariance matrix of the residual vector, $S_e(0)$, is obtained as

$$S_e(0) = E\{e(t)e^T(t)\} = \sum_{g=0}^{n} \sum_{j=0}^{n} F_g S_q(g-j) F_j^T \qquad (6.3)$$

Here $S_q(g-j)$ is the $(g-j)$–shift covariance matrix of the noise vector $\delta q(t)$.

If the noise is white (uncorrelated in time) then $S_q(g-j)=0$ for all $j \neq g$ and Eq. (6.3) simplifies to

$$S_e(0) = \sum_{g=0}^{n} F_g S_q(0) F_g^T \qquad (6.4)$$

If, further, the elements of the noise vector are uncorrelated with each other then $S_q(0)$ is a diagonal matrix and the elements of $S_e(0)$ may be computed as

$$s_{e,ir}(0) = \sum_{j=1}^{k+m} s_{q,jj}(0) \sum_{g=0}^{n} f_{g,ij} f_{g,rj} \qquad (6.5)$$

Here $s_{e,ir}$, $s_{q,jj}$ and $f_{g,ij}$ are elements of S_e, S_q and F_g, respectively.

If the noise covariance matrices and the plant model are known then the residual covariances for the fault–free syatem may be computed using Eq. (6.3), (6.4) or (6.5), whichever the case be. The individual thresholds are then established as multiples of the fault–free standard deviation of the respective residual (square–root of the resp. diagonal element in $S_e(0)$):

$$\eta_i = c\sqrt{s_{e,ii}(0)} \qquad (6.6)$$

where η_i is the threshold for the i–th residual. The multiplier c depends on the desired false alarm rate and the type of the distribution. If the noises are normally distributed, so are the residuals. Then e.g. a 1% false alarm rate (0.5% at either tail of the distribution) implies $c=2.6$ while a 0.1% false alarm rate implies $c=3.2$.

Note that the procedure described above relies entirely on the diagonal elements of the residual covariance matrix. It may be possible to devise a more sophisticated testing procedure that utilizes the additional information represented by the off–diagonal elements.

Note also that the residuals are correlated in time, that is, $S_e (g-j) \neq 0$ for some $g \neq j$. This fact is irrelevant for threshold selection for instantaneous tests but will be taken into account in residual filtering.

6.2. Failure Sensitivity

While the test thresholds for each parity equation are determined by the noise, with no respect to the failure sizes, the question arises what size of failures will then drive the residuals to their respective thresholds. Since in the presence of a failure any residual is the combined effect of this failure and noise, it is more accurate to ask what size of failure will drive the mean of the residual's distribution to the threshold or, equivalently, what is the critical size of a failure that results in a residual equal to its threshold if the instantaneous value of the combined noise is zero.

A sensitivity measure along these lines, called *"triggering limit"*, was introduced in earlier work [9]. Obviously, the triggering limit depends on the threshold and on the "gain" of the parity equation relative to the failure. In a dynamic environment, further clarifications are necessary: one may assume that the failure acts as a step function and then the gain may be defined by its steady state value or maximum transient. This leads, for example, to the following definition for the triggering limit ρ_{ij}:

$$\rho_{ij} = \frac{\eta_i}{|f_{ij}(1)|} \tag{6.7}$$

Here η_i is the threshold for the i–th residual and

$$f_{ij}(1) = [f_{ij}(z)]_{z=1} \tag{6.8}$$

is the steady state gain belonging to the j–th failure in the i–th equation. Note that the triggering limit, in general, is different for each failure within a given parity equation and also in each equation for a given failure.

The computation of missed detection probabilities, conditioned on failure size, now follows naturally. Define $p_{ei}(e|0)$ as the probability density function of $e_i(t)$ with no failures and $p_{ei}(e|\Delta q_j)$ as the same with a Δq_j

fault present. Obviously, the faulty density is a shifted version of the fault–free function:

$$p_{ei}(e|\Delta q_j) = p_{ei}(e - \Delta q_j f_{ij}(1)|0) \tag{6.9}$$

Now the probability of missed detection of a failure Δq_j, assuming a symmetrical fault–free distribution with thresholds $\pm \eta_i$ is (Fig. 2.)

$$\beta_i(\Delta q_j) = \int_{-\eta_i}^{+\eta_i} p_{ei}(e|\Delta q_j)\,de \tag{6.10}$$

For a fault equaling its triggering limit, the probability of missed detection is about 50% while for one twice its triggering limit it is half the probability of false alarm.

An alternative and perhaps more practical sensitivity measure can be obtained by defining, as part of the problem specification, the nominal value Δq_j^o for each failure variable that should, under zero instantaneous noise, bring the given residual to its threshold. Assuming again a step function failure, a steady state *sensitivity number* λ_{ij} may be computed as

$$\lambda_{ij} = \frac{\eta_i}{|\Delta q_j^o f_{ij}(1)|} \tag{6.11}$$

Ideally, each sensitivity number should be 1. If the threshold is adjusted by filtering, the sensitivity numbers change accordingly. The necessary adjustment for any given equation may be determined by its maximum or average sensitivity number.

Filtering does not influence the ratio of sensitivity numbers within a parity equation. The maximum ratio will be called the *"sensitivity conditon"* κ_i of the equation

$$\kappa_i = \frac{\max_j\{\lambda_{ij}\}}{\min_j\{\lambda_{ij}\}} \tag{6.12}$$

The sensitivity condition is the most important sensitivity measure of a parity equation. Ideally it should be 1; if it is much larger then there is an obvious

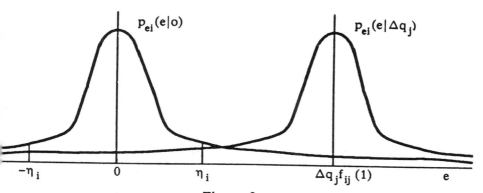

Figure 2.
Probability of missed detection

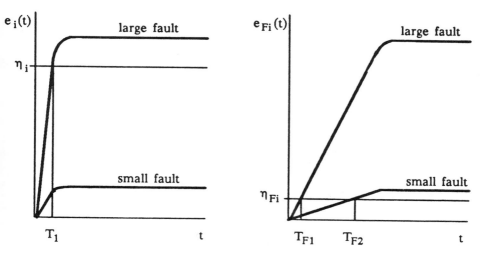

Figure 3.
Detection delays without and with residual filtering

imbalance among the different failures in the equation. If the threshold is adjusted, by filtering, to satisfy the failure with the smallest sensitivity number, failures with larger sensitivity numbers have to exceed their nominal value significantly to trigger the test. On the other hand, if the threshold is adjusted to the largest sensitivity number, failures with smaller sensitivity numbers will trigger the test even if their size is a fraction of the nominal value, causing false alarms and seriously hindering failure isolation. Since the sensitivity condition of any given parity equation can not be influenced, equations with excessive condition measure should be identified and eliminated from the design.

Note that tying failure sensitivity measures to steady state failure gains is not always the best approach to follow. The step response of parity equations with respect to certain failures may have peaks and may even tend to zero. In particular, if the plant is integrating then the $h(z)$ polynomial (the coefficient of $\Delta y(t)$) has zero steady state gain. In such cases, failure sensitivity should be defined using some appropriate transient failure gain.

6.3. Filtering the Residuals: Moving Average

Failure sensitivity can be improved by the individual averaging of the residuals. The average of $e_i(t)$ over an N sample long moving window, $e_{Ni}(t)$, is obtained as

$$e_{Ni}(t) = \frac{1}{N} \sum_{r=0}^{N-1} e_i(t-r) \tag{6.13}$$

Obviously, the mean of the average is the same as that of the residual while its variance is reduced. The length of the sample is selected to achieve a desired reduction of the variance and, thus, of the test threshold.

Recall that $E\{e_i(t-g)e_i(t-r)\}\neq0$ for some $g\neq r$, that is, the residuals are correlated in time. Therefore, the computation of the variance of the moving average is not trivial, though always possible if the noise statistics are known. In the following, only the simplest case will be treated, when the noise is white and the elements of the noise vector are uncorrelated:

$$E\{\delta q_j(t-g)\delta q_j(t-r)\} = 0, \text{ for all } j, \text{ if } g\neq r, \tag{6.14-a}$$

$$E\{\delta q_j (t-g)\delta q_v (t-r)\} = 0, \text{ for all } g,r, \text{ if } j\neq v \tag{6.14-b}$$

Express the residual, in accordance with Eq. (6.2), as

$$e_i(t) = \sum_{j=1}^{m+k} \sum_{g=0}^{n} f_{g,ij}\delta q_j (t-g) \tag{6.15}$$

Then the moving average can be written as

$$e_{Ni} (t) = \sum_{j=1}^{m+k} e_{Ni,j}(t) \tag{6.16}$$

where

$$e_{Ni,j} (t) = \frac{1}{N} \sum_{r=0}^{N-1} \sum_{g=0}^{n} f_{g,ij}\, \delta q_j (t-g-r) \tag{6.17}$$

It follows from Eq. (6.14-b) that the $e_{Ni,j}(t)$ components of the average are uncorrelated, that is,

$$E\{e_{Ni,j}(t)e_{Ni,v}(t)\} = 0 \quad \text{if } j\neq v \tag{6.18}$$

Therefore, the variance of the average may be written as

$$E\{e_{Ni}^2(t)\} = \sum_{j=1}^{m+k} E\{e_{Ni,j}^2(t)\} \tag{6.19}$$

It remains to compute the variance of the $e_{Ni,j}(t)$ components:

$$E\{e_{Ni,j}^2 (t)\} = \frac{1}{N^2} \sum_{r=0}^{N-1} \sum_{w=0}^{N-1} \sum_{g=0}^{n} \sum_{v=0}^{n} f_{g,ij}f_{v,ij}E\{\delta q_j(t-r-g)\delta q_j(t-w-v)\} \tag{6.20}$$

According to (6.14-a), this simplifies to

$$E\{e_{Ni,j}^2 (t)\} = \frac{1}{N^2} \sum_{g=0}^{n} \sum_{v=0}^{n} c_{gv}\, f_{g,ij}\, f_{v,ij}\, E\{\delta q_j^2 (t)\} \tag{6.21}$$

Here c_{gv} is a non-negative interger, indicating, for each (g,v) pair, the number of (r,w) combinations that make $t-r-g=t-w-v$. It can be shown that

$$c_{gv} = \begin{cases} N - |g - v| & \text{if } |g - v| < N \\ 0 & \text{otherwise} \end{cases} \tag{6.22}$$

The moving average filter, of course, is a dynamic element. Its discrete transfer function is

$$M_N(z) = \frac{1}{N} \sum_{r=0}^{N-1} z^{-r} = \frac{1 - z^{-N}}{N(1 - z^{-1})} \tag{6.23}$$

where the rational form is valid only for $|z|<1$. With this, the response of the filtered residual to additive failures is written as

$$e_{Ni}(t) = \sum_{j=1}^{k+m} M_N(z) f_{ij}(z) \Delta q_j(t) \tag{6.24}$$

Filtering of the residuals results in smaller test thresholds and longer transients. The latter, however, usually leads to "delayed detection" only if the failure is so small that, without filtering, it would not be detected at all. For larger failure sizes, the lower thresholds usually compensate for the increased transient delay (Fig. 3.).

A word of caution is in order here concerning averaging and, in general, low-pass filtering of the residuals. If the failure transfer polynomial $f_{ij}(z)$ has a monotonous step response then the failure effect on the residual is the highest in steady state. Lowering the thresholds by filtering always results in improved failure sensitivity in this case. However, if the failure response has a significant peak, then the highest effect occurs during the transient; smoothing this peak by low-pass filtering may outweigh the threshold reduction.

6.4. Filtering the Residuals: Recursive Filter

An alternative approach to residual filtering implies the use of some discrete recursive low-pass filter, characterized by a discrete transfer function $\Gamma(z)$. The filter is so designed that is has unit steady state gain and, under given noise statistics and parity equation coefficients, results in the desired no-failure variance for the filtered residual, reduced relative to the same residual variance without filtering.

The filtered residual $e_{Fi}(t)$, with no failure and modeling error present, is

$$e_{Fi}(t) = \sum_{j=1}^{m+k} \Gamma(z) f_{ij}(z) \delta q_j(t) = \sum_{j=1}^{m+k} e_{Fi,j}(t) \tag{6.25}$$

If Eq. (6.14-b) holds then the variance of the filtered residual may be written, similarly to Eq. (6.19), as

$$E\{e_{Fi}^2(t)\} = \sum_{j=1}^{k+m} E\{e_{Fi,j}^2(t)\} \tag{6.26}$$

The component variances $E\{e_{Fi,j}^2(t)\}$ may be computed using the usual frequency domain techniques or directly in the time-domain. In the following, we will outline a time-domain scheme, applied to a second order filter and white noise sequences. The extension of the scheme to higher order filters is straightforward. For colored noise, the fictitious transfer function generating the given noise from a white source is also to be incorporated. The proof of the scheme involves somewhat lengthy but straightforward manipulations with auto- and cross-correlations (variances) of discrete signals and will be omitted here.

Consider a second order autoregressive filter with unit steady state gain

$$\Gamma(z) = \frac{1 + \gamma_1 + \gamma_2}{1 + \gamma_1 z^{-1} + \gamma_2 z^{-2}} \tag{6.27}$$

and recall that

$$f_{ij}(z) = f_{o,ij} + f_{1,ij} z^{-1} + \ldots + f_{n,ij} z^{-n} \tag{6.28}$$

Now construct the following matrices:

$$
FF_{ij} = \begin{bmatrix} f_{o,ij} & f_{1,ij} & \cdots & f_{n-1,ij} & f_{n,ij} \\ f_{1,ij} & f_{2,ij} & \cdots & f_{n,ij} & 0 \\ \cdot & & & & \\ \cdot & & & & \\ \cdot & & & & \\ f_{n,ij} & 0 & \cdots & 0 & 0 \end{bmatrix}
\qquad
ff_{ij} = \begin{bmatrix} f_{o,ij} \\ f_{1,ij} \\ \cdot \\ \cdot \\ \cdot \\ f_{n,ij} \end{bmatrix}
\qquad (6.29)
$$

$$
\Gamma_L = \begin{bmatrix} 0 & 0 & 0 & 0 & \cdots & 0 & 0 \\ \gamma_1 & 0 & 0 & 0 & \cdots & 0 & 0 \\ \gamma_2 & \gamma_1 & 0 & 0 & \cdots & 0 & 0 \\ 0 & \gamma_2 & \gamma_1 & 0 & \cdots & 0 & 0 \\ \cdot & & & & & & \\ \cdot & & & & & & \\ \cdot & & & & & & \\ 0 & 0 & 0 & 0 & \cdots & \gamma_1 & 0 \end{bmatrix} \begin{matrix} 1 \\ 2 \\ 3 \\ 4 \\ \\ \\ \\ n+1 \end{matrix}
\qquad
\Gamma_U = \begin{bmatrix} 0 & \gamma_1 & \gamma_2 & 0 & \cdots & 0 \\ 0 & \gamma_2 & 0 & 0 & \cdots & 0 \\ 0 & 0 & 0 & 0 & \cdots & 0 \\ 0 & 0 & 0 & 0 & \cdots & 0 \\ \cdot & & & & & \\ \cdot & & & & & \\ \cdot & & & & & \\ 0 & 0 & 0 & 0 & \cdots & 0 \end{bmatrix}
$$

$$
\begin{matrix} 1 & 2 & 3 & 4 & \cdots & n & n+1 \end{matrix} \qquad\qquad \begin{matrix} 1 & 2 & 3 & 4 & \cdots & n+1 \end{matrix}
$$

$$(6.30)$$

Then the component variances may be obtained as

$$
E\{e_{Fi,j}^2\} = (1 + \gamma_1 + \gamma_2)^2 [1 \quad 0 \; \cdots \; 0] [I + \Gamma_L + \Gamma_U]^{-1} FF_{ij} [I + \Gamma_L]^{-1} ff_{ij} E\{\delta \, q_j^2(t)\}
$$

$$(6.31)$$

Designing a filter means finding the filter parameters (in the second order case γ_1 and γ_2) given the noise variances, parity equation coefficients and the desired variance of the filtered residual. The design problem is not completely defined, even in the second order case, since the variance condition provides only a single equation for two (or more) parameters. As additional conditions, the transient behavior (in response to sudden failures) of the parity equation—filter system may be considered. The latter may be analyzed using Eq. (6.25), with the noise inputs replaced with the respective failure inputs.

Equation (6.31), together with Eqs. (6.29), (6.30) and (6.26), are much too complex to allow for an explicit solution for the filter parameters. Instead, a numerical solution is needed that implies the repeated evaluation of the filtered residual variance, for different filter parameters, in some search procedure, until the desired variance is obtained. The transient behavior of the different filter variants may also be taken into account in the search.

The design problem may be made more tractable by imposing some restriction on the filter. For example, choosing the Butterworth structure for the second order filter, the two filter coefficients become functions of a single parameter (the cutoff frequency) and the design simplifies to finding this parameter. With a single parameter, of course, there is little or no room for bringing the transient response into the design.

7. MODELING ERROR ROBUSTNESS

Dynamic plant models are inevitably subject to modeling errors. This may be due to linearization, lower order approximation of higher order systems or simply parameter errors. As shown in Section 2, modeling errors contribute some additional terms to the residuals (see Eq. (2.14) and (3.1)) and thus may interfere with the detection and isolation of failures on the basis of the mathematical model.

By robustness of a parity equation or of an entire model we mean its insensitivity to modeling errors. Such robustness may be defined with the worst case error of all the model parameters in mind [6] or with the assumption that there is a discrete set of uncertain models [14]. We find the notion of "*partial robustness*" with respect to single parameters more appealing, from a practical point of view. Partial robustness measures relative to parameters of some underlying model will be described below. Such underlying models, as introduced in Section 5, may be state space models or first principle characterizations of the plant.

As pointed out earlier (in Sections 2. and 5.), modeling errors appear in the parity equations the same way as parametric failures. Therefore, our treatment of modeling error robustness will be based on the results in Section 5. concerning parametric failures, in particular Equations (5.1) through (5.7). Now $\Delta \mathbf{d}$ will represent the uncertaity (modeling error) associated with the underlying parameters **d**. Some sensitivity concepts described in Sub-section 6.2. will also be utilized.

Partial robustness relates a single residual to a single underlying parameter. A direct measure of such robustness may be defined as the residual caused by the selected modeling error, if there are no failures and the instantaneous noise is zero, divided by the test threshold. Alternatively, the limit value of the parameter error that moves the center of the residual distribution to the threshold may be used as the measure of partial robustness [9]. In eiter case, the robustness measure depends on the test thresholds and on the operating point. For dynamic systems, the $n+1$ most recent values of the plant variables need to be specified (n is the order of the system); it may suffice to assume that they are constant.

Recall Equation (5.3), describing the i-th residual in response to Δ**d** (in this case, the modeling errors of the underlying parameters), with all faults and noises zero

$$e_i(t|\Delta d) = \sum_{j=1}^{k+m} \Delta f_{ij}(z,\Delta d)\,\tilde{q}_j(t) \tag{7.1}$$

Introduce

$$\Delta d^g = \{0,\ ...,\ 0,\ \Delta d_g,\ 0,\ ...\} \tag{7.2}$$

to describe the situation when all but one (the g-th) of the underlying parameters are accurate. Then the direct measure of partial robustness with respect to this parameter is obtained as

$$\mu_i(t,\Delta d_g) = \frac{|e_i(t,\Delta d^g)|}{\eta_i} \tag{7.3}$$

where η_i is the test threshold for $e_i(t)$. If the relationship between the underlying model parameters and the parity equation coefficients is linear (or linearized) then Eq. (5.5) holds and

$$e_i(t|\Delta d^g) = \tilde{p}_{ig}(t)\Delta d_g \tag{7.4}$$

where $\tilde{p}_{ig}(t)$ is defined in Eq. (5.6) and (5.4). With this, the direct robustness measure simplifies to

$$\mu_i(t, \Delta d_g) = \frac{|\tilde{p}_{ig}(t)| \Delta d_g}{\eta_i} \tag{7.5}$$

The limit value of Δd_g that triggers the test on $e_i(t)$ may be written as

$$\tau_{ig}(t) = \underset{\Delta d_g}{\text{sol}}\{|e_i(t, \Delta d^g)| = \eta_i\} \tag{7.6}$$

where sol{...} denotes the solution of the equation in the brackets. For a non-linear relationship, the solution has to be obtained numerically. For linear (linearized) relationships

$$\tau_{ig}(t) = \frac{\eta_i}{|\tilde{p}_{ig}(t)|} \tag{7.7}$$

Obviously, if the parameter uncertainty affects the residual linearly, then the two measures are equivalent. Notice also the clear dependence of both robustness measures on the operating point (via $\tilde{p}_{ig}(t)$).

For a model consisting of several parity equations and with a number of underlying parameters selected, the robustness situation is described by a matrix of partial robustness measures. A model is as robust relative to a selected underlying parameter as its least robust equation.

8. GENERATING DIAGNOSTIC MODELS [9]

As described earlier, a diagnostic model is a consistent set of parity equations. For isolability, the model should be in the approporiate column canonical structure. There is a combinatorial multitude of models satisfying this requirement. The selection can be made on the basis of a combination of performance measures, including the
− sensitivity condition
− maximum (or average) sensitivity number
− partial robustness measures.

The sensitivity condition and partial robustness measures characterize single parity equations. When the equations form a diagnostic model, the poorest performer determines the model performance. The maximum (or average) sensitivity number, by specifying the necessary threshold adjustment for the design of residual filters, influences the detection delays. In a diagnostic model, the isolation decision is held up until the slowest equation detects a threshold passing (meanwhile, the failure signature is undefined). Therefore, equations with smaller and/or uniform delays should be preferred in the design.

Some of the above measures may be used to reduce the set of useful equations by setting limits on the acceptable performance while others may serve as a basis for optimization. The larger the set of parity equations the more flexibility the designer has in selecting the diagnostic model.

A reasonable design strategy is to set limits on the sensitivity condition (and perhaps the maximum sensitivity number) and eliminate the outlying equations, then perform optimization, constrained to satisfy the structural requirements, over the reduced set with respect to different partial robustness measures. This leads to a family of diagnostic models, each column canonical and within the prescribed sensitivity limits, and each one optimally robust, within those constraints, relative to a selected parameter uncertainty.

As the first step of model design, the complete set of parity equations is to be generated. To simplify the design problem, only row–canonical equation structures are considered. For m outputs and k inputs, there are

$$V = \binom{m+k}{m-1} \tag{8.1}$$

such structures. However, as mentioned before, zeroes and linear dependences in the system matrices may render some of the parity equations unattainable and thus irrelevant for model generation. Even so, apart from trivially small systems, the attainable set may be quite extensive (for example, for $m=4$ and $k=6$, $V=120$; but for $m=6$ and $k=8$, $V=2184$).

Subsequently, the selected performance measures are computed for each equation. If limits are set for some of the measures then the set is reduced by eliminating the outlying equations. The remaining set is then ordered according to the measure chosen as a basis for optimization (if there is a number of such measures, the same number of ordered sets is formed).

Optimization for a selected performance measure within the framework of isolability requires a search over the available (reduced) set of equations under the constraint of column–canonical model structure. Two approaches will be discussed below. One implies a global search over the set; this guarantees the optimality of the model but its computational cost may be prohibitive. The other approach utilizes a local search procedure; this is computationally attractive but may lead to a local optimum.

When forming column–canonical models of row–canonical equations, the number v of equations in the model and the number r of zeroes per column have to satisfy the following relationships:

$$v\ (m-1) = r\ (m+k)$$

$$\binom{v}{r} \geq m+k \tag{8.2}$$

Here the equation simply expresses that the total number of zeroes in the rows is identical with that in the columns while the inequality guarantees that a sufficient number of different columns may be created. While there are many possible solutions, the one yielding the simplest diagnostic model is to be sought.

The numerical coefficients of the parity equations are needed for the performance computations but not for the subsequent search. The memory requirements of the algorithm may be reduced significantly if the coefficients are not stored, only the equation structure and the performance measures.

8.1. Global Search Procedure

The global search procedure operates on the ordered set of equations starting with the best performers. The first v equations are selected and checked for column stucture. If this subset is column canonical, it represents the best model. Otherwise the investigated subset is expanded to include the first $v+1$ equations. All the possible v–tuples are formed and checked for column structure; if any of them qualifies, it is declared as the best model. Otherwise the subset is expanded to $v+2$, $v+3$, etc. equations, until the first column–canonical v–tuple is found.

Some saving in the search results from the observation that the v–tuples to check always contain the last equation (since all possibilities without the last equation have been exhausted before). Thus in the stage when the investigated subset contains $v+i$ equations, the number of v–tuples to check is

$\binom{v+i-1}{v-1}$ while the total number of v–tuples checked up to and including

this i–th stage is $\binom{v+i}{v}$. This number may grow extremely large if a column–

canonical model is not found early in the search. Though usually there is a combinatorial number of suitable structures, since the equations are ordered according to performance measures (that depend on numerical coefficients), their distribution over the set of equations does not follow any predictable rule. Therefore there is no guarantee that the first suitable model would be found sufficiently early. (For example, with $m=4$ and $k=6$, covering one quarter of the equation set would require the checking of about 3E7 model structures, that is quite feasible to handle. With $m=6$ and $k=8$, going over 2% of the list implies only about 2E6 models but covering 10% implies 4E21, that is obviously not tractable by present–day computer technology.)

8.2. Local Search Procedure

The local search procedure starts from an arbitrary coloumn–canonical model and uses a series of systematic row replacements to improve its performance while maintaining the column–canonical structure.

The procedure utilizes the concept of *structural distance* between row–canonical equations. Interchanging a 0 and a 1 in an equation structure with $m-1$ zeroes results in another structure with the same number of zeroes. The minimum number of such 0–1 interchanges needed to bring one structure into another is called the structural distance between the two equations. For example, the structural distance between 111100 and 111010 is one while between 111100 and 001111 it is two. The distance can be conveniently obtained by halving the number of 1's in the exlusive OR of the two boolean patterns. The maximum possible distance in a system with k inputs and m outputs is $min[k+1,m-1]$. For a set of equations, a distance matrix may be set up, at least conceptually, showing the structural distances between all possible pairs.

Given any column–canonical initial model, the equation with the poorest performance measure is selected for replacement. In the basic form of the algorithm, only equations with *distance* 1 from the original equation will be considered as substitutes; there are $(k+1)x(m-1)$ such potential substitutes for any equation. The search starts with the distance 1 substitute that has the best performance. The candidate substitute equation disqualifies if it is alrerady a member of the model or if its perormance is worse that that of the equation to be replaced.

Replacing a single equation would spoil the column–canonical structure; a complementary replacement is necessary to restore it. There may be several equations in the initial model that, if appropriately replaced, restore the structure. For each of these, a distance 1 substitute is sought, with the best possible performance, and the one with the best overall performance is then chosen. A candidate for complementary substitute disqualifies if it is already in the model or if its performance is worse than that of the equation originally replaced (i.e. of the worst equation in the initial model). If no suitable complementary substitute is found, the original replacement has to be modified, moving to the substitute equation with the next best performance.

Once the equation with the poorest performance has been successfully replaced, the procedure is repeated for the worst equation in the resulting model, and so on. The search stops if no distance 1 replacement (with distance 1 complementary replacement) can be found for the worst equation in the latest model.

Since the search is limited to distance 1 replacements, it may easily stop in a local optimum. Convergence toward the global optimum can be improved by allowing distance 2, 3, etc. replacements, preferably once all the distance 1 replacements have been exhausted. (In fact, after a higher distance replacement, a series of distance 1 steps may become possible again.) This, however, requires a significantly more complex algorithm; especially the complementary replacements become rather complicated.

The column–canonical initial model needed to start the search may be generated according to some arbitrary symmetrical pattern. If it turns out to contain equations that are unattainable for the given system, other symmetrical patterns may be attempted. Alternatively, the unattainable equations may be assigned some very poor (arbitrary) performance measures, thus making them the first candidates for replacement.

8.3. Further Design Considerations

The procedure to design isolable diagnostic models with implied sensitivity and robustness considerations is rather complex and can not be completely automated. At some points, the human designer has to interact with the design program and make certain decisions. These require a good understanding of the meaning of different design parameters and of the implications of modifying them.

One of the decisions is related to the truncation values of the different performance measures. If truncation does not leave a sufficiently large equation pool for model selection, then the designer may want to relax the truncation values or otherwise modify the specification (e.g. by relaxing some nominal failure sizes).

A deeper issue concerns the trade–off between sensitivity and detection/isolation delay and implies such considerations as the selection of the sampling interval and residual filtering. With the discretized model and the noise situation given, the residual variances follow; these, with a selected false alarm ratio, determine the test thresholds. However, the choice of the sampling interval influences the parameters of the discretized model and has a significant effect on the thresholds and sensitivities. Increasing the sampling interval, in general, leads to smaller residual variances and better sensitivity numbers (while leaving the sensitivity conditions uneffected), at the expense of slower reaction to failures. Filtering of the residuals is another way of improving failure sensitivity, again at the expense of delayed detection/isolation. While a longer sampling interval affects the entire model, residual filters are designed separately for each parity equation, tailored to the conditions of the particular equation. However, filtering does not distinguish between failure and modeling error, unless they are in different frequency ranges; improvements in failure sensitivity are usually accompanied by the deterioration of modeling error robustness.

The partial robustness design concept described in this paper has, of course, certain limitations. Any diagnostic model is only optimized for some selected modeling uncertainty in a selected operating point. The behavior of models under different circumstances may be analyzed and the thresholds increased, if necessary, to improve modeling error robustness; this, however, would adversely effect failure sensitivity. An important implication of isolable model structures is that they provide a certain kind of inherent robustness against modeling error effect as well, in that it is extremely unlikely that modeling errors would lead to any valid failure signature and thus be mis-interpreted as a particular additive failure. With multiple parallel diagnostic models, if designed with realistic modeling error situations in mind, there is a good chance that one or more of the models is also sufficiently shielded from modeling error effects and thus provides a clear failure indication at any given time. The combination of incomplete and/or contradictory evidence from parallel models has been the subject of a related research effort [16].

As pointed out earlier, the global search for the best diagnostic model (under a selected performance measure) may easily become computationally intractable because of the combinatorial nature of the problem. The local search procedure introduced in this paper is computationally efficient and places practically no limitation on the size of systems that can be dealt with, but it does not guarantee the optimality of the solution. As in many other engineering problems, one may have to settle here for *acceptability* (instead of optimality); if the designer is not satisfied with the performance measures of a locally optimal model, a new local search may be initiated from a different arbitrary initial set of equations. The local search procedure may be improved, at the expense of increased computational complexity, by allowing higher distance replacements or by including some annealing technique.

9. DESIGN EXAMPLES

9.1. Car Engine Model

The design procedure will be first demonstrated on the linearized model of a car engine [17]. The model has two inputs and four outputs as listed in Table 1. The system is described by an 8th order continuous-time state-space model.

The design specifications are listed in Table 1; percentages refer to nominal operating values. The noises are assumed to be white with zero mean and uncorrelated with each other.

TABLE 1.

		noise st. dev.	nominal failure
INPUTS	$u_1(t)$: throttle comm.	0.10%	1%
	$u_2(t)$: fuel rate	0.25%	1%
OUTPUTS	$y_1(t)$: torque	0.50%	2%
	$y_2(t)$: air/fuel	0.50%	2%
	$y_3(t)$: in. manif. press.	0.20%	1%
	$y_4(t)$: throttle pos.	0.10%	1%

As uncertain underlying parameters, the elements of the **B** matrix in the state equation have been chosen. This choice is supported by the common understanding that under linearization the model gains may be at error if we depart from the nominal operating point. Robustness measures are computed with the assumption that (1) the entire first column (2) the entire second column is subject to a 10% modeling error. Both measures are evaluated in a temporary operating point, 10% away from the nominal values of the plant inputs.

A discretized state–space model, obtained with 80 ms sampling interval, served to generate the orthogonal parity equations. There are 20 possible orthogonal equations; their structure is shown in Table 2, together with their sensitivity condition κ_i and maximal sensitivity number λ_{imax}.

TABLE 2.

u_1	u_2	y_1	y_2	y_3	y_4	κ_i	λ_{imax}
0	0	0	1	1	1	1.28	0.27
0	0	1	1	0	1	1.77	3.01
0	1	0	1	0	1	1.79	1.34
1	1	0	0	1	0	1.85	0.52
0	1	1	0	1	0	2.09	0.96
1	1	0	0	0	1	3.00	0.34
1	0	1	0	0	1	3.06	0.59
0	0	1	0	1	1	3.12	0.57
1	0	0	1	0	1	3.37	1.22
0	1	0	0	1	1	3.51	0.39
0	1	0	1	1	0	4.49	2.13
0	0	1	1	1	0	4.52	4.86
1	1	0	1	0	0	5.36	2.75
1	0	1	1	0	0	5.41	6.14
1	0	0	0	1	1	6.51	0.36
1	0	0	1	1	0	8.32	2.13
1	1	1	0	0	0	28.6	6.43
1	0	1	0	1	0	34.2	7.76
0	1	1	0	0	1	87.5	5.00
0	1	1	1	0	0	155	97.9

The equations are ordered according to their sensitivity condition; the set is truncated at 10. The max sensitivity numbers of the remaining 16 equations indicate that residual filtering is not necessary.

The smallest isolable model has to contain 4 equations. The best model for sensitivity condition, obtained by global search, is shown in Table 3, including all four performance measures. As it turns out, this model is also optimal for the max sensitivity number and for the first robustness measure μ_{i1} (relative to uncertainty in the first column of the B matrix).

TABLE 3.

pattern						κ_i	λ_{imax}	μ_{i1}	μ_{i2}
1	1	0	1	0	0	1.77	3.01	0.51	0.33
1	0	1	0	1	0	2.09	0.96	1.04	0
0	0	0	1	1	1	3.00	0.33	0	0
0	1	1	0	0	1	4.49	2.13	0	0.47

Finally, Tables 4 and 5 show simulated on–line experiments obtained with the above diagnostic model, with no momentary noise. In Table 4, a 4% actuator error step was applied to the fuel–rate control; in Table 5 a 1% sensor error step to the throttle position. In both cases, the signature corresponding to the respective column of the model's incidence matrix appeared after a brief parity equation transient.

TABLE 4.

0	0	0	1	1	1	1
0	0	0	0	0	0	0
0	0	0	0	0	0	0
0	0	1	1	1	1	1

TABLE 5.

0	0	0	0	0	0	0
0	0	0	0	0	0	0
0	0	1	1	1	1	1
0	1	1	1	1	1	1

9.2. Distillation Column Model [9]

As a more extensive design example, a distillation column model [18] was chosen. The column has 9 plates. The flow rate of the reflux and of the vapor stream are controlled, based on the liquid composition at the second and ninth plate. The composition and the flow rate of the feed are unmeasured disturbances.

The plant is described by a linear state–space model, obtained by linearization around a nominal steady state operating point. The model has the following properties:

11 state variables, namely
    distillate composition
    9 plate compositions
    bottom product composition

2 controlled inputs, namely
    $u_1(t)$: reflux flow rate
    $u_2(t)$: vapor stream flow rate

2 disturbance inputs, namely
    $u_3(t)$: feed stream composition
    $u_4(t)$: feed stream flow rate

2 outputs, namely
    $y_1(t)$: distillate composition
    $y_2(t)$: bottom product composition

All variables in the linear model are changes relative to the nominal steady state operating point. The main characteristics of the nominal operating point are summarized in Table 6.

TABLE 6.

Steady state operating point.

Flow rate, feed		10.00	mol/min
	distillate	4.35	mol/min
	bottom product	5.65	mol/min
	vapor stream	17.40	mol/min
	reflux	13.05	mol/min
Liquid composition, feed		0.54036	
	condenser	0.90000	
	2nd plate	0.72211	
	4th plate	0.52804	
	7th plate	0.32005	
	9th plate	0.16772	
	reboiler	0.10400	

The eigenvalues of the continuous–time state–space model are all real and range from 0.02 1/min to 3.3 1/min, corresponding to time–constants between 50 min and 0.3 min. The model was discretized using an 8 min sampling interval. Shorter sampling intervals would lead to poor sensitivity performance of the parity equations and, in general, to numerical difficulties.

With $k=4$ inputs and $m=2$ outputs, there is a total of $\binom{6}{1} = 6$ row-canonical equations, that is a trivially simple system. Two of the plate compositions are sensed and used for control; considering these as additional outputs would lead to a system with $k=4$ inputs and $m=4$ outputs, resulting in a total of $\binom{8}{3} = 56$ row-canonical equations. Such a selection of parity equations may still be insufficient to allow a full-blown design for a combination of sensitivity and robustness measures. To demonstrate the potentials of the methodology, let us include two more state-variables as outputs. Thus there is a total of

6 outputs, namely

$y_1(t)$: composition of the distillate
$y_2(t)$: liquid composition on the 2nd plate
$y_3(t)$: liquid composition on the 4th plate
$y_4(t)$: liquid composition on the 7th plate
$y_5(t)$: liquid composition on the 9th plate
$y_6(t)$: composition of the bottom product

Now with $k=4$ inputs and $m=6$ outputs, the full row-canonical set contains $\binom{10}{5} = 252$ parity equations. The smallest column-canonical model that may be formed of these equations consists of $v=6$ rows (equations) with $r=3$ zeroes per column.

The system is subjected to sensor, actuator and plant noises, the latter acting additively on the state variables. All noises are assumed to be normal, white (uncorrelated in time) with zero mean and independent of each other. Their variances were established (guessed) on the basis of the character and nominal value of the variables. The nominal detectable failure sizes were specified using similar considerations. The noise variances and nominal failure sizes are shown in Table 7. Of course, the validity of the selected values may be argued; they might be placed on more solid ground utilizing a detailed process knowledge and noise measurements.

TABLE 7.

		noise variance	nominal failure
all state-variables		1E-8	–
inputs	u_1	6.76E-4	0.75
	u_2	11.70E-4	0.75
	u_3	–	0.02
	u_4	–	0.50
outputs	y_1	4E-6	0.025
	y_2	4E-6	0.025
	y_3	3E-6	0.025
	y_4	2E-6	0.010
	y_5	1E-6	0.010
	y_6	1E-6	0.010

All units conform with Table 6.

The test thresholds are set at four times the standard deviation of the respective residual.

As uncertain underlying parameters, the two columns of the **B** matrix that belong to the control inputs have been chosen. One full column at a time is assumed to have modeling error, the same 10% for each element. Robustness is investigated at the temporary operating point characterized by

$$u_1(t) = 1.30 \qquad u_2(t) = 1.74$$

that corresponds to a 10% departure from the nominal steady state operating point.

The design program generates the full row–canonical set of 252 parity equations. Also, for each equation, the following measures are computed:

– test threshold η_i

– sensitivity condition κ_i

– maximum sensitivity number λ_{imax}

– robustness measure μ_{i1} relative to the first column of **B**

– robustness measure μ_{i2} relative to the second column of **B**

The sensitivity condition κ_i for the 252 equations ranges from 1.35 to 1.4E8. The extremely high sensitivity conditions apparently signify unattainable equations; excess zeroes are checked numerically in the transformation procedure and computational inaccuracies may cause some zero coefficients seem nonzero. The maximum sensitivity number λ_{imax} ranges from 1.23 to

8E8; the extremely high numbers appear in the equations that have very high condition numbers as well. Choosing 15 as the maximum acceptable sensitivity condition reduces the number of useful equations to 97, with a λ_{imax} range of 1.23 to 115. Eliminating also the equations for which $\lambda_{imax} > 25$ leaves 72 equations.

A sensitivity number exceeding 1 indicates that the concerned fault, occuring at its nominal size, would not trigger the test on the concerned residual. In order to reduce the thresholds, the residuals are filtered. The new thresholds are selected as

$$\eta_{Fi} = \eta_i \sqrt{\frac{\kappa_i}{\lambda_{imax}}} = \frac{\eta_i}{\sqrt{\lambda_{imax} \lambda_{imin}}}$$

This way, the minimal and maximal responses to nominal size failures are symmetrical relative to the threshold in any equation.

Residual filtering does not change the sensitivity condition but will modify the robustness measures. Therefore, before optimization for robustness, the robustness measures μ_{Fi1}, μ_{Fi2} for the filtered residuals were computed, only for the 72 equations with $\kappa_i < 15$ and $\lambda_{imax} < 25$. The filtered robustness measures range from 1.6E-8 to 0.538 relative to uncertainties in the first column of the B matrix and from 1.3E-5 to 0.541 relative to those in the second column. (The reader should be reminded here that these are inverse robustness measures; the smaller the number, the better the performance.) The extremely small but non-zero values are due to computational inaccuracies; for any equation that does not contain the first (second) input variable, the robustness measure relative to the first (second) column should be zero. The truly non-zero values span from 0.055 to 0.538 and from 0.065 to 0.541, respectively.

Both a global and a local search was performed, the latter initiating from an arbitrary column canonical model, over the 72 equations with the sensitivity condition κ_i as performance index. The results are shown in Tables 8. and 9. As it can be seen, the sensitivity condition of the worst equuation is almost twice as large in (this particular) local optimum than in the globally optimal model. However, the global optimum took about an hour of main-frame computer time to find while the local optimum was found in about a minute.

A local search was performed with the first robustness measure μ_{Fi1} as performance index, starting from the model globally optimal for the sensitivity condition. The resulting model is shown in Table 10. The evolution of the robustness measures over the 4 steps of the local search procedure is demonstrated in Table 11.

Local search was also attempted, using the same initial model as above, for the second robustness measure μ_{Fi2} and the max. sensitivity number λ_{imax} but no improvement could be achieved. However, with a different (arbitrary) initial model, the local search procedure yielded an improved model for λ_{imax} as shown in Table 12.

As the final step of the deign procedure, the equation coefficients for the optimal diagnostic models are computed. The coefficients are polynomials in the shift operator. The maximum degree of the polynomials depends on the order of the system and the number of outputs in the particular equation. For the 11th order system at hand with one to five outputs in the equations this maximum degree ranges from 7 to 11. The polynomials obtained by computation are generally of lower degree than these maxima.

TABLE 8.
Globally optimal model for sensitivity condition κ_i
($\lambda_{imax} < 25$)

outputs	structure inputs	κ_i	λ_{imax}	μ_{Fi1}	μ_{Fi2}
0 0 0 1 1 0	1 1 1 0	1.841	9.352	0.169	0.171
0 0 1 0 0 1	1 0 1 1	3.285	9.139	0.139	0
0 1 0 1 0 0	1 1 0 1	4.119	5.504	0.265	0.114
1 1 1 1 1 0	0 0 0 0	4.550	5.439	0	0
1 0 1 0 1 1	0 0 1 0	5.950	6.338	0	0
1 1 0 0 0 1	0 1 0 1	5.967	1.843	0	0.095

TABLE 9.

Locally optimal model for sensitivity condition κ_i
($\lambda_{imax} < 25$)

structure outputs	inputs	κ_i	λ_{imax}	μ_{Fi1}	μ_{Fi2}
0 1 1 1 0 0	1 0 0 1	1.819	1.572	0.175	0
0 0 0 1 1 0	1 1 1 0	1.841	9.352	0.169	0.171
0 1 0 0 0 1	1 1 0 1	3.126	1.579	0.098	0.135
1 0 0 0 1 1	0 1 0 1	5.112	4.958	0	0.103
1 0 1 1 0 1	0 0 1 0	8.055	2.914	0	0
1 1 1 0 1 0	0 0 1 0	10.650	5.504	0	0

TABLE 10.

Locally optimal model for robustness measure μ_{Fi1}
($\kappa_i < 15$; $\lambda_{imax} < 25$)

structure outputs	inputs	κ_i	λ_{imax}	μ_{Fi1}	μ_{Fi2}
1 1 1 1 1 0	0 0 0 0	4.550	5.439	0	0
1 0 0 1 1 0	0 1 0 1	3.837	7.058	0	0.179
0 1 1 0 1 1	0 0 1 0	6.526	2.154	0	0
1 0 1 0 0 0	1 0 1 1	13.200	23.190	0.055	0
0 1 0 0 0 1	1 1 0 1	3.126	1.579	0.098	0.135
0 0 0 1 0 1	1 1 1 0	2.142	13.620	0.157	0.159

TABLE 11.

The evolution of the robustness measure μ_{Fi1} over the local search

equation		initial	1st step	2nd step	3rd step	4th step
	1	0	0	0	0	0
	2	0	0	0	0	0
	3	0	0	0	0	0
	4	0.139	0.139	0.055	0.055	0.055
	5	0.169	0.169	0.169	0.098	0.098
	6	0.265	0.199	0.172	0.169	0.157

TABLE 12.

Locally optimal model for max. sensitivity number λ_{imax}

($\lambda_{imax} < 25$)

structure outputs	inputs	κ_i	λ_{imax}	μ_{Fi1}	μ_{Fi2}
0 1 1 1 0 0	1 0 0 1	1.819	1.572	0.175	0
0 0 1 0 1 0	1 1 0 1	2.550	1.990	0.109	0.325
1 1 1 0 1 0	0 0 1 0	10.650	5.504	0	0
1 0 0 1 0 1	0 1 0 1	5.746	6.377	0	0.218
0 1 0 0 0 1	1 1 1 0	6.275	6.557	0.434	0.437
1 0 0 1 1 1	0 0 1 0	7.691	6.728	0	0

10. REFERENCES

1. A. S. Willsky: A survey of design methods for failure detection in dynamic systems. *Automatica, 12,* 601–611 (1976).

2. J. L. Tylee: On-line failure detection in nuclear power plant instrumentetion. *IEEE Transactions on Automatic Control,* AC–28, 406–415 (1983).

3. A. Emami–Naeini, M. M. Akhter and S. M. Rock (1988). Effect of model uncertainty on failure detection: the threshold selector. *IEEE Transactions on Automatic Control, AC–33,* 1106–1115 (1988).

4. P. M. Frank: Evaluation of analytical redundancy for fault diagnosis in dynamic systems. *Preprints of the IFAC Symposium on Advanced Information Processing in Automatic Control* (Nancy, France, 1989), I.7–I.19.

5. J. C. Deckert, M. N. Desai, J. J. Deyst and A. S. Willsky: F–8 DFBW sensor failure identification using analytical redundancy. *IEEE Transactions on Automatic Control, AC–22,* 795–803 (1977).

6. E. Y. Chow and A. S. Willsky: Analytical redundancy and the design of robust failure detection systems. *IEEE Transactions on Automatic Control, AC–29,* 603–614 (1984).

7. J. Gertler and D. Singer: Augmented models for statistical fault isolation in complex dynamic systems. *Proceedings of the American Control Conference* (Boston, MA, 1985), 317–322.

8. J. Gertler and D. Singer: A new structural framework for parity equation based failure detection and isolation. *Automatica, 26,* to appear (1990).

9. J. Gertler and Q. Luo: Robust isolable models for failure diagnosis. *AICHE Journal, 31,* 1856–1868 (1989).

10. Y. Ben–Haim: An algorithm for failure location in a complex network. *Nuclear Science and Engineering, 75,* 191–199 (1980).

11. M. A. Massoumnia, G. C. Verghese and A. S. Willsky: Failure detection and identification. *IEEE Transactions on Automatic Control,* AC–34, 316–321 (1989).

12. G. G. Leininger: Model degradation effects on sensor failure detection. *Proceedings of the American Control Conference,* (Blacksburg, VA, 1981), Paper FP–3a.

13. D. T. Horak: Failure detection in dynamic systems with modeling errors. *Journal of Guidance, Control and Dynamics*, *11*, 508–516 (1988).

14. X. C. Lou, A. S. Willsky and G. C. Verghese: Optimally robust redundancy relations for failure detection in uncertain systems. *IEEE Transactions on Automatic Control*, AC–22, 333–344 (1986).

15. J. Gertler: Survey of model based failure detection and isolation in complex plants. *IEEE Control Systems Magazine*, *8*, No. 6, 3–11 (1988).

16. J. Gertler and K. Anderson: An evidential reasoning extension to model based failure diagnosis. *Proceedings of the IEEE Conference on Intelligent Control Systems* (Albany, NY, 1989), 520–525.

17. E. Kamei, H. Namba, K. Osaki and M. Ohba: Application of reduced order model to automotive engine control systems. *Proceedings of the American Control Conference* (Minneapolis, MN, 1987), 1815–1820.

18. T. Takamatsu, I. Hashimoto and Y. Nakai: A geometric approach to multivariable control system design of a distillation column. *Automatica*, *15*, 387–402 (1979).

APPENDIX

In the following, we are presenting the proof of the parity equation degree conditions described in Eqs. (4.9) and (4.10). For simplicity, the ^ sign will be omitted from the model matrices/coefficients.

Consider the discrete linear dynamic system described by its state equations:

$$x(t) = A\ x(t-1) + B\ u(t) \tag{A.1}$$

$$y(t) = C\ x(t) + D\ u(t) \tag{A.2}$$

where $u(t) \in R^k$ are the system inputs, $y(t) \in R^m$ are the system outputs and $x(t) \in R^{n_s}$ are the state variables; n_s is the order of the underlying system. The primary set of parity equations (cf. Eqs. (4.1) – (4.6)) is

$$e^*(t) = F^*(z)\ \tilde{q}(t) = [G^*(z)\ |\ -H^*(z)] \begin{bmatrix} \tilde{u}(t) \\ \tilde{y}(t) \end{bmatrix} \tag{A.3}$$

where the $G^*(z)$ and $H^*(z)$ matrices are related to the state–space model matrices as

$$H^*(z) = Det[I - Az^{-1}]\ I \tag{A.4}$$

$$G^*(z) = C\ Adj[I - Az^{-1}]\ B + Det[I - Az^{-1}]\ D \tag{A.5}$$

Lemma 1.

A. Consider a system described by Eqs. (A.1) and (A.2), with $m = n_s$ and C invertible. A set of parity equations can be obtained as

$$e^0(t) = [G^0(z)\ |\ -H^0(z)] \begin{bmatrix} \tilde{u}(t) \\ \tilde{y}(t) \end{bmatrix} \tag{A.6}$$

with

$$Deg[G^0(z)] = Deg[H^0(z)] = 1 \tag{A.7}$$

B. Consider the same system with $D = 0$. Then (A.6) holds with

$$Deg[G^0(z)] = 0, \qquad Deg[H^0(z)] = 1 \tag{A.8}$$

and $G^0(z)$ upper triangular.

Proof

A. Obtain $\mathbf{F}^0(z)$ by transformation as

$$\mathbf{F}^0(z) = [\text{Det}(\mathbf{I}-\mathbf{A}z^{-1})]^{-1} \; (\mathbf{I}-\mathbf{A}z^{-1}) \; \mathbf{C}^{-1} \; \mathbf{F}^*(z) \tag{A.9}$$

Then with Eqs. (A.4) and (A.5)

$$\mathbf{F}^0(z) = [\mathbf{B} + (\mathbf{I}-\mathbf{A}z^{-1}) \; \mathbf{C}^{-1} \; \mathbf{D} \; | \; -(\mathbf{I}-\mathbf{A}z^{-1}) \; \mathbf{C}^{-1}] \tag{A.10}$$

B. Decompose $\mathbf{B}$ as

$$\mathbf{B} = \mathbf{B}_L\mathbf{B}_U \tag{A.11}$$

where $\mathbf{B}_L$ is lower triangular and invertible and $\mathbf{B}_U$ is upper triangular. Obtain $\mathbf{F}^0(z)$ as

$$\mathbf{F}^0(z) = [\text{Det}(\mathbf{I}-\mathbf{A}z^{-1})]^{-1} \; \mathbf{B}_L^{-1} \; (\mathbf{I}-\mathbf{A}z^{-1}) \; \mathbf{C}^{-1} \; \mathbf{F}^*(z) \tag{A.12}$$

yielding

$$\mathbf{F}^0(z) = [\mathbf{B}_U \; | \; -\mathbf{B}_L^{-1} \; (\mathbf{I}-\mathbf{A}z^{-1}) \; \mathbf{C}^{-1}] \tag{A.13}$$

Remark. The decomposition (A.11) is not unique.

Remark. The shape of $\mathbf{B}_U$ depends on the value of k relative to m. In particular, $\mathbf{B}_U$ is non–zero in a kxk upper triangle if $k{\leq}m$ and it is zero in an $(m-1)$x$(m-1)$ lower triangle if $k{\geq}m$.

Lemma 2.

Consider a polynomial matrix

$$\mathbf{F}^p(z) = \begin{bmatrix} \mathbf{f}_{1.}^p(z) \\ \vdots \\ \mathbf{f}_{m.}^p(z) \end{bmatrix} \tag{A.14}$$

with rows

$$\mathbf{f}_{i.}^p(z) = [\underset{1}{0} \; \ldots \; \underset{p}{0} \; f_{i,p+1}^p(z) \; \ldots \; f_{i,k+m}^p(z)] \tag{A.15}$$

and assume that $\mathbf{F}^p(z)$ has been obtained from $\mathbf{F}^0(z)$, $\text{Deg}[\mathbf{F}^0(z)]{\leq}1$, by successive eliminations. Then for all rows

$$\text{Deg}[\mathbf{f}_{i.}^p(z)] \leq p+1 \tag{A.16}$$

Proof

The proof uses induction. Starting from $F^0(z)$, rows of $F^1(z)$ are obtained by eliminating a single variable from a pair of rows of $F^0(z)$; the maximum degree of the new rows is 2. Then when obtaining rows of $F^2(z)$ from pairs of rows of $F^1(z)$, the maximum degree of the new rows does not become 4 but, due to cancellations, only 3. Similarly, when rows of $F^p(z)$ are obtained from pairs of rows of $F^{p-1}(z)$, the maximum degree of the new rows does not double relative to the previous set but only increases by 1.

Theorem

A. Consider a system described by Eqs. (A.1) and (A.2), with $m=n_s$ and C invertible. For any attainable parity equation containing $m-1$ zeroes

$$\text{Deg}[f_i.(z)] \leq n_s \qquad\qquad\qquad (A.17)$$

B. Consider the same system with $D=0$. Then for any attainable parity equation containing $m-1$ zeroes

$$\text{Deg}[f_i.(z)] \leq n_s - s + 1 \qquad\qquad\qquad (A.18)$$

where $s \leq m = n_s$ is the number of output variables present in the equation.

Proof

A. Start with $F^0(z)$ as in Eq. (A.10). According to Lemma 1.A, $\text{Deg}[F^0(z)]=1$. Any attainable parity equation, containing $m-1$ zeroes, can be obtained from $F^0(z)$ by $m-1$ successive eliminations, that is,

$$f_i.(z) = f_i^{m-1}(z) \qquad\qquad\qquad (A.19)$$

Then by Lemma 2,

$$\text{Deg}[f_i.(z)] \leq m = n_s \qquad\qquad\qquad (A.20)$$

B. Start with $F^0(z)$ as in Eq. (A.13). According to Lemma 1.B, $\text{Deg}[F^0(z)] \leq 1$ and $G^0(z)$ is upper triangular. Thus $F^0(z)$ already contains zeroes on the input variable side. To generate a parity equation with $m-s$ outputs missing (s outputs present), choose the last $m-s+1$ rows in (A.13). It takes $m-s$ successive eliminations to obtain the new equation, that is

$$f_i.(z) = f_{i.}^{m-s}(z) \tag{A.21}$$

Then with Lemma 2,

$$\text{Deg}[f_i.(z)] \leq m - s + 1 = n_s - s + 1 \tag{A.22}$$

Corollary

A. If $m < n_s$ then the system needs to be augmented with $n_s - m$ linearly independent dummy outputs. For the augmented system, all previous results apply. To obtain parity equations containing $m-1$ zeroes from $F^0(z)$, a total of $n_s - 1$ variables have to be eliminated, including all the dummies. (The dummy outputs, of course, will not appear in the actual model transformation procedure that uses $F^*(z)$.)

B. If $m > n_s$ then the outputs are not linearly independent and C is not invertible. All the above results apply, however, to any subset of n_s linearly independent outputs.

MODEL-PREDICTIVE CONTROL
OF PROCESSES WITH
MANY INPUTS AND OUTPUTS

N. LAWRENCE RICKER

Department of Chemical Engineering, BF-10
University of Washington
Seattle, WA 98195

I.   INTRODUCTION

The historical approach to the design of a complex process in the chemical industry has been to partition it into subsystems that interact as little as possible. The subsystems can then be regulated by human operators with, perhaps, the aid of local (multi-loop) controllers. When this strategy is used, however, communication between the operators is usually inadequate, resulting in poor coordination of the subsystems, *i.e.*, a decision to change the operation of one subsystem may upset several others.

Thus there is a need for *plant-wide control* methods that can manage a number of distributed subsystems, with due consideration given to their interconnections. The advent of computer-based data acquisition and retrieval systems has made such methods feasible, and several examples have appeared in the recent literature. Ruiz *et al.* [1], for example, describe a system for optimal control of a pulp-and-paper production facility, and Downs and Vogel [2] consider the operation of a hypothetical multi-product chemical plant.

The problem is essentially the repeated optimization of a high-order dynamic system with many manipulated and output variables. A key aspect of the optimization problem is the presence of physical constraints on the manipulated variables and, in some cases, safety or product-quality constraints on the output variables. Although other techniques could be used (*e.g.*, "rule-based" methods or "expert systems"), the form of the problem makes it a natural application for Model-Predictive Control (MPC).

The state of the art in MPC has been described in several recent books [3,4,5], which, however, stress applications to relatively small subsystems. The purpose of this article is to show what one must do to use MPC techniques for larger problems. The organization is as follows: *Section II* gives an overview of MPC theory with an emphasis on: 1) the use of linear, state-space models with state estimation for the prediction of process outputs, 2) advantages and disadvantages of the most common quadratic and linear objective functions, 3) methods for efficient formulation and solution of the optimization problem. *Section III* describes the application of MPC to inventory control in a process with 37 states, 8 outputs, 18 manipulated variables, and 7 measured disturbance inputs.

II. MPC THEORY

A generic MPC strategy consists of the following steps, which repeat every T time units, where T is the sampling period:

*   Measure the plant variables. These include *manipulated* variables that can be varied to achieve the process goals, *measured disturbances* that affect the operation of the process but cannot be regulated, and *output* variables for which we have specified objectives (usually given in terms of setpoints or inequality constraints).
*   Use the measurements and a model of the process to estimate the current process state.

- Calculate new settings of the manipulated variables that are optimal with respect to a specified objective function.
- Send the calculated settings to the plant and wait for the beginning of the next cycle. It is assumed here that a zero-order hold latches the manipulated variables at these values for the duration of the sampling period.

The following sections expand on each of these steps.

A. Internal model

The *internal model* serves a dual purpose: it estimates the current state of the process and predicts future plant outputs as a function of past and (anticipated) future inputs and outputs. There are no *a priori* restrictions on the form of this model. It can be a nonlinear dynamic system [6-9], a combination of linear and nonlinear models [10,7], or a linear, time-invariant model. The latter category includes truncated step-response or impulse-response (convolution) models [3, 11, 12-15], and transfer-function or state-space models [3,16-19]. Finally, there is increasing interest in the use of time-varying (adaptive) models in combination with MPC [20-22]

The following development is based on a linear, time-invariant (LTI) model for the following reasons:

- For problems of the scope considered here, it is unlikely that an adequate nonlinear model would be available. Even if it were, the on-line computing effort required to solve the optimal control problem would usually be prohibitive.
- The use of time-varying models of large MIMO plants is too risky from the point of view of robustness.

Ongoing work on numerical methods for nonlinear systems and parameter estimation methods for adaptive schemes may make alternatives to the LTI model more attractive in the future.

1. State-space model form

Although all the usual LTI models (*i.e.*, pulse-response, step-response, transfer-function and state-space) are essentially equivalent, the standard discrete-time state-space formulation is a good choice for a general development of MPC. As shown in section II.A.2, for example, it is a natural starting point for an analysis of state-estimation strategies. Thus let the internal model be written in the form:

$$x(k+1) = \Phi\, x(k) + \Gamma_u\, u(k) + \Gamma_v\, v(k) + \Gamma_w\, w(k) \tag{1}$$

$$y(k) = C\, x(k) + z(k) \tag{2}$$

where the integer k refers to the sampling period beginning at time $t = k \cdot T$, T is the sampling period, x is a vector of n state variables, u is a vector of m manipulated inputs[1], v is a vector of m_v measured but unregulated inputs (*i.e.*, measured disturbances), w is a vector of m_w unmeasured state disturbance inputs, y is a vector of p outputs[2], z is measurement noise, and Φ, Γ_u, Γ_v, Γ_w, and C are constant matrices of appropriate size. Note that the terms involving v(k), w(k) and z(k) allow one to model the plant's disturbance characteristics. They are optional.

From Eqs. (1) and (2) it is easy to show that:

$$\Delta x(k) = \Phi\, \Delta x(k\text{-}1) + \Gamma_u\, \Delta u(k\text{-}1) + \Gamma_v\, \Delta v(k\text{-}1) + \Gamma_w\, \Delta w(k\text{-}1) \tag{3}$$

$$y(k) = C\Phi\, \Delta x(k\text{-}1) + y(k\text{-}1) + C\Gamma_u\, \Delta u(k\text{-}1) +$$
$$C\Gamma_v\, \Delta v(k\text{-}1) + C\Gamma_w\, \Delta w(k\text{-}1) + \Delta z(k) \tag{4}$$

where $\Delta x(k) = x(k) - x(k\text{-}1)$, *etc.* The use of a model written in terms of changes in the manipulated variables leads to a controller that is analogous to the *velocity form* of the PID control law. This has advantages with respect to bumpless transfer from manual to automatic operation, anti-reset windup, *etc.* (see, *e.g.* [23,24]).

---

[1] If these can be measured (as is usually the case), the actual values seen in the plant should be used here rather than the values requested by the controller at the previous sampling period. Otherwise, if the plant were unable to deliver the requested values (*e.g.*, because of an unexpected constraint), the model predictions would be biased.

[2] Normally all p outputs are measured, but this is not necessary. If fact, MPC is occasionally used for "inferential control", in which one or more "secondary" outputs are measured in order to estimate the value of one or more "primary" unmeasured (or infrequently measured) outputs. Control objectives are specified for the primary outputs. In that case it is convenient to order the output variables so that the first p_m are the measured outputs and the remaining $p\text{-}p_m$ are the unmeasured outputs.

It is convenient to augment the original states with the outputs to form the new state-space vector $x_a(k) = [\Delta x(k)^T \ y(k)^T]^T$. Then Eqs. (3) and (4) can be re-written as:

$$x_a(k+1) = \Phi_a x_a(k) + \Gamma_{ua} \Delta u(k) + \Gamma_{va} \Delta v(k) + \Gamma_{wa} \Delta w(k) \tag{5}$$
$$y(k) = C_a x_a(k) + \Delta z(k) \tag{6}$$

where

$$\Phi_a = \begin{bmatrix} \Phi & 0 \\ C\Phi & I \end{bmatrix} \quad \Gamma_{ua} = \begin{bmatrix} \Gamma_u \\ C\Gamma_u \end{bmatrix} \quad \Gamma_{va} = \begin{bmatrix} \Gamma_v \\ C\Gamma_v \end{bmatrix} \quad \Gamma_{wa} = \begin{bmatrix} \Gamma_w \\ C\Gamma_w \end{bmatrix}$$
$$C_a = [\ 0 \quad I \ ]$$

Eqs. (5) and (6) are now in the same form as Eqs. (1) and (2), which is the usual starting point for the development of a state estimator.

2. State estimation

Since the signals $\Delta w(k)$ and $\Delta z(k)$ in Eqs. (5) and (6) are unknown, we use an estimator to account for their effects on the states and the outputs. Most of the state estimation strategies proposed for use in linear MPC can be written in the form:

$$\hat{x}(k+1|k) = \Phi_a \hat{x}(k|k-1) + \Gamma_{ua} \Delta u(k) + \Gamma_{va} \Delta v(k) + K[\tilde{y}(k) - \hat{y}(k|k-1)] \tag{7}$$
$$\hat{y}(k|k-1) = C_a \hat{x}(k|k-1) \tag{8}$$

where $\hat{x}(k+1|k)$ is a prediction[3] of the *augmented* plant state (accounting for the effect of the unmeasured disturbances), $\hat{y}(k|k-1)$ is a prediction of the plant output, $\tilde{y}(k)$ is the measured plant output[4], and K is an estimator gain matrix. During on-line operation, Eqs. (7) and (8) are used at each sampling period to update the state and output

---

[3] The nomenclature $k+1/k$ denotes a prediction of a variable at sampling period $k+1$ based on information available at period k.

[4] It is assumed that there are p_m measured outputs and $p_m = p$. The more general case (with $p_m \leq p$) is easily derived following the approach outlined here.

estimates based on the *measured* values of $\tilde{y}(k)$, $\Delta u(k)$, and $\Delta v(k)$. This is one of the two essential functions of the internal model. (The other is output prediction – see the next section).

In order to use Eq. (7) one must first specify the estimator gain, $\mathbf{K}$. As explained in many texts (*e.g.*, [25], which also covers disturbance modeling), one way to obtain $\mathbf{K}$ is to assume that the signals $\Delta w(k)$ and $\Delta z(k)$ are random-normal inputs with known means and covariances, in which case $\mathbf{K}$ may be interpreted as the steady-state gain of the standard Kalman filter, given by:

$$\mathbf{K} = \Phi_a \mathbf{P} \mathbf{C}_a^T (\mathbf{R}_2 + \mathbf{C}_a \mathbf{P} \mathbf{C}_a^T)^{-1} \tag{9}$$

where $\mathbf{P}$ is the steady-state solution of the Riccati equation:

$$\mathbf{P}(k+1) = \Phi_a \mathbf{P}(k) \Phi_a^T + \mathbf{R}_1 - \Phi_a \mathbf{P}(k) \mathbf{C}_a^T (\mathbf{R}_2 + \mathbf{C}_a \mathbf{P}(k) \mathbf{C}_a^T)^{-1} \mathbf{C}_a \mathbf{P}(k) \Phi_a^T \tag{10}$$

and $\mathbf{R}_1$ and $\mathbf{R}_2$ are the covariance matrices for $\Delta w(k)$ and $\Delta z(k)$:

$$\mathbf{R}_1 = E\{\Gamma_{wa} \Delta w \Delta w^T \Gamma_{wa}^T\} \tag{11}$$

$$\mathbf{R}_2 = E\{\Delta z \Delta z^T\} \tag{12}$$

where $E\{\cdot\}$ is the expectation operator. Details of the solution of Eq. (10) and other aspects of Kalman filtering theory may be found in [25] and other sources. A closely related approach is the use of "disturbance prediction" (see, *e.g.*, [26]).

An alternative is the *specification* of a matrix $\mathbf{K}$ such that the resulting estimator has convenient properties. For example, if we choose

$$\mathbf{K} = \begin{bmatrix} \mathbf{K}_1 \\ \mathbf{K}_2 \end{bmatrix} = \begin{bmatrix} \mathbf{0} \\ \mathbf{I} \end{bmatrix} \tag{13}$$

where $\mathbf{K}_2 = \mathbf{I}_{pxp}$, Eqs. (7) and (8) reduce to the form of the internal model used in DMC [3, pp. 93-95]. This is optimal (in the sense of Kalman filtering) if the measurements are accurate (*i.e.*, R_2 is small relative to R_1) and the state disturbances, Δw, are random steps at the plant output.

Although the DMC estimator is a reasonable first guess for process control applications, other values of K may improve the overall performance of MPC. For example, the development in [19] assumes that the dominant disturbances are random steps at the plant input, so that $\Gamma_{wa} = \Gamma_{ua}$, and $E\{\Delta w \Delta w^T\} = \sigma^2 I$, where σ is an adjustable parameter used to tune the estimator. Then Eq. (11) becomes:

$$R_1 = \sigma^2 \Gamma_{ua} \Gamma_{ua}^T \tag{14}$$

With the additional assumption $R_2 = I_{pxp}$, one can calculate K from Eqs. (9) and (10).

It was shown in [19] that the resulting estimator performs better than the DMC estimator for certain types of ill-conditioned MIMO plants, as well as for processes including integrating elements (such as in level-control applications). Additional examples of the use of state estimators in combination with MPC are given in [16, 18, 27].

3. *Prediction of future outputs – the "prediction horizon"*

Eqs. (7) and (8) can be used to predict plant outputs over a specified *prediction horizon* of N future sampling periods. Suppose we are at the beginning of sampling period k. The signals $\Delta v(k)$ and $\tilde{y}(k)$ have just been measured, and $\hat{x}(k|k-1)$ and $\hat{y}(k|k-1)$ have been calculated *via* Eqs. (7) and (8). We now propose to make a change in the manipulated variable that will apply for the duration of period k. Let this proposed change be designated $\Delta u(k|k)$. Then from Eqs. (7) and (8), the predicted value of the outputs at the beginning of period $k+1$ is:

$$\begin{aligned}
\hat{y}(k+1|k) &= C_a \hat{x}(k+1|k) \\
&= C_a \Phi_a \hat{x}(k|k-1) + C_a \Gamma_{ua} \Delta u(k|k) + C_a \Gamma_{va} \Delta v(k) \\
&\quad + C_a K[\tilde{y}(k) - \hat{y}(k|k-1)]
\end{aligned} \tag{15}$$

In order to use a prediction horizon with $N > 1$, we need to make an assumption regarding future values of the estimator error, $\tilde{y}(k) - \hat{y}(k|k-1)$. Assuming that the estimator is well designed, the expectation of future values of the estimator error is zero. Then the prediction of the outputs at interval $k+2$ is:

$$
\begin{aligned}
\hat{y}(k+2|k) &= C_a\hat{x}(k+2|k) \\
&= C_a\Phi_a\hat{x}(k+1|k) + C_a\Gamma_{ua}\,\Delta u(k+1|k) + C_a\Gamma_{va}\,\Delta v(k+1|k) \\
&= C_a\Phi_a^2\,\hat{x}(k|k-1) + C_a\Phi_a\Gamma_{ua}\,\Delta u(k|k) + C_a\Phi_a\Gamma_{va}\,\Delta v(k) \\
&\quad + C_a\Phi_aK[\tilde{y}(k) - \hat{y}(k|k-1)] \\
&\quad + C_a\Gamma_{ua}\,\Delta u(k+1|k) + C_a\Gamma_{va}\,\Delta v(k+1|k)
\end{aligned}
\tag{16}
$$

where $\Delta u(k+1|k)$ and $\Delta v(k+1|k)$ are the expected future values of the manipulated variables and measured disturbances, respectively. Continuing in this manner for N steps we get the following linear prediction of the future outputs:

$$
\begin{aligned}
\boldsymbol{y}(k) &= S_x\,\hat{x}(k|k-1) + S_u\Delta \boldsymbol{u}(k) + S_v\Delta \boldsymbol{v}(k) + S_y[\tilde{y}(k) - \hat{y}(k|k-1)] \\
&= (S_x - S_yC_a)\hat{x}(k|k-1) + S_u\Delta\boldsymbol{u}(k) + S_v\Delta\boldsymbol{v}(k) + S_y\tilde{y}(k)
\end{aligned}
\tag{17}
$$

where, by definition,

$$
\boldsymbol{y}(k) = \begin{bmatrix} \hat{y}(k+1|k) \\ \hat{y}(k+2|k) \\ \vdots \\ \hat{y}(k+N|k) \end{bmatrix}
\qquad
\Delta\boldsymbol{u}(k) = \begin{bmatrix} \Delta u(k|k) \\ \Delta u(k+1|k) \\ \vdots \\ \Delta u(k+N-1|k) \end{bmatrix}
\qquad
\Delta\boldsymbol{v}(k) = \begin{bmatrix} \Delta v(k) \\ \Delta v(k+1|k) \\ \vdots \\ \Delta v(k+N-1|k) \end{bmatrix}
$$

$$
S_x = \begin{bmatrix} C\Phi & I \\ C(\Phi^2+\Phi) & I \\ \vdots & \vdots \\ C\sum_{i=1}^{N}\Phi^i & I \end{bmatrix} \qquad S_y = \begin{bmatrix} K_2 \\ C\Phi K_1+K_2 \\ \vdots \\ \sum_{i=2}^{N} C\Phi^{i-1}K_1+K_2 \end{bmatrix}
$$

$$
S_u = \begin{bmatrix} C\Gamma_u & 0 & \cdots & 0 \\ C(\Phi+I)\Gamma_u & C\Gamma_u & \ddots & \vdots \\ \vdots & \vdots & \ddots & 0 \\ C(\sum_{i=1}^{N}\Phi^{i-1})\Gamma_u & C(\sum_{i=1}^{N-1}\Phi^{i-1})\Gamma_u & \cdots & C\Gamma_u \end{bmatrix}
$$

$$
S_v = \begin{bmatrix} C\Gamma_v & 0 & \cdots & 0 \\ C(\Phi+I)\Gamma_v & C\Gamma_v & \ddots & \vdots \\ \vdots & \vdots & \ddots & 0 \\ C(\sum_{i=1}^{N}\Phi^{i-1})\Gamma_v & C(\sum_{i=1}^{N-1}\Phi^{i-1})\Gamma_v & \cdots & C\Gamma_v \end{bmatrix}
$$

Note that for the definition of S_y, the estimator gain matrix has been partitioned as in Eq. (13).

In order to use Eq. (17) we need to specify the future values of the measured disturbances, $\Delta v(k)$. The usual approach (*e.g.*, in DMC) is to *assume* that $\Delta v(k+1|k) = \Delta v(k+2|k) = \ldots = \Delta v(k+N-1|k) = 0$, *i.e.*, future values of v will be equal to the present value. In that case, Eq. (17) becomes:

$$\boldsymbol{y}(k) = (\mathbf{S}_x - \mathbf{S}_y\mathbf{C}_a)\hat{\mathbf{x}}(k|k-1) + \mathbf{S}_u\Delta\boldsymbol{u}(k) + \mathbf{S}_{v0}\Delta v(k) + \mathbf{S}_y\tilde{\mathbf{y}}(k) \qquad (18)$$

with

$$\mathbf{S}_{v0} = \begin{bmatrix} \mathbf{C}\Gamma_v \\[2mm] \mathbf{C}(\Phi+\mathbf{I})\Gamma_v \\[2mm] \vdots \\[2mm] \mathbf{C}(\sum_{i=1}^{N}\Phi^{i-1})\Gamma_v \end{bmatrix}$$

In some situations it is possible to *predict* future variations of these disturbances (see, *e.g.*, section III). Then Eq. 17 should be used for prediction of the plant outputs rather than Eq. 18.

4. *Blocking of the manipulated variables*

As defined in [28], *blocking* of the manipulated variables implies that, *for purposes of output prediction*, the future manipulated variables are assumed to be held constant over "blocks" of time[5]. The duration of each such block is $\kappa_i \cdot T$ where κ_i is an integer ($\kappa_i \geq 1$) and T is the sampling period. Let n_b designate the number of blocks with the restriction that $1 \leq n_b \leq N$, where N is the number of sampling periods in the prediction horizon. The blocking strategy[6] for the prediction horizon is then described by the vector κ, where

$$\kappa = [\ \kappa_1 \ \kappa_2 \ ... \ \kappa_{n_b}\ ] \qquad \text{such that} \qquad \kappa_i \geq 1 \ \text{ and } \ \sum_{i=1}^{n_b}\kappa_i = N.$$

---

[5]   Blocking applies only for the output-prediction calculations. During process operation, the control system actually changes the manipulated variables at each sampling period (see the description of MPC at the beginning of section II).

[6]   For simplicity, the discussion here assumes that the same blocking strategy is used for each of the m manipulated variables. This would not be necessary in general.

For example, consider a hypothetical example with $N=5$, $n_b=2$, $\kappa_1=2$, $\kappa_2=3$. This would constrain the manipulated variables to change at only two points in the prediction horizon – at beginning of the first sampling period and at the beginning of the third, *i.e.*, of the elements in $\Delta \mathbf{u}(k)$, only $\Delta u(k|k)$ and $\Delta u(k+2|k)$ would be non-zero. When doing predictions, we can then eliminate the elements of $\Delta \mathbf{u}(k)$ that are assumed to be zero and the corresponding columns of the $\mathbf{S}_u$ matrix.

5. Relationship of blocking to DMC control horizon

Blocking is a generalization of the specification of a *control horizon* in DMC [3]. Let the duration of the DMC control horizon be M sampling periods, where $M \le N$. The equivalent blocking strategy has $n_b = M$ blocks, with $\kappa_i = 1$ for $i = 1,M-1$, and $\kappa_M = N-M+1$.

There is an obvious advantage to blocking (or the use of $M < N$ in DMC) – it reduces the size of the $\mathbf{S}_u$ matrix and the number of elements of $\Delta \mathbf{u}(k)$. Since the elements of $\Delta \mathbf{u}(k)$ are decision variables in the MPC optimization problem (see section II.C), this is especially useful for large problems of the type considered here.

A less obvious advantage is that blocking allows one to compensate for the effects of non-minimum phase elements in the internal model [28]. For example, if the nominal (stable) internal model has an unstable inverse (which is very common in the discrete-time case), a large value of N and a sufficiently small value of n_b will result in a stable control law. Blocking can also be beneficial for systems with time delays, as described in the next section.

6. Processes with time delays

Suppose that at the beginning of sampling period k the plant is at steady-state. At this instant, we make step changes in *all* of the manipulated variables and predict their effect on the output variables (according to the internal model). Let $k+\tau_i$ be the sampling period at which we first see a change in y_i, the ith plant output[7]. The value

---

7   The form of the internal model used in Eqs. (1) and (2) implies that $\tau_i \ge 1$ (*i.e.*, $\mathbf{D} = \mathbf{0}$ in the standard state-space formulation).

τ_i-1 is termed the *minimum time delay* for y_i. The minimum time delay for each output is, by definition, given by $\tau = [\ \tau_1$-$1 \quad \tau_2$-$1 \quad ... \quad \tau_p$-$1\ ]$.

Recall that the purpose of the prediction step in MPC is to show how proposed adjustments in the manipulated variables would affect the outputs. From this feedback-control viewpoint, if we are at period k it is useless to predict y_i between period k and period $k+\tau_i$-1 – we cannot use the manipulated variables to control y_i during this period. It is better to define a prediction horizon for y_i that begins at sampling period $k+\tau_i$ and ends at sampling period $k+\tau_i+N$-1. The same applies for the other outputs as well.

This change in the definition of the prediction horizon can be accommodated easily in Eqs. (17) and (18) by shifts in the rows of the $\mathbf{y}(k)$ vector and corresponding shifts in the $\mathbf{S_u}$, $\mathbf{S_v}$, $\mathbf{S_x}$, and $\mathbf{S_y}$ matrices [28]. If this is not done, the resulting optimization problem will be ill-posed unless the designer compensates for the presence of the time delays in some other way.

One example of such compensation is used in DMC. Let τ be the *maximum* value of the elements in τ. Then if one defines the control horizon and the prediction horizon such that N-$M \geq \tau$, the resulting unconstrained optimization problem is well-posed from the point of view of the time delays. The explicit shifting of the prediction horizon is recommended over this approach, however, because it automatically defines an appropriate "window" for the application of hard inequality constraints on the output variables (see section II.D).

B . Reference trajectory

Consider a SISO system at the beginning of sampling period $k=0$ whose measured output is $\tilde{y}(0) = 0$. The output setpoint is $r(0) = 1$. From the point of view of setpoint tracking, one would usually like to move the output back to its setpoint immediately, *i.e.*, by the end of the sampling period. This is rarely a realistic goal, however. Some forms of MPC (*e.g.*, IMC [11] and MAC [15]) allow one to specify a more gradual path back to the setpoint. Figure 1 shows three examples of such paths.

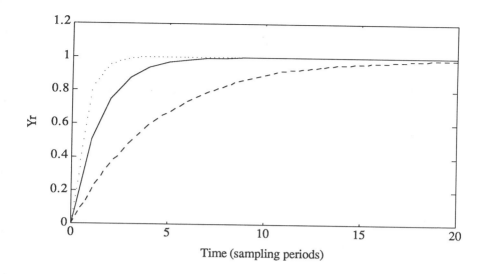

Figure 1. Three example reference trajectories. Dotted line: $\Phi_r = 0.2$.
Solid line: $\Phi_r = 0.5$. Dashed line: $\Phi_r = 0.8$.

The desired path to the setpoint is called the *reference trajectory*. For systems
with multiple outputs, a simple way to define the reference trajectory is as follows:

$$y_r(k+1|k) = r(k+1|k) - \Phi_r [r(k) - \tilde{y}(k)]$$
$$y_r(k+2|k) = r(k+2|k) - \Phi_r^2 [r(k) - \tilde{y}(k)]$$

$$\vdots \quad = \quad \vdots \qquad \vdots$$

$$y_r(k+N|k) = r(k+N|k) - \Phi_r^N [r(k) - \tilde{y}(k)] \tag{19}$$

where $r(k+j|k)$ is the expected value of the setpoint vector j sampling periods in the
future, $y_r(k+j|k)$ is the desired value of the output vector at that time, and Φ_r is a p by
p diagonal matrix with the diagonal elements, ϕ_{ri}, chosen such that $0 \le \phi_{ri} < 1$ for
$i = 1,p$. Eq. (19) can be generalized in the manner of Eqs. (17) and (18) as follows:

$$\mathbf{y}_r(k) = \mathbf{r}(k) - \mathbf{S}_r[\mathbf{r}(k) - \tilde{\mathbf{y}}(k)] \tag{20}$$

where, by definition,

$$\mathbf{y}_r(k) = \begin{bmatrix} y_r(k+1|k) \\ y_r(k+2|k) \\ \vdots \\ y_r(k+N|k) \end{bmatrix} \qquad \mathbf{r}(k) = \begin{bmatrix} r(k+1|k) \\ r(k+2|k) \\ \vdots \\ r(k+N|k) \end{bmatrix} \qquad \mathbf{S}_r = \begin{bmatrix} \Phi_r \\ \Phi_r^2 \\ \vdots \\ \Phi_r^N \end{bmatrix}$$

One can use the ϕ_{ri} values to specify the desired closed-loop response time of the plant. In other words, according to Eq. (20) the ideal servo and regulatory response of the closed-loop system is decoupled – each of the p outputs mimics a first-order system with time a constant $\tau_{ri} = -T/ln(\phi_{ri})$. The use of a "robustness filter" in IMC is a generalization of the above approach [4,11].

C. Objective function

1. Quadratic objective

The most widely used objective function for MPC has the quadratic form:

$$J_q(k) = [\mathbf{y}_r(k) - \mathbf{y}(k)]^T \mathbf{Q}[\mathbf{y}_r(k) - \mathbf{y}(k)] + \mathbf{u}^T(k)\mathbf{R}\mathbf{u}(k) \tag{21}$$

where $\mathbf{Q}$ and $\mathbf{R}$ are weighting matrices[8] that allow one to adjust the performance of the controller. The objective, J_q, is to be minimized with respect to $\mathbf{u}(k)$, subject to the equality constraints, Eqs. (17)[9] and (20), and the following inequalities:

$$\mathbf{u}_{min}(k) \le \mathbf{u}(k) \le \mathbf{u}_{max}(k) \tag{22}$$
$$|\Delta\mathbf{u}(k)| \le \Delta\mathbf{u}_{max}(k) \tag{23}$$
$$\mathbf{y}_{min}(k) \le \mathbf{y}(k) \le \mathbf{y}_{max}(k) \tag{24}$$

---

[8]   These are usually diagonal matrices for the sake of simplicity.
[9]   or Eq. (18) if we assume that measured disturbances will be constant in the future.

which are *optional* bounds on the output variables, the manipulated variables, and the rates of change of the manipulated variables. Note that the bounds may vary from one sampling period to the next as process conditions change. Also, one may include constraints on (unmeasured) model states through an appropriate definition of the process outputs and specification of the bounds in Eq. (24). The variables $\boldsymbol{u}(k)$ and $\Delta\boldsymbol{u}(k)$ are related by:

$$\boldsymbol{u}(k) = \mathbf{R}_\Delta \Delta\boldsymbol{u}(k) + \delta(k) \tag{25}$$

where, by definition,

$$\mathbf{R}_\Delta = \begin{bmatrix} \mathbf{I} & \mathbf{0} & \mathbf{0} & . & \mathbf{0} & \mathbf{0} \\ \mathbf{I} & \mathbf{I} & \mathbf{0} & . & \mathbf{0} & \mathbf{0} \\ \mathbf{I} & \mathbf{I} & \mathbf{I} & . & \mathbf{0} & \mathbf{0} \\ . & . & . & . & . & . \\ \mathbf{I} & \mathbf{I} & \mathbf{I} & . & \mathbf{I} & \mathbf{0} \\ \mathbf{I} & \mathbf{I} & \mathbf{I} & . & \mathbf{I} & \mathbf{I} \end{bmatrix} \qquad \delta(k) = \begin{bmatrix} u(k-1) \\ u(k-1) \\ u(k-1) \\ . \\ u(k-1) \\ u(k-1) \end{bmatrix}$$

Since the objective function is quadratic and the constraints are linear, this form of MPC requires the on-line solution of a quadratic programming (QP) problem at the beginning of each sampling period [3]. General-purpose software for QP is available [29] and specialized methods have also been developed to take advantage of the MPC problem structure [28].

2. Analytical solution of the unconstrained QP problem

An advantage of the quadratic form of MPC is that one can derive an analytical solution for the control law if $\mathbf{Q}$ and $\mathbf{R}$ are specified such that a unique, solution of the unconstrained optimization problem exists[10]. This solution is valid as long as it satisfies the inequality constraints (if any). Then if we use the output prediction of Eq. (18), and assume that future setpoints will be constant (*i.e.*, $r(k+i|k) = r(k)$ for $i=1,N$), we obtain the following expression for the required change in the manipulated variables at each sampling period [19]:

---

[10] A sufficient condition is that $\mathbf{R}$ is positive definite, but such a solution can also exist in many other situations [12,19], *e.g.*, when there are no time delays, $N=n_b$, and $m=p$.

$$\Delta \pmb{u}(k) = \; \pmb{K_r r}(k) + \pmb{K_x \hat{x}}(k|k-1) - \pmb{K_v \Delta v}(k) - \pmb{K_y \tilde{y}}(k) \tag{26}$$

where

$$\pmb{K_r} = \pmb{L}(1 - \pmb{S_r}) \tag{27}$$

$$\pmb{K_x} = \pmb{L}(\pmb{S_y C_a} - \pmb{S_x}) \tag{28}$$

$$\pmb{K_v} = \pmb{L S_v} \tag{29}$$

$$\pmb{K_y} = \pmb{L}(\pmb{S_y} - \pmb{S_r}) \tag{30}$$

$$1 = [\; \pmb{I_{pxp}} \;\; \pmb{I_{pxp}} \;\; \cdots \;\; \pmb{I_{pxp}} \;]^T \tag{31}$$

and $\pmb{L}$ is the first m rows of the matrix $(\pmb{S_u^T Q S_u} + \pmb{R})^{-1} \pmb{S_u^T Q}$.

Let the state variables of the controller be $\pmb{x_c}(k) = \pmb{x_a}(k)$, with the controller inputs and outputs defined as $\pmb{u_c}(k) = [\pmb{r}^T(k) \;\; \pmb{\tilde{y}}^T(k) \;\; \pmb{\Delta v}^T(k)]^T$ and $\pmb{y_c}(k) = \Delta \pmb{u}(k)$. Then the state equations for the controller can be written in the standard form:

$$\pmb{x_c}(k+1) = \pmb{\Phi_c x_c}(k) + \pmb{\Gamma_c u_c}(k) \tag{32}$$

$$\pmb{y_c}(k) = \pmb{C_c x_c}(k) + \pmb{D_c u_c}(k) \tag{33}$$

with the constant coefficient matrices:

$$\pmb{\Phi_c} = \pmb{\Phi_a - KC_a + \Gamma_{ua} K_x} \qquad \pmb{\Gamma_c} = [\pmb{\Gamma_{ua} K_r} \quad \pmb{K - \Gamma_{ua} K_y} \quad \pmb{\Gamma_{va} - \Gamma_{ua} K_v}]$$

$$\pmb{C_c} = \pmb{K_x} \qquad\qquad\qquad \pmb{D_c} = [\quad \pmb{K_r} \qquad \pmb{-K_y} \qquad \pmb{-K_v} \quad]$$

Equations (32) and (33) can be combined with state equations describing the plant[11] for linear simulations and performance/robustness studies (in which case it is most convenient to put the plant equations in the form of Eqs. (5) and (6)).

3.   Linear objective function

An alternative to Eq. (21) is a linear objective [30-32], such as:

---

[11]   The plant model need not be the same as the internal model -- one can simulate the effect of modeling error by making them different. For the unconstrained case one can use the analytical approaches described in [4].

$$J_l(k) = a_r^T \mid y_r(k) - y(k) \mid + a_y^T \, y(k) + a_{\Delta u}^T \mid \Delta u(k) \mid + a_u^T \, u(k) \qquad (34)$$

where J_l is to be minimized subject to the same constraints as for the quadratic case. The a_r, a_y, $a_{\Delta u}$, and a_u vectors are weights (*i.e.*, tuning parameters), analogous to the Q and R matrices in the quadratic objective. Vectors a_r and $a_{\Delta u}$ must be non-negative, but the elements of a_y and a_u are can be any finite real number. Equation (34) allows a great deal of flexibility in the definition of the control objectives[12]. For example, one can force certain outputs to go to setpoints by making the corresponding elements of a_r positive and setting those in a_y equal to zero. One can encourage other outputs to go to their upper or lower bounds by making the corresponding weights in a_r zero, and those in a_y positive (for an upper bound), or negative (for a lower bound). Similarly, one can encourage certain inputs to be minimized or maximized.

A disadvantage of the linear objective is that an analytical solution for the control law is only possible when the number of manipulated variables equals the number of output variables, a_r is strictly positive, all other weights are zero, and all inequalities are inactive. In that case the optimal control law is the *perfect controller*, *i.e.*, the inverse of the internal model [28]. The perfect controller has poor robustness with respect to modeling errors, however, so for all practical purposes we must use the numerical solution, which is a linear programming (LP) problem.

Thus, in contrast to the unconstrained quadratic case, one cannot use standard analytical techniques to determine stability and robustness properties of the closed-loop system. These issues must be addressed *via* a series of simulations. Fortunately, efficient, general-purpose LP packages are widely available, making simulation studies feasible (if not completely satisfying from a theoretical point of view).

---

12   The linear terms involving a_y and a_u could also be added to the quadratic objective, providing essentially the same degree of flexibility. Note, however, that there might not be an unconstrained solution to such a problem.

4. Formulation of the LP problem

As described in [30], to accommodate the absolute value terms in Eq. (34) one can introduce the following slack variables:

$$e_+(k) - e_-(k) = y_r(k) - y(k) \tag{35}$$
$$\Delta u_+(k) - \Delta u_-(k) = \Delta u(k) \tag{36}$$

The objective function can then be re-written in the standard LP form:

$$J_1(k) = a_r^T[e_+(k)+e_-(k)] + a_y^T\, y(k) + a_{\Delta u}^T[\Delta u_+(k)+\Delta u_-(k)] + a_u^T\, u(k) \tag{37}$$

which is to be minimized with respect to the variables ($\Delta u(k)$, $u(k)$, $y(k)$, $e_+(k)$, $\Delta u_+(k)$ and $\Delta u_-(k)$), subject to the constraints, Eqs. (17), (20), (22) to (25), and the following non-negativity conditions:

$$e_+(k) \geq 0 \tag{38a}$$
$$e_-(k) \geq 0 \tag{38b}$$
$$\Delta u_+(k) \geq 0 \tag{38c}$$
$$\Delta u_-(k) \geq 0 \tag{38d}$$

As in the QP case, this LP must be solved at the beginning of each sampling period. The calculated values[13] of $\Delta u(k|k)$ are then sent to the plant and the procedure is repeated at the next sampling period.

5. Multi-objective formulation

It should be mentioned here that it can be difficult to define an effective MPC strategy for a large system using a single objective function such as Eqs. (21) or (34). In other words, if there are many decision variables and control objectives, it may be far from obvious how one should choose the objective-function weights so

---

13   By definition these are the first m values in the $\Delta u(k)$ vector, determined by solution of the LP problem.

as to achieve the performance specifications for the real plant. In that case it may be better to partition the overall optimization problem into group of related sub-problems requiring a coordinated, iterative solution. Such methods are beyond the scope of this article but an introduction to the subject may be found in [33].

D. Recommendations regarding the use of hard constraints

The inequality constraints, Eqs. (22) to (24), are to be satisfied *exactly* in the minimization of the objective function, *i.e.*, they are "hard". It is emphasized that although this can improve overall control-system performance, especially for MIMO systems [3, 28], it introduces some pitfalls that may trap the unwary. First a general observation: *the constraints are written in terms of model estimates, not the real plant variables*. Because of the inevitable presence of unexpected disturbances and errors in measurement and modeling, there is *no way* to guarantee that the real variables will always be within a specified region. Now consider the following specific points:

1. Constraints on output variables

Hard constraints on output variables are especially problematic. Unanticipated combinations of active constraints and state trajectories may lead to instabilities in the closed-loop system [34]. Also, modeling errors or unmeasured disturbances can make the on-line LP or QP problem infeasible (see, *e.g.*, [34,35] and section III.B.1.b). Although some progress has been made on analytical methods that can feret out such problems [34], *the best policy is to avoid hard output constraints unless they are essential to the definition of the MPC problem.*

When output constraints must be used, *constraint softening* can make the closed-loop system more stable and prevent infeasibilities [35]. The disadvantage is that this introduces additional controller parameters that must be chosen by trial-and-error. Overall performance suffers if they are specified incorrectly.

Another *ad hoc* approach is the use of *constraint windows*. The idea is to constrain the outputs for only a specified fraction of the prediction horizon – the constraint window. The window can be different for each output variable. For example, if the minimum time delay for output y_i is 3, it would be dangerous to constrain this output for the first 3 sampling periods of the prediction horizon[14]. Unfortunately, although reducing the window size and moving it further out in the prediction horizon usually reduces the chance of an infeasibility, there are no guarantees, and the selection of window sizes must be done by trial and error.

2. Constraints on manipulated variables

The use of simultaneous constraints on the absolute value and rate-of-change of a manipulated variable can also result in an infeasible optimization problem. Suppose, for example, that one has defined a lower bound for a manipulated variable, but the measured value from the plant is, for some reason, below this value. If there is a tight restriction on $|\Delta u|$, it may be impossible to make the manipulated variable satisfy its lower bound within one sampling period.

In practice, measurement errors and other uncertainties also make it difficult to estimate the *true* upper and lower bounds on the manipulated variables. In fact, u_{min} and u_{max} are often time-varying (see, *e.g.*, section III.A). If one must use constraints on $|\Delta u|$ as well as u_{min} and/or u_{max}, logic should be added to check for infeasibilities of the type cited above. The following *ad hoc* fixes are easily implemented:

- Assume that the measurement is incorrect and set the measured value to the lower bound (or the upper bound where appropriate).
- Assume that the bound on the variable was specified incorrectly and reset it such that the current measured condition satisfies the bound.
- Relax the specification on $|\Delta u|_{max}$ to allow the variable to move within the bounds within one sampling period.
- Use constraint softening for either the bounds on Δu or the bounds on u (as for the output constraints).

It is better to avoid the use of such simultaneous constraints, however. If each manipulated variable is constrained by either Eq. (22) or (23), but not both, there will be no problems of the type noted above.

---

14   The use of time-delay factorization, as described in section II.A.5, automatically avoids this problem.

E. Efficiency considerations

It is essential to formulate large MPC problems carefully and code the algorithm in an efficient manner. The following considerations apply both to QP and LP problems.

1. Accounting for problem sparseness.

It is rarely practical to model all the interactions between the manipulated and output variables in a large system. In fact, one should decouple the control problem by a clever choice of the manipulated variables, if possible. Consider, for example, the level control problem shown in Figure 2. The goal is to maintain the levels, $h_1(t)$, $h_2(t)$, and $h_3(t)$, at setpoints. If we select the valve positions as the manipulated variables, then a change in any of the three valve positions (with the other two held constant) will cause a change in all three levels. In other words, the system is fully interacting. It would also be nonlinear because, for example, $q_1 \propto \sqrt{h_1 - h_2}$ (for a given valve position) and the relationship between valve position and flowrate (at constant Δh) is generally nonlinear as well.

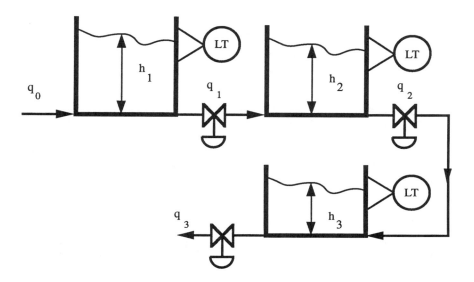

Figure 2: Process schematic for a level-control system.

Now suppose that we install flow control loops for q_1, q_2, and q_3, and use the setpoints of these loops as the manipulated variables for the level-control problem. Also suppose that the flow-control loops provide a fast and accurate servo and regulatory response. Then if we change one of the manipulated variables, holding the others constant, a maximum of 2 levels would be affected. Moreover, the new plant would be linear (unless a valve saturates). In effect, the flow control loops physically decouple and linearize the system.

With respect to MPC, decoupling makes the constraint matrices sparse. In other words, the transfer-function matrix relating the manipulated variables to the plant outputs contains many elements that are essentially[15] equal to zero. In that case the matrix $\mathbf{S}_u$ in Eq. (17), which is rather sparse even for fully interacting systems, would have very few non-zero elements. (For the example problem in Section III, only about 2 % of the elements of $\mathbf{S}_u$ were non-zero). Other matrices, *e.g.*, $\mathbf{R}_\Delta$ in Eq. (25), are equally sparse. The use of software packages such as MINOS [36], which is designed to solve sparse LP, QP, and general NLP problems, can provide huge savings in computer storage and computation time for such problems. Only the non-zero elements of the constraint matrices are stored and manipulated.

2. *Eliminating unnecessary variables and constraints.*

Extraneous variables and constraints should be eliminated from the problem definition. For example, the following rules apply to the LP formulation:

- Recall that by definition, the $\mathbf{y}$ vector includes all p output variables at each of the N sampling periods in the prediction horizon. If a given element of $\mathbf{y}$ is unbounded and has a zero weight in both the $\mathbf{a}_r$ and $\mathbf{a}_y$ vectors, remove this element from $\mathbf{y}$, $\mathbf{e}_+$ and $\mathbf{e}_-$, and delete the corresponding rows in Eqs. (17), (35), and (38a,b).
- Of the remaining elements in $\mathbf{y}$, delete those having zero weights in the $\mathbf{a}_r$ vector from the definition of $\mathbf{e}_+$ and $\mathbf{e}_-$, and remove the corresponding rows from Eqs. (35) and (38a,b). If, for example, $\mathbf{a}_r$ has all elements equal to zero,

---

[15]   If a given manipulated variable has a small effect on an output -- relative to the effects of other manipulated variables and the disturbances -- the relationship is best modeled as a zero-gain transfer function.

meaning that setpoint tracking is not required at all, one should eliminate these variables and equations completely.

- Similarly, if an element of $\Delta\mathbf{u}$ has a zero weight in the $\mathbf{a}_{\Delta u}$ vector, eliminate the corresponding elements of $\Delta\mathbf{u}_+$ and $\Delta\mathbf{u}_-$, and the corresponding row of Eqs. (36) and (38c,d).

- Finally, if an element of $\mathbf{u}$ is unbounded and has a zero weight in the $\mathbf{a}_u$ vector, eliminate this element and the corresponding rows in Eqs. (22) and (25).

It may be possible to formulate additional rules for specific cases. If, for example, it can be shown that a particular constraint can never become active, it should be eliminated along with all variables that appear only in that constraint. The clever use of constraints (*e.g*, the use of constraint windows – see Section II.D) can also have a big impact on the number of variables and constraints that must be retained.

3. "Warm start" at each sampling period.

MINOS and other such packages have provisions that allow one to start the LP or QP iterations for sampling period k from the solution of the optimal control problem at sampling period $k-1$. For large problems, this *warm start* feature usually reduces the computational time for a solution by more than a factor of 5 relative to a "cold start". Note that once the problem has been defined (including the specification of the internal model and the weighting vectors), the number of constraint equations is fixed, as are their coefficients. Then during real-time operation, only the bounds and the right-hand-side elements of the constraint equations vary from one sampling period to the next. Thus a warm start is easily accomplished.

4. Sparse-matrix representation of the internal model.

The internal model (Eqs. (5) and (6)) can be stored in a sparse-matrix format for use in state estimation (see section V for an example).

III. EXAMPLE APPLICATION

The purpose of this section is to illustrate the application of MPC to a typical large-scale system. The proper formulation of such a problem (*i.e.*, selection of the objective function and the controlled and manipulated variables) is crucial. It is hoped that the reasoning leading to the formulation used here will serve as a guide for others. It should be emphasized, however, that certain aspects are problem-specific. For example, the control objectives suggest the use of a linear rather than a quadratic objective function. This is not meant to imply that a LP formulation is always to be preferred over a QP formulation.

A . Problem description

Many cities in the U.S. are faced with the problem of *combined sewer overflows* (CSOs). Combined sewers carry both sewage and storm runoff. During large storms the runoff can exceed the capacity of the collection system, at which point the mixture of storm drainage and sewage must be released to the environment untreated.

One way to reduce CSOs is to construct separated storm water and sewage conduits and/or storage basins to handle peak loads. In most cases, however, the cost is prohibitive. An attractive alternative is real-time control of regulating gates and pumps to take maximum advantage of storage volumes that already exist in the collection system.

Figure 3 shows a schematic of a single sewer pipe. The flowrate entering at the upstream end is either controlled by an upstream regulator or is an unregulated inflow (from storm runoff or a sanitary sewer). Unregulated inflows are also distributed along the length of the pipe. The flowrate leaving the section can be controlled by an automated gate or, in some cases, a pump with a regulated discharge rate. This flowrate has an upper bound determined by one of the following factors:

•   The maximum flow through the gate or pump depends on the total upstream and downstream heads. These vary with time.

- There may be a constraint imposed by the downstream element in the collection system. If, for example, the downstream element is a treatment plant, there is usually an upper limit on the treatment plant inflow.

There is a liquid-level sensor at the downstream end near the flow regulation point. If a certain level is reached, sewage either overflows a fixed-elevation weir or a second gate opens automatically to release the sewage to the overflow outlet.

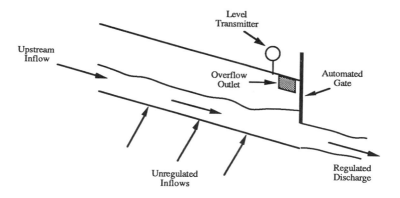

Figure 3: Schematic of a single combined-sewer pipe

If the collection system were only a single pipe, the optimal strategy would be clear: set the downstream flowrate at its maximum value at all times. In reality, however, there are many such sections connected in a complex network. During a given storm, certain parts of the network will be subjected to heavier loads than others. An optimal strategy for minimization of *total* overflows must consider the conditions throughout the network.

1. The plant

Papageorgiou [37] studied the application of a form of optimal control to the hypothetical combined sewer system shown in Figure 4.   There are 2 treatment plants (fed by q_6 and q_8), 8 locations where sewage can accumulate (volumes V_1 to

V_8, represented by the triangles in Figure 4), and 10 controllable flowrates (q_1 to q_8, p_1 and p_2). See Table I for upper bounds on these variables. All have lower bounds at zero.

Variable	Upper Bound	Variable	Upper Bound
V_1	100,000 m^3	q_1	35 m^3/s
V_2	50,000 m^3	q_2	20 m^3/s
V_3	5,000 m^3	q_3	30 m^3/s
V_4	70,000 m^3	q_4	20 m^3/s
V_5	50,000 m^3	q_5	25 m^3/s
V_6	50,000 m^3	q_6	17 m^3/s
V_7	40,000 m^3	q_7	9 m^3/s
V_8	100,000 m^3	q_8	5 m^3/s
		p_1	10 m^3/s
		p_2	13 m^3/s

Table I: Upper bounds for variables in Papageorgiou's problem

The two numbered circles in Fig. 4 represent flow division points. One flowrate leaving each division point is controllable. The other is determined by the difference between the combined inflows and the controlled outflow, *i.e.*, there is no accumulation of mass at these points.

There are 8 combined-sewer overflows (CSOs) labeled s_1 to s_8. The overall control objective is to minimize a weighted sum of these CSOs[16]. There are also 7 inflow disturbances (d_1 to d_7)[17] and 10 long conduits represented by pure time delays – the numbered rectangles in Figure 4. The periods of delay in the ten conduits were 4, 3, 1, 1, 1, 2, 3, 1, 1, and 12, respectively. Papageorgiou's

---

[16] Note, however, that s_5 is not released to the environment, so there is no penalty if s_5 is non-zero. Also, Papageorgiou did not include s_1, which is used here to avoid problems due to infeasible LPs for the simulations with measurement error in section III.B.2. It turns out that $s_1 = 0$ at all times, so the inclusion of this variable has no effect on the CSOs.

[17] Papageorgiou's original problem had 8 disturbance inputs, but the last 2 could be summed with no change in the results. That was the approach taken here.

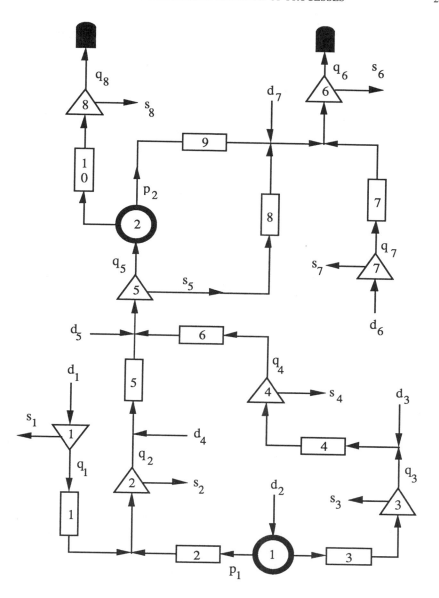

Figure 4: Papageorgiou's CSO network [37]

sampling period was 20 minutes, which was used here as well.

2. Internal model

The internal model of this plant can be derived theoretically [37]. A mass balance around each storage volume in the system gives:

$$V_i(k+1) = V_i(k) + T [d_j(k) + q_i^0(k) - q_i(k) - s_i(k)]$$
(39)

where V_i is the volume of liquid in the ith storage volume (m^3), d_j is the disturbance inflow (m^3/s) that directly influences V_i, q_i^0 is the flowrate entering V_i from an upstream pipeline section (m^3/s), q_i is the regulated flowrate leaving V_i (m^3/s), and s_i is the overflow rate from V_i (m^3/s). We can also relate the flowrates entering and leaving a pipeline section (which is represented by a time-delay element as noted previously):

$$q_i(k) = q_i^0(k-\tau_{di})$$
(40)

where q_i is the flowrate leaving the ith pipeline section, q_i^0 is the flowrate entering the ith pipeline section, and τ_{di} is the number of sampling periods of time delay in the ith pipeline section (an integer, ≥ 1). Note that when $\tau_{di} > 1$, Eq. (40) expands to τ_{di} state equations [25, pp. 40-42]. The entire model has 37 states – 8 storage volumes and 29 delay states. The coefficient matrices for the model appear in section V.A.

In most cases a theoretical model of the system is unavailable and/or too costly to develop. Experiments must then be carried out in order to identify the model. This is often the most time-consuming step in the installation of an MPC system. Methods for model identification and estimation of linear models are discussed in many texts and articles, such as [38] and [39].

3. Choice of manipulated and output variables for MPC

For this application, it is best to define the manipulated variables, $\mathbf{u}(k)$, to be the set of all regulated flowrates, q_i, and overflow rates, s_i. *The overflows are not manipulated variables from the point of view of the physical process, however.* They are the result of the given inflows and the regulated flows, *i.e.*, output variables. Although it would be possible to formulate a model from this point of view, it would be inherently nonlinear (since the overflows are zero until the liquid levels reach a pre-determined point). The designation of the overflows as decision variables in the LP avoids this problem. The output variables, $\mathbf{y}(k)$, are the liquid volumes held in each storage element.

Since the primary control objective is to minimize the sum of the overflows, the elements of $\mathbf{a_u}$ corresponding to the s_i terms are given a large positive weight in the objective function. Accordingly, they will stay at their lower bounds (*i.e.*, zero) unless a constraint forces one or more of them to be positive. This happens whenever one or more of the liquid volumes would otherwise exceed its specified upper bound.

As discussed in section II.D.1, hard constraints on outputs cause problems in most cases. The upper limits on the liquid volumes are an exception (fortunately, since they are essential to the definition of the MPC objective). The overflows, s_i, provide a natural "safety valve", since they have no upper bounds and no effect on any variable other than the liquid holdup in the storage volume they drain. Thus it is always possible to satisfy the upper bounds on the liquid volumes. The lower bounds on the liquid volumes are another matter, as discussed in section III.B.1.b.

The calculated s_i values are not sent to the plant. Only the desired values of the regulated flowrates, q_i, are sent to the plant, to be implemented by local flow controllers. The measured values of the regulated flowrates and the overflows from the most recent sampling interval are fed back to the MPC system (along with the measured liquid levels) in order to update the state of the internal model using Eqs. (7) and (8).

4. Time-delay factorization and reference trajectory for MPC

According to the model and the problem definition, each output variable can be influenced by at least one manipulated variable within a single sampling period.

Thus time-delay factorization is not required ($\kappa_i = 0$ for all outputs). A fast reference trajectory ($\Phi_r = 0$) was used in all simulations, but a slower reference trajectory ($\Phi_r > 0$) would provide better robustness [4].

5. State estimation

For the estimator design, one could use the standard DMC assumption for this problem ($K_1 = 0$, $K_2 = I$). Since the process contains integrating elements, however, errors in flow measurements would cause steady-state offset in the output predictions [19]. We therefore consider other possibilities.

An unusual feature of this problem is that the volume states are measured and are decoupled (because the flowrates leaving each volume element are manipulated, *i.e.*, independent of the levels). Thus the measurement of a volume at a given location affects the estimate of the true volume at that location, but it has no influence on the estimates of volumes at other locations. This simplifies the design of **K** since we can consider the estimation of the p volume states to be independent of each other.

Consider Eq. (39) written in the following form:

$$x_i(k+1) = x_i(k) + T[v_j(k) - u_i(k)] \tag{41}$$

where $x_i(k)$ is the volume of liquid in the ith reservoir ($i = 1, p$), $v_j(k)$ is the measured net inflow, and $u_i(k)$ is the measured, regulated, net outflow. Let $\tilde{y}_i(k)$ be the measurement of the volume in the ith reservoir. We can augment Eq. (41) with the output, $y_i(k)$, in the form of Eqs. (5) and (6):

$$x_a(k+1) = \Phi_a x_a(k) + \Gamma_{ua}\Delta u_i(k) + \Gamma_{va}\Delta v_i(k) + \Gamma_{wa}\Delta w_i(k) \tag{42}$$
$$y_i(k) = C_a x_a(k) + \Delta z_i(k) \tag{43}$$

where in this case, $x_a(k) = [\Delta x_i(k) \ y_i(k)]^T$, $\Delta w_i(k)$ is an unmeasured flowrate disturbance (or an error in flowrate measurement), $\Delta z_i(k)$ is an error in volume measurement, and

$$\Phi_a = \begin{bmatrix} 1 & 0 \\ 1 & 1 \end{bmatrix} \quad \Gamma_{ua} = \begin{bmatrix} -T \\ -T \end{bmatrix} \quad \Gamma_{va} = \Gamma_{wa} = \begin{bmatrix} T \\ T \end{bmatrix} \quad C_a = [\ 0 \quad 1\ ]$$

Then we can design an estimator in the form of Eqs. (7) and (8) using the method discussed in section II.A.2. As the parameter σ goes to infinity (meaning that the volume measurements are accurate relative to the flowrate measurements), the estimator gain goes to $\mathbf{K} = [1 \quad 2]^T$. As σ goes to zero, both elements of $\mathbf{K}$ go to zero[18]. For example, for $\sigma = 1.0$, $\mathbf{K} = [\; 0.4805 \quad 1.2496 \;]^T$.

Suppose the state equations of the internal model are grouped with the p equations of the form of Eq. (41) followed by the delay equations. Generalizing the above results, the estimator for this problem takes the form:

$$\mathbf{K} = [\mathbf{K}_{1v}^T \quad \mathbf{K}_{1d}^T \quad \mathbf{K}_2^T]^T \tag{44}$$

where $\mathbf{K}_{1v}$ and $\mathbf{K}_2$ are p by p diagonal matrices, and $\mathbf{K}_{1d} = \mathbf{0}$ (*i.e.*, no correction to the delayed-flowrate states). For example, if we assume $\sigma = 1.0$ for all p volumes, then $\mathbf{K}_{1v} = 0.4805\, \mathbf{I}_{pxp}$ and $\mathbf{K}_2 = 1.2496\, \mathbf{I}_{pxp}$. Note that the effect of errors in the delayed-flowrate states on the estimated volumes are accounted for in the Δw term in Eq. (42). Since the delayed-flowrate states are updated by measurements and are not defined as plant outputs, it is reasonable to choose $\mathbf{K}_{1d} = \mathbf{0}$ as suggested above.

It would also be possible to make other assumptions regarding the state and measurement noise terms and solve the Kalman filter problem for the full $\mathbf{K}$ matrix. Section III.B.2 shows how two alternatives for $\mathbf{K}$ affect the performance of MPC when there is "input uncertainty", *i.e.*, errors in the measurements of $u(k)$ and $v(k)$.

6. Choice of objective function

The overriding goal of the control system is to minimize the weighted sum of the overflows, s_i. This goal can be incorporated easily in a linear objective function in the form of Eq. (34). Auxiliary goals, such as the desire to empty the storage volumes at the end of a storm, can also be included (details appear in section III.B.1). Thus Eq. (34) was used rather than Eq. (21). Patry [40] has proposed a

---

[18]   Performance at this limiting value is poor since the resulting estimator has a pole on the unit circle.

similar formulation for CSO problems. A side-benefit is that the resulting LP problem is easily solved on-line (see next section).

One could also use a combination of quadratic and linear terms. This may have certain advantages (*e.g.*, an unconstrained solution might then exist). The resulting QP problem would probably require more computational effort, however. Future studies of this example problem will examine such tradeoffs in more detail.

7. *MPC implementation*

A function was written in Matlab[19] to build the plant model. The estimator gain was also designed in Matlab using functions from the MPC Toolbox [19] and the Control Systems Toolbox. Finally, a Matlab function similar to "MPCQP" in the MPC Toolbox used, as input data, the plant model and specified values of $\mathbf{K}$, N, n_b, the weighting vectors, upper and lower bounds, *etc.* It generated an ASCII data file containing the problem definition in the form[20] required by MINOS. For simulations, MINOS was modified to run under the control of a Fortran main program. After initializing common areas in memory, *etc.*, this main program called MINOS, which then set up and solved successive LP problems until a specified number of sampling periods was simulated.

All MINOS calculations were done on a Macintosh II using Language Systems MPW Fortran. Time requirements depended strongly on the choice of N, n_b, and the number of inequality constraints, as one would expect. For $N = 25$, the largest value used, the LP problem for the first sampling period required about 5 minutes, but subsequent problems required an average of only 20 seconds. Since no blocking was used, the number of decision variables to be optimized was typically 900, with about 500 equality constraints, *i.e.*, it was a rather large LP problem.

---

19   The MathWorks, Inc.
20   The problem definition file is in the standard MPS format, used by many mathematical programming codes.

B. Results

Figure 5 shows the disturbance inflows (d_1 to d_7) and their sum for the storm considered by Papageorgiou [37]. The total treatment plant capacity is 22 m^3/s, which is indicated as a dashed line on the graph of the total inflows in Fig. 5. The treatment plant capacity is exceeded for a period of about 6 hours. During this time, sewage must either accumulate in the 8 holdup volumes in the system or it must overflow.

The strategy leading to the minimum possible total CSO for this case would be: 1) fill the entire volume of the system while keeping the flows to the treatment plants at their maximum values, then 2) allow overflows only to the extent that the inflows continued to exceed the treatment plant capacity. *Note that this assumes that liquid transport between storage volumes is unconstrained (which is not really the case).* The corresponding minimum possible CSO value is easily calculated. For this storm it is about 233200 m^3. In reality, the non-uniform distribution and intensity of the rainfall, combined with the restrictions imposed by upper bounds on flowrates, prevents this from being achieved, but it provides a useful point of reference.

1. Performance with a perfect internal model and predictable inflows

Let us first consider the performance of MPC when: 1) The model used in MPC is the same as that used to represent the plant – there is no modeling error, 2) all inputs – including the disturbance inflows – are measured without error, and 3) the disturbance inflows are predicted perfectly, *i.e.*, the future disturbances in the vector $\boldsymbol{v}(k)$, Eq. (17), are known exactly. This results in the best possible MPC performance within the limitations of the physical constraints and the choice of the design parameters, N, $\mathbf{a}_y$, $\mathbf{a}_u$, *etc.*

The following parameter values were used for all simulations (unless noted otherwise):

* There were hard lower bounds at zero for all decision variables (including the plant outputs). The hard upper bounds given in Table I were also used. There were no bounds on $|\Delta\boldsymbol{u}|$.

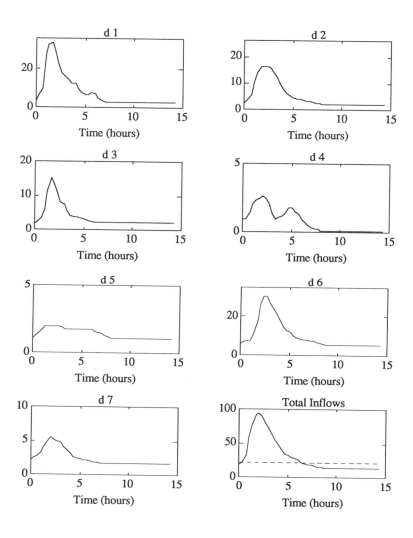

Figure 5: Disturbance inflows (m³/s) for the example application

- $n_b = N$, with N varied as noted in the text for each case.
- $\tau = [\ 0\ \ 0\ \ ...\ \ 0\ ]$, and $\Phi_r = 0$.
- $a_r = a_{\Delta u} = r = 0$.
- The DMC estimator was used but this choice had no effect since there was no model error.
- $a_y = 1 \times 10^{-6} [I_{8 \times 8}\ \ I_{8 \times 8}\ \ ...\ \ I_{8 \times 8}]^T [1\ \ \ 1\ \ \ 1\ \ \ 1\ \ \ 0.5\ \ \ 0.1\ \ \ 0.3\ \ \ 0.1]^T$.
  The positive weights on y encourage the controller to empty the storage volumes at the end of a storm. The volumes near the treatment plant are to be emptied last, all other things being equal – hence the relatively small weights on these volumes.
- a_u was determined from the following MATLAB statements:

```
uwt=T*[0 0 0 00 0 0 1 1 1 1 0 1 1 1 0 0];
uwt=ones(nb,1)*uwt;
uwt=diag(.005*[200:-5:(200+5*(1-nb))])*uwt;
```

where T is the sampling period (1200 seconds) and nb is the number of blocks, which is equal to N in this case as mentioned above. Note that the only non-zero values in the first line are those corresponding to the overflows s_1 to s_8, *but excluding s_5* (since it is not released to the environment). The overall effect of the above statements is to produce a matrix with n_b rows and m columns, the rows of which are the elements of a_u. The first row has the values shown in the first MATLAB statement, and the remaining rows are scaled linearly such that the last is smaller in absolute value than the first. This penalizes near-term overflows more heavily that those further out in the prediction horizon, reflecting the large uncertainties in the prediction of overflows in the distant future, encouraging the optimal controller to postpone overflows as long as possible, all other things being equal.

The system was initially at steady-state and a total of 45 sampling periods (15 hours) were simulated.

a.   Effect of length of prediction horizon (Case 1). The choice of the prediction horizon, N, had a strong effect on the results. Figures 6, 7, and 8 compare the strategy used by the controller for $N = 10$ (dashed lines) and $N = 25$ (solid lines). Variables p_2 and s_1 to s_4 were zero at all times and are not shown. Variable s_7 was identical for the 2 cases (see Fig. 8). As shown in the first column of Table II, the total CSO was nearly constant at about 542,000 with $N \leq 10$ and at about 275,000 with $N \geq 15$. The discontinuity between $N = 10$ and $N = 15$ is caused by the long transport delay of 12 sampling periods between diversion element #2 and storage element V_8 (see Fig. 6). If the prediction horizon is less than this delay, the controller sets q_5 to its maximum value during periods of high loading (Fig. 7), not realizing that this will cause an overflow in s_8 (Fig. 8). Once N is above this threshold value, further increases have very little effect. Note that for large N the total CSOs were only about 18% above the reference value of 233200 m^3.

N	Case 1	Case 2	Case 3
5	542000	601000	524000
10	576000	629000	577000
15	273000	274000	319000
20	275000	274000	272000
25	274000	270000	276000

Table II: Effect of prediction horizon on total CSOs (in m^3).

b.   Relaxation of hard constraints on output variables (Case 2). As mentioned previously, the requirement that $V_i \geq 0$ for $i = 1,8$ can cause problems in the LP. There are conditions under which the linear model predicts that a negative value of q_i is required to keep a storage volume from going negative. Consider, for example, volume V_3 in Figure 4. Suppose that at a given sampling period, k, $p_1(k) = 5$ and we estimate that $d_2(k)$ is 10. Although this is impossible from a physical point of view, there is nothing in the linear model formulation to prevent it

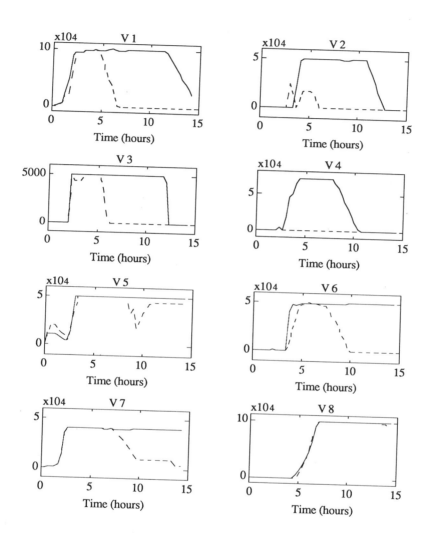

Figure 6: Storage (in m³) *vs* time for 2 prediction horizons:
1)   N = 25 sampling periods (solid line)
2)   N = 10 sampling periods (dashed line)

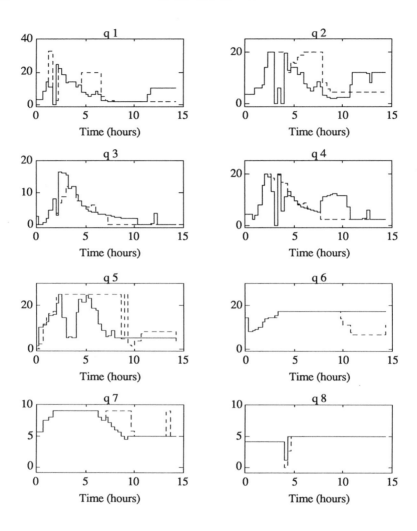

Figure 7:   Outlet flowrate (in m³/s) *vs* time for 2 prediction horizons:
1)   N = 25 sampling periods (solid line)
2)   N = 10 sampling periods (dashed line)

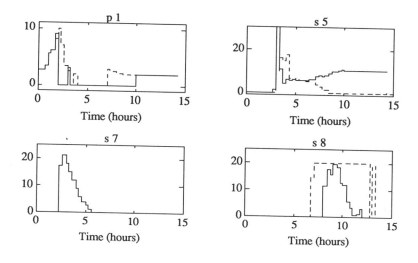

Figure 8: Non-zero diversion and overflow rates (in m^3) *vs* time
for 2 prediction horizons:
1)   25 sampling periods (solid line)
2)   10 sampling periods (dashed line)

from happening. In that case, the "flowrate" of the stream entering delay element #3
is -5. If V_3 were close to zero at that time, even setting $q_3 = 0$ would not prevent the
predicted value of V_3 from going negative. The result would be an infeasible LP.

One way to avoid this is to add logic to the internal model to guarantee that
the variables are always consistent with known physical laws. For this problem, as
long as the measurements are accurate we simply need to make sure that the flowrates
leaving the two flow diversion elements are ≥ 0 at all times. In the previous
situation, for example, we could assume that our prediction of d_2 was incorrect and
set it to 5 instead. Also, to enforce non-negativity over the entire prediction horizon,
these two flowrates would have to be defined as auxiliary output variables with hard
lower bounds at zero.

Even for this relatively simple problem, however, the situation becomes very
complicated as soon as we allow the possibility of errors in the measurements of the
input and output variables (as in section III.B.2). We would then need to devise a

systematic method for checking and adjusting all the measured flowrates, which is difficult at best and is an *ad hoc* solution to the problem in any case.

It is thus a good idea to examine critically the need for the hard lower bounds on the V_i values. Qualitatively, we can anticipate that if a V_i value becomes negative (as calculated by the model), the main effect will be a bias in the prediction of future CSOs. The degree to which a V_i value is allowed to go negative can be controlled by the design of the estimator gain, **K**. Since other difficulties will arise in practice (*e.g.*, poor predictions of the inflows – see next section – and measurement errors – see section III.B.2), the bias caused by a negative V_i may be relatively unimportant. Whether this indeed the case can be determined by simulation.

The results in the second column of Table II are from simulations in which the lower bounds on the outputs have been removed but all other conditions are identical to those for Case 1[21,22]. Although the resulting CSOs are worse for small values of *N*, they are essentially identical to those in Case 1 for *N > 15*. Thus for this problem it does not seem to be critical to retain the lower bounds on the V_i values. This point is reconsidered in section III.B.2.

c.   Effect of inaccurate disturbance forecasting (Case 3). It is also important to see how accurately one must forecast future inflows in order to get good performance. Reasonable forecasts can be made over a horizon of 0-1 hours (since this is influenced mainly by rainfall intensity, which can be measured), but beyond that the accuracy will depend on predictions of future rainfall, which may be very poor. The results in the third column in Table II were from simulations identical to those in Case 1 except that *future disturbance inputs were predicted to be the same as the current disturbance* (which was assumed to be measured perfectly). In other words, Eq. (18) was used for output prediction in place of Eq. (17). Therefore, during periods of high inflow the model over-estimated future inflows. At the

---

21   Note that for *all* simulations the flowrates and volumes in the plant obeyed the applicable physical laws and hence were always non-negative. Thus in the *output prediction* phase, the internal model occasionally predicted that a negative value would be inevitable but this did not actually happen. For the *state estimation* phase of the simulations of section III.B.1, the true values of the plant inputs and outputs were fed back to the estimator, so the state estimates were always perfect. Use of inaccurate measurements (section III.B.2) introduced errors in the state estimates, however.

22   Another option would have been to "soften" these output constraints. Selective softening might have been especially effective in this case, but this was not tested.

beginning of the storm, on the other hand, the near-term inflows were under-estimated.

This had suprisingly little influence on the results (compare columns 1 and 3 in Table II). The reasons for this are as follows:

- Under-estimation of future inflows at the beginning of the storm was not a serious error because the storage volumes started at zero (empty). There was sufficient capacity to accommodate the unanticipated sudden increase in the inflows.

- Once the inflows had increased, future inflows were over-estimated, resulting in a conservative strategy in which the controller kept the flows to the treatment plants at their upper limits while accumulating sewage throughout the system. (This might not work in general if, for example, certain inflows were over-estimated while others were under-estimated due to the appearance of a new, localized storm front).

2. Effect of measurement errors

Flowrate measurements in sewer systems are notoriously inaccurate. Here we consider the potential effect of such measurement errors. The setup for the simulations was the same as in section III.B.1, except that measurements from the simulated "plant" were corrupted by a constant multiple that ranged between 0.2 and 2.0. The multiples used for each measured variable are given in section V.B. This constitutes an example of "input error" as defined in [4], and is expected to be a primary cause of model error in CSO minimization applications.

The introduction of measurement errors has two important effects. First, it becomes very difficult to avoid predictions of negative flowrates and volumes from the internal model (see section III.B.1.b). Therefore, the lower bounds on the V_i output variables were deleted. Second, errors in the state estimates are inevitable (even when the model equations are exact), so the choice of the estimator gain, $\mathbf{K}$, becomes important, in contrast to the previous cases.

Figures 9 to 12 compare the use of the standard DMC estimator, Eq. (13), with the modified estimator described in section III.A.5, where $\mathbf{K}_{1v} = \mathbf{I}_{8x8}$ and

$K_2 = 2 \cdot I_{8 \times 8}$. The prediction horizon was $N = 20$ for all cases. Other MPC parameters were the same as those given in section III.B.1.

a.   Performance with DMC estimator (Case 4). Figure 10 compares the actual liquid volumes (solid lines) to those predicted by Eqs. (7) and (8) (dotted lines) using the DMC value of K and the corrupted measurements. In all cases there is a steady-state offset in the prediction at the end of the simulation, and in some cases it is large. The estimate of V_3, for example, is negative for most of the simulation. This bias causes a small overflow from storage element #3 (see Fig. 12) whereas s_3 was zero in all other cases. A more serious problem is the poor prediction of V_5 and V_6. The controller does not use these volumes effectively (Fig. 10), which ultimately results in a larger-than-necessary overflow from V_8 (see s_8 in Fig. 12). The total CSO for this case was 377,000 m^3, well above the value of 274,000 obtained for $N = 20$ in the perfect measurement cases (see Table I).

b.   Performance with the alternative estimator (Case 5). Figure 9 compares the estimated outputs (dotted lines) to the true outputs (solid lines) for the alternate estimator (see introduction to section III.B.2). Although the estimates are "noisy" in some cases – especially for V_2, they follow the trends of the true outputs and there is no steady-state offset. Consequently, this version of the MPC strategy uses a different set of adjustments in the manipulated variables than in Case 4 (compare solid and dashed lines in Fig. 11). In particular, V_5 and V_6 are used much more effectively (compare Figs. 9 and 10). The total CSO for this case was 332,000 m^3, a significant improvement over Case 4.

IV.   CONCLUSIONS

It has been demonstrated that MPC can be effective when applied to systems with many (constrained) input and output variables. For the example problem, MPC was surprisingly robust with respect to errors in the prediction of future disturbance inputs, and proper selection of the estimator gain matrix made it insensitive to large errors at the plant input as well. More general types of modeling errors were not

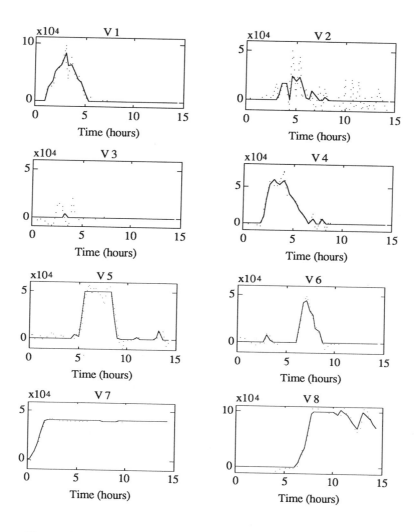

Figure 9:   Actual storage volumes (solid lines) *vs* output
estimates (dotted lines) from Eqs. (7) and (8) with the
alternative estimator.

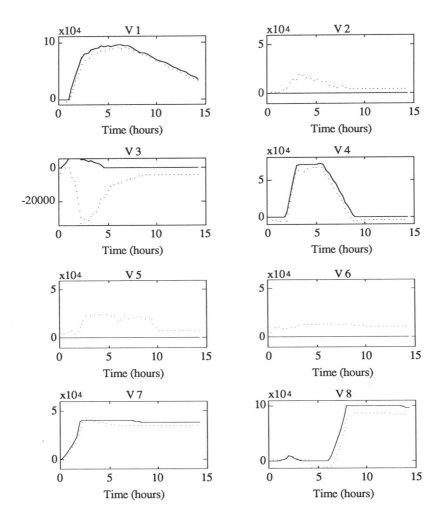

Figure 10:   Actual storage volumes (solid lines) *vs* output estimates
(dotted lines) using Eqs. (7) and (8) with the DMC
estimator.

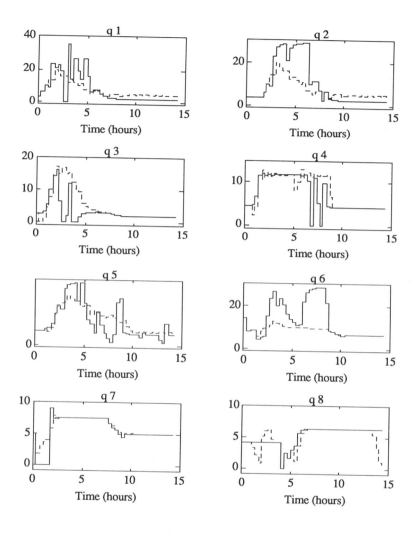

Figure 11: Adjustments in regulated flowrates in the presence of measurement error using the alternative estimator (solid lines) and the DMC estimator (dashed lines).

202

N. LAWRENCE RICKER

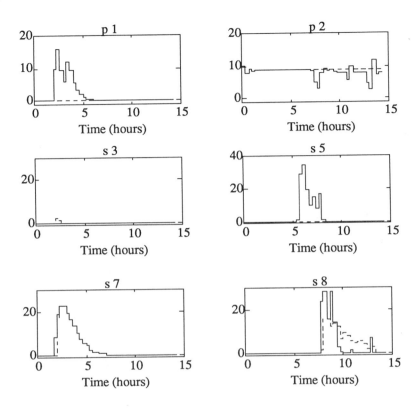

Figure 12: Adjustments in p_1 and p_2, and non-zero overflows in the presence of measurement error using the alternative estimator (solid lines) and the DMC estimator (dashed lines).

considered[23], but experience with smaller-scale systems [3, 4, 41] suggests that this should not be a serious problem. The MPC parameter having the largest effect on performance was the prediction horizon. If it was less than the maximum time delay in the various pipeline sections, overflows increased dramatically.

The example also suggests that large problems require special attention to the definition of the internal model, objective function, and the constraints.

---

23 This will be the subject of future work. A test of MPC in a real CSO minimization application is planned, as are simulation studies using a nonlinear representation of the plant.

Shortcomings in any of these areas can impair both efficiency and performance. For example, one may need to neglect certain (relatively unimportant) process interactions in order to obtain a model that is sparse enough for identification and on-line use. Identification of model parameters was not required in the example, but is the weak link in many MPC applications. If sufficient attention is given to these points, it is expected that MPC will be used more and more frequently in plant-wide control applications.

V.   ADDITIONAL DETAILS FOR THE EXAMPLE PROBLEM

A. Sparse-matrix version of the internal model used in the example problem

The same convention found in MINOS [36] is used to store each of the state-space coefficient matrices in the internal model. The following Fortran variables are needed to store a sparse matrix with N_r rows, N_c columns, and N_e non-zero elements:

$A(N_e)$ is a REAL vector that holds the *non-zero* elements of the original state-space matrix. Elements are stored column-wise, *i.e.*, all non-zero elements of the first column of the original matrix, followed by those from the second column, *etc.*

$HA(N_e)$ is an INTEGER vector that specifies the row (in the original state-space matrix) in which each corresponding element of A is located.

$KA(N_c)$ is an INTEGER vector that points to location in A and HA corresponding to the *first* non-zero element in each column of the original matrix, where N_c is the number of columns in the original matrix. If KA(i)=0, this signifies that all elements of the ith column of the original matrix were zero.

Given the values of N_c, N_r, N_e, A, HA, and KA, one can re-construct the original matrix. Special routines for matrix multiplication, *etc.* can be written that operate on matrices in this format.

For the example problem, the model matrices are as follows:

Φ_a: $N_r=N_c=47$, $N_e=66$, and

KA =

1	3	5	7	9	11	13	15	17	18	19	20	22	23	24
26	28	30	32	33	35	36	37	39	41	43	44	45	46	47
48	49	50	51	52	53	55	57	58	59	60	61	62	63	64
65	66													

HA =

1	38	2	39	3	40	4	41	5	42	6	43	7	44	8
45	10	11	12	2	39	14	15	2	39	3	40	4	41	5
42	20	5	42	22	23	6	43	6	43	6	43	27	28	29
30	31	32	33	34	35	36	37	47	8	45	38	39	40	41
42	43	44	45	46	47									

A =

```
1.000000E+00   1.000000E+00   1.000000E+00   1.000000E+00   1.000000E+00
1.000000E+00   1.000000E+00   1.000000E+00   1.000000E+00   1.000000E+00
1.000000E+00   1.000000E+00   1.000000E+00   1.000000E+00   1.000000E+00
1.000000E+00   1.000000E+00   1.000000E+00   1.000000E+00   1.200000E+03
1.200000E+03   1.000000E+00   1.000000E+00   1.200000E+03   1.200000E+03
1.200000E+03   1.200000E+03   1.200000E+03   1.200000E+03   1.200000E+03
1.200000E+03   1.000000E+00   1.200000E+03   1.200000E+03   1.000000E+00
1.000000E+00   1.200000E+03   1.200000E+03   1.200000E+03   1.200000E+03
1.200000E+03   1.200000E+03   1.000000E+00   1.000000E+00   1.000000E+00
1.000000E+00   1.000000E+00   1.000000E+00   1.000000E+00   1.000000E+00
1.000000E+00   1.000000E+00   1.000000E+00   1.000000E+00   1.200000E+03
1.200000E+03   1.000000E+00   1.000000E+00   1.000000E+00   1.000000E+00
1.000000E+00   1.000000E+00   1.000000E+00   1.000000E+00   1.000000E+00
1.000000E+00
```

Γ_a: $N_r=47$, $N_c=25$, $N_e=56$, and

KA =

1	4	7	10	13	16	18	21	23	25	27	29	31	34	36
38	40	44	45	47	49	50	51	53	55	57				

HA =

1	9	38	2	18	39	3	17	40	4	19	41	5	26	42
6	43	7	21	44	8	45	1	38	2	39	3	40	4	41
5	24	42	6	43	7	44	8	45	13	16	25	46	26	1
38	16	46	17	18	5	42	7	44	6	43				

A =

```
-1.200000E+03   1.000000E+00  -1.200000E+03  -1.200000E+03   1.000000E+00
-1.200000E+03  -1.200000E+03   1.000000E+00  -1.200000E+03  -1.200000E+03
 1.000000E+00  -1.200000E+03  -1.200000E+03   1.000000E+00  -1.200000E+03
-1.200000E+03  -1.200000E+03  -1.200000E+03   1.000000E+00  -1.200000E+03
-1.200000E+03  -1.200000E+03  -1.200000E+03  -1.200000E+03  -1.200000E+03
-1.200000E+03  -1.200000E+03  -1.200000E+03  -1.200000E+03  -1.200000E+03
-1.200000E+03   1.000000E+00  -1.200000E+03  -1.200000E+03  -1.200000E+03
-1.200000E+03  -1.200000E+03  -1.200000E+03  -1.200000E+03   1.000000E+00
-1.000000E+00   1.000000E+00  -1.000000E+00  -1.000000E+00   1.200000E+03
 1.200000E+03   1.000000E+00   1.000000E+00   1.000000E+00   1.000000E+00
 1.200000E+03   1.200000E+03   1.200000E+03   1.200000E+03   1.200000E+03
 1.200000E+03
```

C_a: $N_r=8, N_c=47, N_e=8,$ and

KA =

```
0    0    0    0    0    0    0    0    0    0    0    0    0    0    0
0    0    0    0    0    0    0    0    0    0    0    0    0    0    0
0    0    0    0    0    0    0    1    2    3    4    5    6    7    8
```

HA =

```
1    2    3    4    5    6    7    8
```

A =

```
1.000000E+00   1.000000E+00   1.000000E+00   1.000000E+00   1.000000E+00
1.000000E+00   1.000000E+00   1.000000E+00
```

The definitions of the state, input, and output vectors were as follows:

$$\mathbf{x}^T = [\ V_1 \quad V_2 \quad ... \quad V_8 \quad x_{11} \quad ... \quad x_{1,n1} \quad x_{21} \quad ... \quad x_{2,n2} \quad ... \quad x_{10,n10}\ ]$$
$$\mathbf{u}^T = [\ q_1 \quad q_2 \quad ... \quad q_8 \quad s_1 \quad s_2 \quad ... \quad s_8 \quad p_1 \quad p_2\ ]$$
$$\mathbf{v}^T = [\ d_1 \quad d_2 \quad ... \quad d_7\ ]$$
$$\mathbf{y}^T = [\ V_1 \quad V_2 \quad ... \quad V_8\ ]$$

where x_{ij} is the jth delay state for pipeline element i.

B . Measurement error factors used in Cases 4 and 5 of the example problem

To simulate measurement error, the *true* values of $u(k)$ from the plant were multiplied by the following factors (one for each of the 18 manipulated variables as defined in the previous section):

$$e_u^T = [1.6\ \ 0.7\ \ 1.9\ \ 1.7\ \ 0.4\ \ 0.6\ \ 1.2\ \ 0.8\ \ 0.8\ \ 1.8$$
$$2.0\ \ 1.8\ \ 0.7\ \ 1.3\ \ 0.5\ \ 0.3\ \ 0.4\ \ 1.5]$$

Similarly, the following values multiplied the elements of $v(k)$:

$$e_v^T = [1.0\ \ 0.1\ \ 0.3\ \ 0.2\ \ 0.5\ \ 0.7\ \ 0.7]$$

VI. ACKNOWLEDGEMENTS

Professor Manfred Morari (Caltech) provided a draft of a forthcoming book, a chapter of which was very influential in the development of the MPC theory in sections II.A.1 to II.A.3. I would also like to thank Dr. Z. Vitasovic (Metro, Seattle WA) for background and practical details on the CSO minimization problem discussed in section III.

VII. REFERENCES

1.   Ruiz, J.; Muratore, E.; Ayral, A.; Durand, D. "OPTIMILL: Optimal management of pulp mill production departments and storage tanks" *6th Int. IFAC/IFIP/IMEKO Conf. on Instrumentation and Automation in the Paper, Rubber, Plastics and Polymerization Industries*, Akron, OH, Oct. 27-9, 1986.

2.   Downs, J.; Vogel, E. F. "Application of Model Predictive Control to a Production Scheduling Problem", paper presented at the Annual AIChE Meeting, Washington DC, Nov. 29, 1988.

3. Prett, D. M.; Garcia, C. E. *Fundamental Process Control*, Butterworths: Stoneham, MA; 1988.

4. Morari, M.; Zafiriou, E. *Robust Process Control*, Prentice-Hall: Englewood Cliffs, N. J.; 1989.

5. McAvoy, T. J.; Arkun, Y.; Zafiriou, E., eds. *Model-based pr;ocess control – proceedings of the 1988 IFAC workshop*, Pergamon Press: Oxford, 1989.

6. Eaton, J. W.; Rawlings, J. B. "Feedback control of chemical processes using on-line optimization techniques", Annual AIChE Meeting, Washington, D.C., 1988.

7. Lee, P. L.; Sullivan, G. R. "Generic Model Control – theory and applications", pp. 111-120 in [5].

8. Garcia, C. E. "Quadratic Dynamic Matrix Control of nonlinear processes" paper presented at the AIChE Annual Meeting, San Francisco, 1984.

9. Parrish, J.R.; Brosilow, C. B. "Nonlinear inferential control", *AIChE J.* **1988**, *34*, 633-644.

10. Georgiou, A.; Georgakis, C.; Luyben, W. "Nonlinear Dynamic Matrix Control for high-purity distillation columns", *AIChE J.* **1988**, *34*, 1287-1298.

11. Garcia, C. E.; Morari, M. "Internal model control 1. A unifying review and some new results" *Ind. Eng. Chem. Process Des. Dev.* **1982**, *21*, 308-323.

12. Garcia, C. E.; Morari, M. "Internal model control 2. Design procedure for multivariable systems" *Ind. Eng. Chem. Process Des. Dev.* **1985**, *24*, 472-484.

13. Marchetti, J. L.; Mellichamp, D. A.; Seborg, D. E. "Predictive control based on discrete convolution models, *Ind. Eng. Chem. Process Des. Dev.* **1982**, *22*, 488-495.

14. Richalet, J.; Rault, A. ; Testud, J. L.; Papon, J. "Model predictive heuristic control: applications to industrial processes" *Automatica* **1978**, *14*, 413-428.

15. Rouhani, R.; Mehra, R. K. "Model Algorithmic Control (MAC); basic theoretical properties" *Automatica* **1982**, *18*, 401-414.

16. Navratil, J. P.; Lim, K. Y.; Fisher, D. G. "Disturbance Feedback in Model Predictive Control Systems", pp. 63-68 in [5].

17. Li, S.; Lim, K. Y.; Fisher D. G. "A State Space Formulation for Model Predictive Control". *AIChE J.* **1989**, *35*, 241-249.

18. Marquis, P.; Broustail, J.P. "SMOC, a bridge between State Space and Model Predictive Controllers: Application to the automation of a hydrotreating unit", pp. 37-46 in [5].

19. Ricker, N. L. "Model Predictive Control with State Estimation". *Ind. Eng. Chem. Res.* **1989** (submitted).

20. Clarke, D. W.; Mohtadi, C.; Tuffs, P. S. "Generalized Predictive Control Part I. The Basic Algorithm". *Automatica* **1987**, *23*, 137-148.

21. De Keyser, R. M. C.; Van de Velde, Ph. G. A.; Dumortier, F. A. G. "A comparative study of sSelf-adaptive long-range predictive control methods". *Automatica* **1988**, *24*, 149-163.

22. Shen, G-C.; Lee, W-K. "Multivariable adaptive inferential control", *Ind. Eng. Chem. Res.* **1988**, *27*, 1863-1872.

23. Seborg, D. E.; Edgar, T. F.; Mellichamp, D. A. *Process Dynamics and Control*, p. 196, Wiley: New York, 1989.

24. Astrom, K. J. "Integrator windup and how to avoid it" *Proc. Am. Control Conf.* **1989**, *2*, 1693-1698.

25. Astrom, K. J.; Wittenmark, B. *Computer Controlled Systems Theory and Design*; Ch. 11, Prentice-Hall: Englewood Cliffs, N.J., 1984.

26. Wellons, M. C.; Edgar, T. F. "The Generalized Analytical Predictor", *Ind. Eng. Chem. Res.* **1987**, *26*, 1523-1536.

27. Walgame, K. S.; Fisher, D. G.; Shah, S. L. "Control of processes with noise and time delays" *AIChE J.* **1989**, *35*, 213-222.

28. Ricker, N. L. "The use of quadratic programming for constrained Internal Model Control". *Ind. Eng. Chem. Process Des. Dev.* **1985**, *24*, 925-936.

29. Edgar, T. F.; Himmelblau, D. M. *Optimization of Chemical Processes*, Ch. 8, McGraw-Hill: New York, 1988.

30. Chang, T. S.; Seborg, D. E. "A linear programming approach for multivariable feedback control with inequality constraints" *Int. J. Control* **1983**, *37(3)*, 583-597.

31. Morshedi, A. M.; Cutler, C. R.; Skrovanek, T. A. "Optimal solution of Dynamic Matrix Control with linear programming techniques (LDMC)" *Proceedings Am. Control Conf.*, **1985**, 199-208.

32. Brosilow, C. B.; Zhao, G. Q. "A linear programming approach to constrained multivariable process control" *Advances in Control and Dynamic Systems* **1986**, *24*.

33. Geering, H. P.; Mansour, M. *Large-scale systems: Theory and applications 1986* , Vol. 1&2, (IFAC Proceedings Series), Pergamon Press: Oxford, (1987).

34. Zafiriou, E. "Robust Model-predictive Control of Processes with Hard Constraints", paper presented at the Annual AIChE Meeting, Washington, DC, Nov., 1988.

35. Ricker, N. L.; Subrahmanian, T.; Sim, T. "Case studies of model-predictive control in pulp and paper production", pp. 13-22 in [5].

36. Murtagh, B. A.; Saunders, M. A. "MINOS 5.1 Users' Guide", *Technical Report SOL 83-20R*, Stanford University, 1987.

37. Papageorgiou, M. "Automatic control strategies for combined sewer systems" *J. Environ. Eng.* **1983**, *109(6)*, 1385-1402.

38. Ljung, L. *System identification – theory for the user* Prentice-Hall: Englewood Cliffs, NJ, 1987.

39. Prett, D. M.; Skrovanek, T. A.; Pollard, J. F. "Process identification – past, present, future" *Shell Process Control Workshop*, D. M. Prett and M. Morari, eds., Butterworths: Stoneham, MA, 79-104(1987).

40. Patry, G. G. "A linear programming model for the control of combined sewer systems with off-line storage facilities" *Can. Water Res. J.* **1983**, *8(1)*, 83-105.

41. Ricker, N. L.; Sim, T.; Cheng, C-M. "Predictive control of a multieffect evaporation system" *Proc. Am. Control Conf.* **1986**, *1*, 355-359.

ROBUST ESTIMATION THEORY
FOR BAD DATA DIAGNOSTICS
IN ELECTRIC POWER SYSTEMS

L. MILI   **V. PHANIRAJ**

Virginia Polytechnic Institute and
State University, Blacksburg, VA 24061

P. J. ROUSSEEUW

Vrije Universiteit Brussel
Pleinlaan 2, B-1050 Brussels

I. INTRODUCTION

Electric power systems consist of lines and transformers which interconnect large centralized synchronous generators and decentralized loads. Since only a small amount of electric energy can be stored, the production must meet the continuous random fluctuation of energy demand. As a result, they become increasingly complex and interdependent. Their on–line operation is performed through hierarchical multilevel control centers whose role is to minimize the cost of the generation while maintaining the system in a normal and secure operating state. A trade–off between economy and security is achieved through security functions; at their heart there is the state estimation function. It is responsible for providing a complete, coherent, and reliable data base from a collection of switch and breaker status data and measurements.

In power systems, experience has shown that discordant measurements, which are referred to as outliers, appear at each state estimation run; typically some of them are temporary ones. For instance, VanSlyck and Allemong [1] have found that during 13 years of state estimation experience at the American Electric Power, only 5 days were free of outliers. Moreover, they noticed that it appeared on the average a percentage of 1 to 2% outliers among the measurements. Several sources of outliers have been reported in the literature [1–6]. Major sources are

(i) model approximations, including the statistical error modelling inadequacies;

(ii) rounding errors, which affect all the metered values;

(iii) gross measurement errors, which are due to device failures, incorrect wiring, non–instantaneous meter scan, bad calibration or transducer bias drift with the weather, time or temperature;

(iv) parameter errors, which corrupt the values of the π–equivalent circuit elements;

(v) and topology errors, which are induced by status data errors.

Model approximations generally give rise to temporary small or medium size outliers whereas "hard" failures are prone to induce permanent large size outliers.

Up to now, three lines of research have been carried out for solving the power system state estimation problem. The first line was initiated by Schweppe *et al.* [7]. They proposed to use the Weighted Least Squares (WLS) estimator in conjunction with some bad data detection and identification rules in order to overcome its lack of robustness against gross errors. For the detection step, Handschin *et al.* [8] recommended the use of statistical tests applied to the weighted or the normalized residuals. As for the identification step, it was thought of as an extension of the detection stage through cycles of measurement eliminations and residual tests. Along this line, the most advanced methods are (i) the combinatorial optimization method of Monticelli *et al.* [9] which finds the minimal set of eliminated measurements which make the detection tests non–significant, (ii) the geometric approach of Clements and Davis [10], (iii) the Hypothesis Testing Identification (HTI) method developed by Mili *et al.* [11–14], from an idea proposed by Xiang *et al.* [15]. The HTI method consists of eliminating (either explicitly or through a linearized formula) sets of suspected measurements and applying a hypothesis test to their residuals, the so–called measurement error estimates. The main weaknesses of all these approaches results from the fact that they make use of detection tests based on the least squares residuals, which are prone to the masking effect of multiple bad data (see the comparative study performed by Handschin *et al.* [8] and updated by Mili *et al.* [12]).

A second line of research advocated by Merrill and Schweppe [16] and Handschin *et al.* [8], among others, consists of applying the M–estimators to power systems. These estimators downweight the measurements having large residuals. A third line of research was suggested by Irving *et al.* [17] and by Kotiuga and Vidyasagar [18]. They proposed to use instead the Least Absolute Value (LAV) estimator. Unfortunately, as reported recently in the statistical literature [3,19], the M–estimators, including the LAV estimator, become unreliable in presence of gross errors affecting leverage points. These are measurement points which are distant from the point cloud in the space associated with the observation matrix, the so–called factor space of the regression [3,19]. Such points are numerous in power system regression models; usually, they appear in clusters. Indeed, these are power measurements associated with relatively short lines or power injections on buses having several incident lines.

Recently a new family of robust estimators has been devised. They are called high breakdown point estimators to indicate that they are able to cope with a large fraction of bad leverage points. A comprehensive review is given by Rousseeuw and Leroy in [19]. These estimators stem from the robust estimation theory initiated by Huber [20] and Hampel [21], following the alarm raised by Tukey [22], among others, about the dramatic lack of robustness of the least squares estimator against the outliers. This paper is devoted to one of them, the Least Median of Squares (LMS) estimator, developed by Rousseeuw [23] from an idea proposed by Hampel [24]. The resampling technique advocated by Rousseeuw and Leroy [19] will be extended to large–scale nonlinear regression models. Fast decoupled LMS state estimation along with observability techniques will be proposed; they make this estimator applicable to large–scale systems. Moreover, concepts of fundamental set and of local breakdown will be defined. Based on these concepts, meter placement methodologies which are optimal for bad data identification will be proposed.

The paper is organized as follows. Section II gives some elements of the robust estimation theory. In particular the approach based on the influence function and the breakdown point will be reviewed. Section III defines the concept of the leverage point in regression and reviews the estimation methods proposed so far for power system applications. Section IV is devoted to the LMS estimator and its implementation in power system non–linear models. Some simulation results performed on the IEEE 14–bus system are reported in Section V. Section VI defines the local breakdown concept and shows through an example how it can be used for developing meter placement strategies which are optimal for bad data identification.

II. ELEMENTS OF ROBUST ESTIMATION THEORY

A. BRIEF HISTORY

The *parametric estimation theory* devised by Fisher in the early 1920's assumes the a priori knowledge of the family of the probability distributions of the measurement errors, the so–called parametric model. On this ground, Fisher developed the class of the maximum likelihood estimators which are optimal, i.e. unbiased with minimum asymptotic variance, when the postulated parametric model is exact. Moreover, he proposed several concepts such as consistency, efficiency, sufficiency, which assess the performance of an estimator under the assumptions.

In the early sixties, the need for new estimation theories different from the Fisherian one became more and more urgent. This resulted from the increasing use of the computer in all scientific and industrial fields in conjunction with the widespread use of the Least Squares (LS) estimator for data processing. This estimator is generally applied without any verification beforehand of the validity of the Gaussian assumption, and above all, without any protection against the presence of outliers.

In practice, it is usually the Gaussian model which is assumed (generally implicitly), and hence the least squares estimator , which is the maximum likelihood estimator for this particular model. In fact, the exclusive usage of this estimator in practice stems mainly from its straightforward implementation. This is indeed an advantage over other possible candidates in the pre–computer age. It is no longer true nowadays. Unfortunately, this simplicity has a price, which is the great vulnerability of this estimator to outliers. Indeed, as shown by Tukey [22] in an important communication, this estimator loses its efficiency very rapidly as soon as the tails of the probability distribution become a little longer. In particular, he showed that the arithmetic mean (the LS estimator in the one–dimensional case) becomes less efficient than the sample median (the LAV estimator) as soon as the fraction of contamination ϵ of the Gaussian mixture model, $G = (1-\epsilon)N(0,1) + \epsilon N(0,9)$, exceeds 8%. This result seems to contradict the Gauss–Markov theorem which states that the LS estimator is the Best Linear Unbiased Estimator (BLUE) for a general class of symmetric distributions with a finite variance. But all linear estimators exhibit poor performances under departure from the Gaussian model. Moreover, the LS estimator is biased for asymmetric distributions, which are by far the most common distributions in the real world. This is supported by the fact that any finite sample drawn from a symmetric distribution appears to be asymmetric. In fact, its bias can be carried over all bounds by the actions of a single outlier. The least squares estimator is said to be non–robust.

Following the growing alarm raised by Pearson and Chandra Sekar [25], Box [26], Tukey [22], Anscombe [27], Wilcoxon and then Hodges and Lehman [28] among others took an opposite position to the Fisherian one and devised the *non–parametric estimation theory*. Indeed, as suggested by its name, this theory requires no a priori knowledge of the probability distribution of the measurement errors. Here, only mild but nevertheless strict assumptions such as the continuity of the distributions or the independence of the measurements are postulated.

Taking a middle position, Huber [20] initiated the *robust estimation theory*. He started from the idea that usually in practice, a reasonable approximate model can be postulated, a model which fits not all the data points but only the majority ones. For instance, on the grounds of the central limit theorem, one can claim that the majority of the data follow a Gaussian distribution; the remaining ones are the outliers whose influence has to be downweighted in some way. In this spirit, Huber [29] defined the concept of robustness as being "synonymous to insensitivity to small departure from the assumptions". Hence, the objective is to build classes of estimators which are not optimal under the assumptions but whose bias and variance remain bounded when the assumptions are not fulfilled. In this framework, Huber [20] initiated the minimax approach of robustness and developed the class of M–estimators (M– as in generalized maximum likelihood estimators), which have been reinvented in the electric power field by Merrill and Schweppe [16] under the name of non–quadratic criteria.

Following the theoretical lines drawn by Huber, Hampel [21] introduced several concepts which complement the Fisherian ones in the sense that they assess the robustness of an estimator under departures from the assumptions. Roughly speaking, an estimator is regarded as a system for which we want to analyze the stability of the outputs, i.e. the bias and variance of the estimates, when some deviations affect the inputs, i.e. the assumptions. Hampel [30,31] proposed to analyze the robustness of an estimator in three different ways :

(i)   The qualitative robustness assesses the effect of small deviations from the assumptions;

(ii)   The global robustness, which is quantified by the breakdown point, determines the largest deviation that an estimator can handle;

(iii)   The local robustness, which is measured by means of the influence and change–of–variance functions, analyzes the effect of infinitesimal deviations on the bias and variance, respectively.

B. ROBUSTNESS CONCEPTS

For the sake of simplicity, we will define the basic robustness concepts in the one–dimensional case. The multidimensional case will be considered later.

1. STATEMENT OF AN ESTIMATION PROBLEM OF LOCATION

We intend to estimate a parameter of location θ from a sample of m real–valued observations $\{z_1, \cdots, z_m\}$ assumed to be independent and identically distributed (i.i.d.) according to a cumulative distribution function (c.d.f.) G. Since G is generally unknown, we approximate it by F. Let $\hat{\theta}_m(z_1, \cdots, z_m)$ be an estimator of the parameter θ processed from the m measurements. Let $\mathscr{L}_G(\hat{\theta}_m)$ and $\mathscr{L}_F(\hat{\theta}_m)$ be the cumulative distribution function of $\hat{\theta}_m$ associated with the true c.d.f. G and the assumed c.d.f. F, respectively.

2. FUNCTIONAL FORM OF AN ESTIMATOR

We may associate with the sample $\{z_1, \ldots, z_m\}$ an empirical c.d.f. given by

$$G_m(u) = \frac{1}{m} \sum_{i=1}^{m} \Delta(u - z_i), \qquad (1)$$

where $\Delta(u-z_i)$ is the unit step which is equal to 0 for $u < z_i$ and to 1 for $u \geq z_i$. It is the point mass 1 in z_i. Note that for i.i.d. observations, G_m will tend to G for increasing m by virtue of the Glivenko–Cantelli theorem [32]. By noting that

$$\int_{-\infty}^{+\infty} u \, d\Delta(u-z_i) = z_i, \qquad (2)$$

we may replace the sample by G_m and the estimator by a function of G_m, which yields

$$\hat{\theta}_m(z_1, \cdots, z_m) = \hat{\theta}_m(G_m). \qquad (3)$$

If this function remains the same for all m and G_m, it is said to be a functional.

Definition. An estimator $\hat{\theta}_m(G_m)$ is a functional if we have

$\hat{\theta}_m(G_m) = \hat{\theta}(G_m)$ for all m and G_m.

3. ASSUMPTIONS REGARDING THE ESTIMATORS

In the next sections, we will make the following assumptions:

1) We will consider only estimators which may be replaced at least asymptotically (m → +∞) by a functional. For example, the asymptotic functional form of

(i) the arithmetic mean is : $\hat{\theta}(G) = \int_{-\infty}^{+\infty} u \, dG(u);$

(ii) the sample median is : $\hat{\theta}(G) = G^{-1}(\frac{1}{2});$

(iii) the α–trimmed mean is: $\hat{\theta}(G) = \frac{1}{1-2\alpha} \int_{\alpha}^{1-\alpha} G^{-1}(u) \, du.$

Recall that the sample median of a sample $\{z_1, \cdots, z_m\}$ is obtained by first ordering the observations by increasing values, $\{z_{1:m}, z_{2:m}, \cdots, z_{m:m}\}$, and then by taking the νth observation, $z_{\nu:m}$, if m is odd, or the mid–point $(z_{\nu:m} + z_{\nu+1:m})/2$, if m is even, where $\nu = [\frac{m}{2}] + 1$. The sample median is a point situated exactly at the center of the sample. As for the α–trimmed mean, it is equal to the arithmetic mean of the sub–sample obtained after having disregarded the $[\alpha \, m]$ largest and the $[\alpha \, m]$ smallest observations. Here [x] denotes the integer part of the real number x.

2) We will assume in addition that the estimators are Fisher consistent. This means that the estimator $\hat{\theta}(F_m)$ converges to the true value

θ for increasing m at F. Formally, we have $\hat{\theta}(F) = \theta$.

4. QUALITATIVE ROBUSTNESS

Now we are able to give the following

Definition. A sequence of estimators $\{\hat{\theta}_m; \; m \geq 1\}$ is said to be qualitatively robust at F if a small deviation between F and G yields a small deviation between $\mathscr{L}_F(\hat{\theta}_m)$ and $\mathscr{L}_G(\hat{\theta}_m)$, for all sample size m.

We see that this definition expresses the equicontinuity of the cumulative distribution functions of $\hat{\theta}_m$ with respect to m (Hampel *et al.*, 1986). In other words, if G is in the close neighborhood $\mathscr{N}(F)$ of F, then $\mathscr{L}_G(\hat{\theta}_m)$ will remain in the close neighborhood $\mathscr{N}(\mathscr{L}_F)$ of $\mathscr{L}_F(\hat{\theta}_m)$ (see Fig. 1).

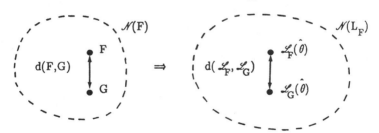

Figure 1. Neighborhoods of F and $\mathscr{L}_F(\hat{\theta})$ in the probability distribution space.

Formally we have:

$\forall \delta > 0, \exists \epsilon > 0$ such that $\forall m$ and $\forall G \in \mathscr{N}(F)$, then

$$d(F,G) < \epsilon \implies d(\mathscr{L}_F, \mathscr{L}_G) < \delta, \qquad (4)$$

where $d(F,G)$ denote the distance between F and G . Here, several metrics may be considered for measuring such a distance. According to Hampel [30],

the most appropriate metric which is able to account for the various types
of errors, i.e. rounding errors, grouping errors and gross errors, is the
Prohorof distance [29].

In practice, we may assess the qualitative robustness of an estimator
by simply verifying the continuity of this estimator at F. This statement is
based on the following theorem given by Hampel [30]:

*Hampel's theorem. A sequence of estimators $\{\hat{\theta}_m \; ; m \geq 1\}$ which is a
continuous sequence of continuous functions $\hat{\theta}_m$ of the observations at
F, is qualitatively robust at this c.d.f..*

Another important theorem given by Hampel shows the tight link between
the continuity of the functional $\hat{\theta}(F_m)$ at all F and the qualitative
robustness property of T. In particular, if we consider the contaminated
model $G = (1-\epsilon) F + \epsilon H$, where H may be any c.d.f., and ϵ is the fraction
of contamination which ranges from 0 to 1 , then the qualitative robustness
implies the continuity of the asymptotic estimator $\hat{\theta}(G)$, and thereby of the
asymptotic bias $b = |\hat{\theta}(G) - \hat{\theta}(F)|$ at $\epsilon = 0$. Note that here we define the
asymptotic bias of the estimator $\hat{\theta}(G)$ as being equal to the absolute value
of the difference between the estimated value $\hat{\theta}(G)$ and the true value $\hat{\theta}(F)$
$= \theta$. The finite sample bias will be $b_m = |\hat{\theta}(G_m) - \hat{\theta}(F_m)|$.

5. GLOBAL ROBUSTNESS : THE BREAKDOWN POINT

Qualitative robustness implies that the maximum possible bias
remains bounded for a positive fraction of contamination ϵ, i.e. for at least
one outlier over m measurements. At this point, the question that arises is
to find the maximum fraction ϵ^* of arbitrarily large gross errors that a
robust estimator can deal with.

A more formal definition is as follows [19,33]. Consider a sample of m
good data points, $Z = \{z_1, \cdots, z_m\}$. Let $\hat{\theta}_m$ be the value estimated from Z.

Now consider all possible corrupted sample Z' that are obtained by replacing any f of the good data by arbitrary large values, i.e. the fraction of contamination $\epsilon = f/m$. Let $\hat{\theta}'_m$ be the estimate processed from Z' and

$$b_{max} = \sup | \hat{\theta}_m - \hat{\theta}'_m | \qquad (5)$$

be the maximum bias caused by the contamination. Then the breakdown point $\epsilon*$ of the estimator $\hat{\theta}_m$ at the sample Z is given by

$$\epsilon* = \max \{ \epsilon = f/m \; ; \; b_{max} \text{ is finite} \}. \qquad (6)$$

In this definition one can choose which observations are replaced, as well as the magnitude of the outliers, in the least favorable way. Note that this definition involves no probability distributions. This means that the breakdown point is independent of the distribution of the good data points. Since robust estimators rely upon the majority of the data, and reject partially or totally the minority ones, the largest number of bad data that any estimator can handle is equal to half the number of redundant observations [19,23], i.e.

$$f_{max} = [(m-n)/2], \qquad (7)$$

which yields

$$\epsilon*_{max} = [(m-n)/2]/m, \qquad (8)$$

where n is the number of parameters (or variables) to be estimated and the notation [x] denotes the integer part of x. The asymptotic value (for $m \to +\infty$) of $\epsilon*_{max}$ is 50 % . Estimators with large value of $\epsilon*$ are called high breakdown point estimators.

6. THE BREAKDOWN POINT OF SOME ESTIMATORS OF LOCATION

Let us give the breakdown point of some estimators of location. A comprehensive account can be found in the Princeton robustness study reported in Andrews *et al.* [34] where more than 68 estimators have been analyzed. Another interesting reference is Hampel [35]. Here, we will consider only the three estimators previously mentioned in Section II.B.3.

The arithmetic mean constitutes the first borderline case. Indeed, this estimator is the most efficient estimator at the Gaussian distribution. However, it is not qualitatively robust and has a breakdown point of zero since it can be carried over all bounds by the action of a single outlier. In fact the lack of robustness is the characteristic of any WLS estimator.

Another borderline case is given by the sample median. This estimator has only a relative asymptotic efficiency of 63.5 % at the Gaussian model; but it is the most robust estimator with a maximum breakdown of $\epsilon^*_{max} = [(m - 1)/2]/m$; here n is equal to 1. Indeed, we have to replace at least half of the observations by outlying values in order to be certain that the middle observation is among them. Note that the outliers are chosen in the least favorable way, all in the same side of the original sample.

An intermediate case is provided by the α–trimmed mean. Indeed, it becomes the arithmetic mean for $\alpha = 0$ and the sample median for $\alpha = \epsilon^*_{max}$. Its breakdown point is obviously equal to α.

7. THE INFLUENCE FUNCTION

The local robustness of an estimator at a given probability distribution of the measurement errors is assessed through two functions: the Influence Function (IF) and the Change–of–Variance Function (CVF). They measure the effect of a single outlier on the bias and variance, respectively.

a. *Finite sample IF in the unidimensional case*

The finite sample influence function $IF(z ; \hat{\theta}_m, F)$ of an estimator of location $\hat{\theta}_m$ at the distribution function F, associated with m observations $\{z_1, ..., z_{m-1}, z\}$ is equal to the difference between the estimates $\hat{\theta}_m$ and

$\hat{\theta}_{m-1}$, processed with and without the observation z, and divided by the fraction of contamination $\epsilon = 1/m$. Here $\{z_1, \cdots, z_{m-1}\}$ follows F and z takes on all real values. Formally

$$IF(z; \hat{\theta}_m, F) = m \left\{ \hat{\theta}_m(z_1, \cdots, z_{m-1}, z) - \hat{\theta}_{m-1}(z_1, \cdots, z_{m-1}) \right\}. \qquad (9)$$

Several local robustness measures can be derived from the influence function. The most important of them is

b. *The gross error sensitivity*

It is the supremum of the absolute value of IF taken over all z,

$$\gamma^* = \sup_z |IF(z; \hat{\theta}_m, F)|. \qquad (10)$$

Its importance stems from its link with the maximum bias of the estimator. Indeed, if we define the bias b as just being the absolute value of the deviation of the estimates as it was stated previously, we get

$$b = |\hat{\theta}_m - \hat{\theta}_{m-1}| \simeq \epsilon |IF|, \qquad (11)$$

and

$$b_{max} \simeq \epsilon \gamma^*. \qquad (12)$$

Hence a well–behaved robust estimator has a bounded gross error sensitivity, namely a bounded bias. Such an estimator is said to be B–robust (B – as bias). On the other hand, it is clear that bounding γ^* constitutes a critical step in the design of a new class of robust estimators. The other local robustness measure is

c. *The local shift sensitivity*

It measures the effect of small fluctuations in the observations, due to rounding errors for instance. As noted by Hampel [31], this is equivalent to moving an observation from point z to a close point y. Its effect is equal to the difference between the values of IF at these two points. The local shift sensitivity $\lambda*$ is a standardized version of the supremum taken over all $y \neq z$ of this difference, which is given by

$$\lambda* = \sup_{y \neq z} \frac{|IF(y;\hat{\theta}_m,F) - IF(z;\hat{\theta}_m,F)|}{|y - z|}. \tag{13}$$

It is the largest slope of all straight lines intersecting IF in two distinct points [36]. When IF is differentiable for all z, the straight lines will be the tangents to IF. In that case, we have

$$\lambda* = \sup_z |\frac{\partial \; IF(z;\hat{\theta}_m,F)}{\partial z}|. \tag{14}$$

d. Asymptotic influence function

Now, let the number m of the observations increase to infinity. In the same time, the fraction of contamination, $\epsilon = 1/m$, will decrease to zero; it becomes an infinitesimal quantity. In that case, it can be shown that under certain regularity conditions [3], the IF will be equal to the directional derivative of the functional form of the estimator, $\hat{\theta}(G)$, in the direction of G, with G being the contaminated model

$$G = (1 - \epsilon)F - \epsilon \, \Delta_z, \tag{15}$$

where Δ_z is a short notation for $\Delta(u - z)$; it is the c.d.f. which puts a mass 1 at z. Formally, we have

$$IF(z;\hat{\theta},F) = \frac{\partial \; \hat{\theta}(G)}{\partial \epsilon}\Big|_{\epsilon=0} = \lim_{\epsilon \to 0} \frac{\hat{\theta}(G) - \hat{\theta}(F)}{\epsilon}. \tag{16}$$

We see that the IF stems from a linearization of the estimator $\hat{\theta}(G)$ at ϵ

equal to 0. It describes the local behavior of $\hat{\theta}$ in the close neighborhood of the assumed model F.

The IF is related to the asymptotic bias b(ϵ) (see Fig. 2) through

$$b(\epsilon) = |\hat{\theta}(G) - \hat{\theta}(F)| \simeq \epsilon \, |IF(z;\hat{\theta},F)|, \qquad (17)$$

and to the asymptotic variance $V(\hat{\theta},F)$ through

$$V(\hat{\theta},F) = \int IF^2(z;\hat{\theta},F) \, dF(z). \qquad (18)$$

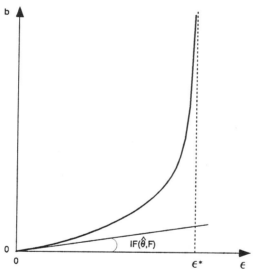

Figure 2. Asymptotic bias b versus ϵ. (Taken from [3]).

Figure 2 shows the link in a schematic way between the various robustness concepts defined previously. As suggested by Huber [29] and by Hampel [3], the qualitative robustness implies the continuity of the maximum possible asymptotic bias b(ϵ) at $\epsilon = 0$, the gross error sensitivity is the slope of its tangent at $\epsilon = 0$ for well–behaved estimators, and the breakdown point is the abscissa of its first asymptote.

8. THE IF OF SOME ESTIMATORS OF LOCATION

Figure 3 sketches the asymptotic IF at the Gaussian distribution of the three estimators of location defined in Section II.B.3. We see that the

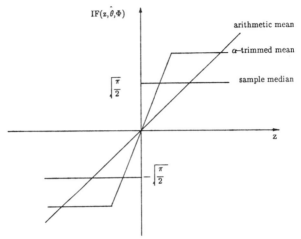

Figure 3. IF of some estimators of location at the Gaussian distribution.

IF of the arithmetic mean is a straight line with slope 1. It indicates that the larger the value of a measurement, the greater its influence. Therefore, it has an infinite gross error sensitivity; this estimator is not robust. On the other hand, $\gamma*$ is bounded for the α − trimmed mean and for the sample median. In fact the latter estimator has the smallest value of $\gamma*$ that an estimator may have ($\gamma* = \sqrt{\pi/2}$ at the standard Gaussian distribution).

Concerning the local shift sensitivity, it is equal to one for the arithmetic mean and to infinity for the sample median. In fact, the arithmetic mean has the smallest possible value of $\lambda*$ that an estimator may have. In this respect, it is the most resistant estimator to rounding errors. On the other hand, because the sample median relies only on some central observations, it is very sensitive to such errors. Note that from a practical point of view, robustness against gross errors is by far more crucial than robustness against rounding errors.

9. HAMPEL'S OPTIMALITY CRITERION

In the light of this comment, a question arises. Can we find an estimator with both a minimum asymptotic bias and a minimum asymptotic variance at a given probability distribution F ? Unfortunately, the answer is in the negative since these two requirements conflict: the lower the gross error sensitivity, the larger the asymptotic variance and vice versa. Therefore in practice, we have to find a trade–off between robustness and efficiency. For instance, the arithmetic mean is the most efficient estimator at the Gaussian distribution but it is not robust whereas the sample median has a poor efficiency at this distribution, but it is the most robust estimator. This dilemma may be solved through the following

Hampel's optimality criterion. It consists of minimizing the asymptotic variance under the constraint that the gross error sensitivity is smaller than a given threshold.

Estimators which satisfies such a criterion are called optimally B–robust.

10. THE CHANGE–OF–VARIANCE FUNCTION

Until now, we have analyzed the local robustness of the asymptotic bias of an estimator by means of the influence function. But what about the local robustness of the asymptotic variance? This can be done through the Change–of–Variance Function (CVF). It is equal to the first derivative of the asymptotic variance with respect to ϵ at $\epsilon = 0$:

$$CVF(z;\hat{\theta},F) = \frac{\partial \ V(\hat{\theta},G)}{\partial \ \epsilon}\bigg|_{\epsilon=0}, \qquad (19)$$

with

$$G = (1-\epsilon) \ F + \epsilon \ (\tfrac{1}{2} \Delta_z - \tfrac{1}{2} \Delta_{-z}). \qquad (20)$$

It expresses the sensitivity of the asymptotic variance with respect to the fraction of contamination ϵ.

Similar to the bias analysis, the variance analysis comprises the determination of the quantity κ^*, the counterpart of γ^*, which is called the change–of–variance sensitivity. This quantity is related to the maximal asymptotic variance via

$$\sup_{G \in \mathcal{N}(F)} V(\hat{\theta},G) \simeq V(\hat{\theta},F) \exp\left(\epsilon \, \kappa^*(\hat{\theta},F)\right). \tag{21}$$

It is apparent that if κ^* is bounded, then the asymptotic variance will be bounded. An estimator with such a property is said to be V–robust. On the other hand, when the variance $V(\hat{\theta},F)$ associated with the assumed probability distribution F is minimal under the constraint that κ^* is less than a given threshold, then the estimator is said to be optimal V–robust (the optimality criterion of Hampel).

Concerning the global robustness of the variance, it is quantified by the breakdown point ϵ^{**} ; it is the fraction of contamination for which the variance remains bounded [29].

III. POWER SYSTEM STATE ESTIMATION : STATE–OF–THE–ART

A. ROLE OF THE STATE ESTIMATION FUNCTION

From the static state estimation standpoint, an electric power system consists of a meshed high–voltage transmission network composed of lines and transformers connecting buses at which are injected the electric power supplied by synchronous generators or absorbed by the distribution system feeding the consumers. This system is supervised by control centers which are provided with computer facilities; the latter are fed with information collected in real–time from the system. These include the status of the circuit breakers and a collection of measurements. The status of all breakers leads to the real time topology of the system. As for the measurements, they are presently collected during cyclic scans which last from 1 to 10 seconds. They have to be sufficient in number and evenly distributed across the network so that the observability of the system is ensured. These information are in turn processed through the state estimation function and displayed to operators which send controls to the various power plants and substations.

The validity of the collected data is first assessed through simple plausibility checks. Here, all the measurements which are beyond the physical limitations of the equipments, or which conflict with breaker status, or which do not satisfy Kirchhoff's laws, are labelled as false and as such rejected from the data base. Non obvious gross errors have to be identified through the state estimation function. Its role is threefold. First, it aims at providing an estimate for all the metered and unmetered electrical quantities of the network. Second, it has to filter out the small errors which unavoidably corrupt any telemetered value in order to compute estimates which satisfy the Kirchhoff's laws. Its final objective is to clear out the data base from all gross errors.

Static state estimation is based on the following assumptions. The first assumption is that the system is balanced and is sinusoidal three phase, which leads to a single phase π–equivalent circuit model of lines and transformers. The second assumption is that the parameters of this circuit such as resistances, reactances and susceptances are exactly known and that the real–time topology of the network can be infered from the available status data of breakers and isolators. The third assumption is that the measurements with which the system is equipped are taken simultaneously so that they constitute a "true" snapshot of the state of the power system. The fourth assumption is that they are corrupted by an error vector, e, of independent Gaussian random variables, e_i, with zero means

$$e \sim N(0, R), \qquad (22)$$

and known covariance matrix

$$R = \text{diag}(\sigma_i^2). \qquad (23)$$

Under these assumptions, the purpose is to estimate the state vector x containing the voltage magnitudes V and phase angles θ at all buses except one bus taken as a reference for the phase angles. It is called the slack bus and its angle is arbitrarily set to zero. Hence, the n–dimensional state vector x is given by

$$x = [\theta_1, \theta_2, \cdots, \theta_{N-1}, V_1, \cdots, V_N]^T, \tag{24}$$

where N is the number of buses of the system and $n = 2N - 1$. The vector x is related to the m–dimensional measurement vector z through the following nonlinear model

$$z = h(x) + e, \tag{25}$$

where $h(\cdot)$ is a nonlinear vector–valued function.

We assume in the following that there exists a certain level of redundancy in the measurements, i.e. the global redundancy $\eta = m/n$ is larger than one. The measurements are presently telemetered values on power flows and power injections as well as on voltage magnitudes. Moreover, they are assumed to be distributed across the system so that the observability of the system is ensured. This means that the Jacobian matrix is of full rank [37,38], implying that the state estimation problem can be solved.

Up to now three approaches have been proposed. These are

(i)   the weighted least squares estimator provided with some bad data rejection rules based on the weighted or normalized residuals;

(ii)   the least absolute value estimator which minimizes the L_1–norm;

(iii)   the M–estimators using the weighted residuals and the generalized M–estimators based on the normalized residuals.

In order to be able to analyze the robustness of these estimators, let us define beforehand the very important concept of leverage point in the regression case.

B. LEVERAGE POINTS IN POWER SYSTEMS

Let $H = \partial h(x)/\partial x$ be the (m x n) Jacobian matrix and

$$\ell_i = [\ell_{i1}, \cdots, \ell_{im}]^T \tag{26}$$

be the ith row of $(R^{-1/2}H)$. Each vector ℓ_i is associated with a particular measurement and defines a point in the factor space of the state estimation. Hampel [39] showed that the total influence function IT of a regression M–estimator $\hat{x}$ at F, including the WLS and LAV estimator, is equal to the product of the scalar influence function of the residuals, IR, and the vector–valued influence of position in the factor space, IP. Note that IR is identical to IF in the unidimensional case. Formally

$$IT(z, \hat{x}, F, \ell, \Lambda) = IR(z, \hat{x}, F) \ IP(\ell, \Lambda), \tag{27}$$

where

$$IP(\ell, \Lambda) = (E[\ell \ell^T])^{-1} \ell, \tag{28}$$

and where $E[\ell \ell^T]$ is the covariance matrix of the vector ℓ, which is assumed to have zero mean. Therefore it is apparent that potentially influential measurements are not only those far from the others in the z direction (vertical outlier) but also those distant in the factor space, the so–called leverage points. The associated measurements have a great influence on the M–estimators, including the WLS and the LAV estimator, as we will see later on. The residuals of such measurements remain small (zero values for the LAV) even if they are corrupted by gross errors. Therefore, they are not detected by hypothesis tests based on the weighted residuals.

Leverage points in power system state estimation result from measurements of the following types [40] :

1) line flows and bus injections associated with lines that are relatively short compared to the others;

2) bus injections at nodes that have a large number of incident lines.

The entries of the Jacobian matrix associated with such measurements become large.

As a result, we may assert that any actual power system model possesses many leverage points, often in a cluster pattern.

C. THE WEIGHTED LEAST SQUARES ESTIMATOR

The WLS estimator was the first estimator to have been applied to power systems. Initiated by Schweppe *et al.* [7], it is the most widely used method in the real–time environment of a control center. Research has been focused on the implementation aspect (see [41–44], among others) as well as on its robustification against gross errors by means of some bad data rejection rules [8,12,44].

1. DEFINITION

This estimator minimizes the sum of the squares of the weighted residuals, $r_{Wi} = r_i/\sigma_i$, yielding the criterion

$$J(x) = \sum_{i=1}^{m} r_{Wi}^2 = [z - h(x)]^T R^{-1} [z - h(x)]. \quad (29)$$

A necessary condition for $J(\hat{x})$ to be minimum is that $\hat{x}$ makes the derivative of the above criterion equal to zero, namely

$$H^T(\hat{x})R^{-1}[z - h(\hat{x})] = H^T(\hat{x}) R^{-1}r = 0, \quad (30)$$

where

$$r = z - h(\hat{x}) \quad (31)$$

is the residual vector.

The nonlinear equation (30) is usually solved through the Gauss–Newton iterative algorithm [7,45] given by

$$G(\mathbf{x_k})(\mathbf{x_{k+1}} - \mathbf{x_k}) = \mathbf{H}^T(\mathbf{x_k})\mathbf{R}^{-1}[\mathbf{z} - \mathbf{h}(\mathbf{x_k})], \qquad (32)$$

where $\mathbf{x_k}$ is the value of $\mathbf{x}$ at the k-th iteration, and

$$G(\mathbf{x_k}) = \mathbf{H}^T(\mathbf{x_k})\mathbf{R}^{-1}\mathbf{H}(\mathbf{x_k}) \qquad (33)$$

is the information matrix.

2. RESIDUAL SENSITIVITY ANALYSIS

Sensitivity analysis makes use of the linearization of $\mathbf{h}(\mathbf{x})$ about $\hat{\mathbf{x}}$ given by

$$\mathbf{h}(\mathbf{x}) \simeq \mathbf{h}(\hat{\mathbf{x}}) + \mathbf{H}(\hat{\mathbf{x}})(\mathbf{x} - \hat{\mathbf{x}}). \qquad (34)$$

Substituting Eqs. (25) and (34) into (30) and (31) yields

$$\mathbf{r} = \mathbf{h}(\mathbf{x}) + \mathbf{e} - \mathbf{h}(\hat{\mathbf{x}}) \simeq \mathbf{e} + \mathbf{H}(\hat{\mathbf{x}})(\mathbf{x} - \hat{\mathbf{x}}) \qquad (35)$$

and

$$\mathbf{H}^T(\hat{\mathbf{x}})\mathbf{R}^{-1}[\mathbf{z} - \mathbf{h}(\mathbf{x}) + \mathbf{H}(\hat{\mathbf{x}})(\mathbf{x} - \hat{\mathbf{x}})] = 0. \qquad (36)$$

Hence,

$$\mathbf{r} = \mathbf{W}\,\mathbf{e}, \qquad (37)$$

where

$$\mathbf{W} = \mathbf{I} - \mathbf{H}\,(\mathbf{H}^T\mathbf{R}^{-1}\mathbf{H})^{-1}\,\mathbf{H}^T\,\mathbf{R}^{-1} \qquad (38)$$

is the (m x m) residual sensitivity matrix.

Under the Gaussian assumption, $\mathbf{e} \sim N(0,\mathbf{R})$, the residual vector has zero mean,

$$E[\mathbf{r}] = \mathbf{W}\,E[\mathbf{e}] = 0, \qquad (39)$$

and a covariance matrix given by

$$\mathrm{Cov}(\mathbf{r}) = E[\mathbf{r}\,\mathbf{r}^T] = \mathbf{W}\,\mathbf{R}\,\mathbf{W} = \mathbf{W}\,\mathbf{W}\,\mathbf{R} = \mathbf{W}\,\mathbf{R}. \qquad (40)$$

3. THE THREE CASES OF BAD DATA

For the sake of clarity, it is relevant to distinguish the following three cases of bad data:

(i)   A single bad data among the measurements.

(ii) Multiple non–interacting bad data; this case occurs when there exist several gross errors corrupting measurements with uncorrelated residuals, namely whose correlation coefficient

$$\rho_{ij} = \frac{\mathrm{Cov}(r_i, r_j)}{\sqrt{\mathrm{Var}(r_i)\,\mathrm{Var}(r_j)}} = \frac{\sigma_j\,W_{ij}}{\sigma_i\sqrt{W_{ii}W_{ij}}} \qquad (41)$$

are close to zero ($W_{ij} \simeq 0$). This property is shared by all measurements which are far away electrically and/or topologically from each other. Note that multiple noninteracting bad data can be regarded as the superposition of cases of a single bad measurement.

(iii) Multiple interacting bad data; these are wrong measurements with correlated residuals ($|W_{ij}| > 0$). They are usually situated in the vicinity of the same buses. Such bad data may mask each other in the sense that that their effects may cancel each other out in the least squares residuals. For instance, let us assume that the errors e_1 and e_2 are rather large whereas the other ones are negligible ($e_i \simeq 0$ for $i = 3, \cdots, m$). Referring to Eq. (37), we may write

$$\begin{cases} r_1 = W_{11}\,e_1 + W_{12}\,e_2\,, \\ r_2 = W_{21}\,e_1 + W_{22}\,e_2\,. \end{cases} \qquad (42)$$

It is apparent that there exist values of e_1 and e_2 that make the residuals small, and hence the bad data undetectable by the hypothesis tests applied to these quantities.

4. BAD DATA DETECTION

It borrows its framework from the Neyman–Pearson testing theory. The objective is to check the validity of the Gaussian assumptions made on the measurement errors by testing the null hypothesis

H_0: absence of bad data,

against the alternative

H_1: presence of bad data.

For that purpose, three statistical tests have been proposed [7,8,12]:

(i)   **The test on $J(\hat{x})$.** Knowing that under H_0 , the criterion

$$J(\hat{x}) = r^T R^{-1} r = e^T W^T R^{-1} W\, e = e^T R^{-1} W\, e \qquad (43)$$

follows a χ_k^2 distribution with $k = m - n$ degrees of freedom (rank$(W) = k$), we choose a threshold $\chi_{k,1-\alpha}^2$ associated with a risk of first kind α and we decide to

– accept H_0 , if $J(\hat{x}) \le \chi_{k,1-\alpha}^2$,
– reject H_0, otherwise.

(ii)   **The test on the residuals.** It consists of comparing the normalized residuals

$$r_{Ni} = \frac{r_i}{\sqrt{Var(r_i)}} = \frac{r_i}{\sigma_i \sqrt{W_{ii}}} \qquad (44)$$

to a fixed threshold $N_{1-\alpha/2}$ associated with a given α risk. More precisely, we conclude the absence of bad data if $|r_{Ni}| \le N_{1-\alpha/2}$, the presence of bad data otherwise. Note that under H_0, the normalized residuals follow a standard Gaussian distribution, $N(0,I)$. In order to avoid the computation of the diagonal entries of the W matrix, we may check the weighted residuals r_{Wi} instead, by comparing them to a given value λ. This amounts to comparing the normalized residuals to

$N_{1-\alpha/2} = \lambda/\sqrt{W_{ii}}$. It is clear that this threshold increases when W_{ii} decreases, namely for the leverage points of the system. On the other hand, when $W_{ii} \simeq 1$, then $r_{Ni} \simeq r_{Wi}$ and $N_{1-\alpha/2} \simeq \lambda$.

Robustness analysis. Neither the J–test nor the r_W–test is robust against even a single bad leverage point. Therefore their breakdown point is zero. On the other hand, the r_N–test is able to reveal the presence of a single bad leverage point, but it is susceptible to the masking effect of multiple bad ones. Indeed, Handschin *et al.* [8] showed that any single bad data, including a bad leverage point, has a normalized residual with the largest absolute value when all the other measurements are free from errors. Unfortunately this nice property does not hold in presence of multiple bad data as demonstrated by Eq. (42) (for more details see [12]). As a result, the r_N–test is qualitatively robust but with a breakdown point ϵ^* of only $1/m$.

5. BAD DATA IDENTIFICATION METHODS

As soon as the detection tests become significant, thus indicating the presence of bad data, we have to identify them. The earliest method [8] and the most widely used one in practice derives from the largest normalized residual property for a single bad data mentioned previously. It consists of performing cycles of measurement eliminations and state re–estimations, at each cycle the measurement with the largest normalized (or weighted) residual is removed. The procedure stops when the detection tests become non–significant.

To overcome the lack of robustness of this approach against multiple bad data, several methodologies have been proposed along this line. The most sophisticated ones are the HTI method developed by Mili *et al.* [11–14], the combinatorial method initiated by Monticelli *et al.* [9], and the geometric approach advocated by Clements and Davis [10]. Basically, these methods consist of eliminating and/or reinserting groups of measurements, and performing hypothesis tests on the residuals either of the processed measurements or of the eliminated ones (the so–called error estimates of the HTI method, called also the predicted residuals by Cook and Weisberg, [46]). These methods belong to the methods known as *outlier diagnostics* in the robust statistical field [47]. Their weaknesses result from the fact that they rely on the detection tests applied to the least squares residuals for deciding whether or not the processed measurements are free from gross errors. As a result, all these methods have at most a breakdown point of $1/m$ when they make use of the normalized residuals, and of zero when they make use of the weighted residuals.

D. THE LEAST ABSOLUTE VALUE ESTIMATOR

To circumvent the lack of robustness of the least squares approach for bad data detection and identification, Irving *et al.* [17] advocated the use of the LAV estimator. This estimator minimizes the L_1– norm given by

$$J(\mathbf{x}) = \sum_{i=1}^{m} |\, r_{Wi}| \, .$$

(45)

The solution is found through linear programming techniques. This approach was further investigated by Kotiuga and Vidyasagar [18] and by Kotiuga [48]. Possible failures of the LAV to reject bad data when it is applied to power systems was mentioned in [49]. The reported simulation results reveal a non–negligible percentage of failure which ranges from 11% to 27% (see discussion of [49]).

Robustness analysis. Since the LAV estimator belongs to the general class of the M–estimators, its robustness will be analyzed in the next Section.

E. THE M–ESTIMATORS

Soon after the development of a weighted least squares procedure for power systems, Merrill and Schweppe [16] tested the class of the

M–estimators and suggested a new M–estimator for bad data suppression, called in the sequel the Merrill–Schweppe estimator. In fact, this application ranks among the earliest use of M–estimators in the regression case. Later on, Handschin *et al.* [8] compared this estimator to the ones proposed by Huber [20] and by Hampel [3,34]. Further investigations were carried out by Muller [50], Couch *et al.* [51], Falcão *et al.* [52,53], Lo *et al.* [54], and by Zhuang and Balasubramanian [55]. A comparative study was reported in Mili *et al.* [12].

Having made this historical summary, let us now define the M–estimators and analyze their robustness via the influence function and the breakdown point.

1. DEFINITION

Huber [56] defines the class of the M–estimators in regression as estimators which minimize the criterion

$$J(\mathbf{x}) = \sum_{i=1}^{m} \rho(\frac{z_i - h_i(\mathbf{x})}{\sigma_i}) = \sum_{i=1}^{m} \rho(r_{Wi}), \tag{46}$$

where ρ is a particular function of the residuals. A solution $\hat{\mathbf{x}}$ may be found by equating to zero the derivative of $J(\mathbf{x})$ with respect to x, namely

$$\sum_{i=1}^{m} \psi(r_{Wi}) \, \ell_i = 0 . \tag{47}$$

Three sub–classes of M–estimators may be defined according to the characteristics of the ρ– and ψ–functions (see Table I and Fig. 4). The first subclass is characterized by convex ρ–functions. Belonging to this class are the least squares estimator, the Huber estimator, and the LAV estimator. The second sub–class is characterized by nonconvex ρ–functions with non–vanishing derivatives ψ. Belonging to this class are the Merrill–Schweppe estimator [16] and the Muller estimator [50]. Finally, the third sub–class is characterized by non–convex ρ–functions with vanishing derivatives ψ. These estimators act in such a way that the influence of residual IR is canceled for measurements with large weighted residuals (but not the total influence). Due to the shape of their ψ–function, they are called redescending M–estimators. Belonging to this subclass are the Hampel estimator, the Huber–type skipped mean, and the WLS estimators provided with some rejection rules defined previously.

In the power system field [8,12], these estimators have been called

o quadratic–tangent for the Huber estimator;

o quadratic–linear for the Muller estimator;

o quadratic–square–root for the Merrill–Schweppe estimator;

o quadratic–constant for the Huber–type skipped mean;

o multiple–segment for the Hampel estimator.

TABLE I. Functions ρ and ψ of some M–estimators.

ESTIMATOR	Domain	$\rho(r_w)$	$\psi(r_w)$
L_1-norm	IR	$\lvert r_w \rvert$	$\text{sign}(r)$
Huber (QT)	$\lvert r_w \rvert \leq b$	$\frac{1}{2} r_w^2$	r_w
	$\lvert r_w \rvert > b$	$b\lvert r_w \rvert - \frac{b^2}{2}$	$b\,\text{sign}(r)$
Muller (QL)	$\lvert r_w \rvert \leq b$	$\frac{1}{2} r_w^2$	r_w
	$\lvert r_w \rvert > b$	$\frac{1}{2} b \lvert r_w \rvert$	$\frac{1}{2} b\,\text{sign}(r)$
Schweppe (QR)	$\lvert r_w \rvert \leq b$	$\frac{1}{2} r_w^2$	r_w
	$\lvert r_w \rvert > b$	$2b^{3/2}\sqrt{\lvert r_w \rvert} - \frac{3}{2}b^2$	$b^{3/2}\dfrac{\text{sign}(r)}{\sqrt{\lvert r_w \rvert}}$
Huber type skipped-mean (QC)	$\lvert r_w \rvert \leq b$	$\frac{1}{2} r_w^2$	r_w
	$\lvert r_w \rvert > b$	$\frac{1}{2} b^2$	0
Hampel (SM)	$\lvert r_w \rvert \leq a$	$\frac{1}{2} r_w^2$	r_w
	$a < \lvert r_w \rvert \leq b$	$a\lvert r_w \rvert - \frac{a^2}{2}$	$a\,\text{sign}(r)$
	$b < \lvert r_w \rvert \leq c$	$a\,\dfrac{c\lvert r_w \rvert - \frac{1}{2} r_w^2}{c-b} - \dfrac{a^2}{2} - \dfrac{ab^2}{2(c-b)}$	$a\,\dfrac{c - \lvert r_w \rvert}{c-b}\,\text{sign}(r)$
	$\lvert r_w \rvert > c$	$\frac{a}{2}(c+b-a)$	0

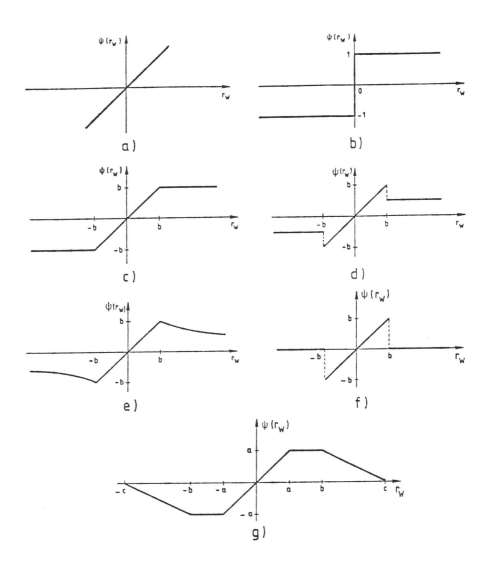

Figure 4. ψ–Function of the following estimators:

a) LS estimator,   d) Muller estimator,

b) LAV estimator,   e) Merrill–Schweppe estimator,

c) Huber estimator,   f) Huber–type skipped mean,

                                  g) Hampel estimator.

Robustness analysis. For this class of estimators, Hampel [3,39] showed that at the Gaussian distribution Φ, the influence of residual is given by

$$\text{IR}(r_w;\hat{\mathbf{x}},\Phi) = \frac{\psi(r_w)}{E[\psi'(r_w)]}, \qquad (48)$$

whereas the influence of position is given by

$$\text{IP}(\boldsymbol{\ell};\hat{\mathbf{x}},\Lambda) = (E[\boldsymbol{\ell}\,\boldsymbol{\ell}^T])^{-1}\,\boldsymbol{\ell}. \qquad (49)$$

Here, Λ is the c.d.f. of the points defined by $\boldsymbol{\ell}$ in the factor space, which are assumed to have zero mean and a covariance matrix $E[\boldsymbol{\ell}\,\boldsymbol{\ell}^T]$. From these equations, it is apparent that when the function ψ is bounded, the influence of residual IR will be bounded, but not the influence of position IP. Indeed, the latter one may become arbitrarily large for increasing $\boldsymbol{\ell}$, i.e. for a single leverage point which is arbitrarily far away from the majority of the point cloud in the factor space. Hence, the product of these two influences, the total influence, will be unbounded. From this statement we conclude that all the regression M–estimators, including the LAV estimator, are not robust and have a zero breakdown point. On the other hand, in the absence of leverage points, for instance when the points are uniformly distributed in the factor space (which is not the case in the power systems regression models), the M–estimators with a bounded function ψ are robust with an asymptotic breakdown point of at most 25% (see Ref. [3,24]).

F. THE GENERALIZED M–ESTIMATORS

Several proposals have been made in order to robustify the M–estimators against bad leverage points by bounding both the influence of residual and of position. Such estimators are referred to as the generalized M–estimators (also called the bounded–influence regression estimators). According to Hill [57], all the proposals may be written in the following form:

$$\sum_{i=1}^{m} u(\ell_i) \, \psi(r_{Wi} \, v(\ell_i)) \, \ell_i = 0 \,, \tag{50}$$

where $u(\ell_i)$ and $v(\ell_i)$ are weighted functions whose role is to downweight the influence of the leverage points of the regression.

The most interesting proposals are those initiated by

(i) Mallows [58], with $v(\ell_i) = 1$; here, the leverage points are downweighted regardless of their residual values. This means that the influence of position IP is bounded separately from the influence of residual IR.

(ii) Schweppe [8], with $v(\ell_i) = 1/u(\ell_i)$; here, the leverage points are downweighted only if their residuals have large values. Unlike the previous case, IP and IR depend on each other in such a way that the total influence is bounded.

Let us analyze in more detail the form advocated by Schweppe in the power system field. The idea is to choose not the same fixed tuning constant b for all the weighted residuals but a tuning constant $b \sqrt{W_{ii}}$, which varies with their standard deviation (see Eq. (40)). In fact, this amounts to compare instead the normalized residuals r_{Ni} to b, namely to set

$$v(\ell_i) = \frac{1}{u(\ell_i)} = \frac{1}{\sqrt{W_{ii}}} \,. \tag{51}$$

Due to the lack of robustness of the diagonal entries of the **W**—matrix against multiple bad leverage points (see the robustness analysis reported in Section III.C.4), research has been carried out in order to find more robust weights. It results in the development of the estimators proposed by Hampel and Krasker [3,39] and by Krasker and Welsch [59], among others. According to Hampel *et al.* [3], Schweppe's form leads to optimal B—robust estimators. Note that although software programs exist which implement such estimators [60,61], they have not been yet applied to power systems.

Robustness analysis. The generalized M–estimators constitute an important improvement with respect to the M–estimators since they are able to bound the influence of the leverage points. Unfortunately, Maronna [62] and Maronna *et al.* [63] showed that the asymptotic breakdown point $\epsilon*$ of such estimators cannot exceed the reciprocal of the number of variables to be estimated, $1/n$. As a result, it tends to vanish for large n. For instance, for a medium size power system with 100 buses, their breakdown point is less than $1/200 = 0.005$. It is nearly zero for larger size systems.

G. THE HIGH BREAKDOWN POINT ESTIMATORS

To overcome these weaknesses, a third family of robust estimators have been initiated. They are called the high breakdown point estimators since they give the right solution even if a large fraction of the redundant observations are bad leverage points, and that regardless of the dimension of the regression model. Devised by Siegel [64], the repeated median was the first estimator to be developed with such a robustness property . It was followed by the least median of squares estimator and the least trimmed squares estimator developed by Rousseeuw [23], and by the class of the S–estimators devised by Rousseeuw and Yohai [65]. The interesting feature of the last 3 estimators is that they are equivariant for linear transformations of the ℓ_i , a property which is not shared by the repeated median estimator .

To circumvent the very poor efficiency of these estimators at the Gaussian distribution, we may use them either for computing a robust starting point for the generalized M–estimators or as diagnostic tools for the identification and the rejection of the outliers; the latter procedure will be followed by a standard least squares estimation for improving the efficiency of the estimates, if this is desired.

IV. LEAST MEDIAN OF SQUARES ESTIMATOR

Among the various high breakdown point estimators previously mentioned, only the LMS will be considered.

A. DEFINITION

In the unidimensional and the simple regression case ($n = 1$ and 2), the LMS estimator minimizes not the sum but the median of the squared residuals, hence its name. In multiple regression, it minimizes the ν–th ordered squared residual, yielding the criterion

$$J(x) = (r_W^2)_{\nu:m} , \qquad (52)$$

where

$$\nu = [\tfrac{m}{2}] + [\tfrac{n+1}{2}] . \qquad (53)$$

Here, the weighted residuals are first squared and then ordered by increasing value:

$$(r_W^2)_{1:m} \leq \cdots \leq (r_W^2)_{m:m}. \qquad (54)$$

This estimator has the following geometrical properties [19]:

(i)   In estimation of location, the LMS is the midpoint of the shortest half of the real–valued observations. The halves are subsets containing ν successive ordered observations, with $\nu = [m/2] + 1$; the shortest half is the half having the smallest distance between its extreme observations, namely with the smallest of the differences

$$z_{\nu:m} - z_{1:m}, \; z_{\nu+1:m} - z_{2:m}, \cdots, \; z_{m:m} - z_{m-\nu+1:m}. \qquad (55)$$

(ii)   In simple regression, the LMS line lies at the middle of the narrowest strip covering half the data (see Fig. 5).

(iii)   In multiple regression, the previous statements are true, except that one has hyperstrips and hyperplanes.

Roughly speaking, theses properties state that the LMS estimator fits only the majority of the data. This means that it can withstand a fraction of outliers equal to the half of the redundant measurements, yielding a breakdown point of ϵ_{max}^*.

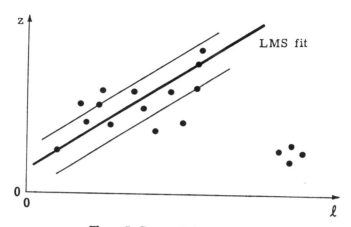

Figure 5. Geometric Interpretation.

B. BAD DATA IDENTIFICATION

Once the robust fit has been computed, one has to identify the outliers. This can be done by comparing the magnitudes of the standardized residuals,

$$r_{Si} = r_{Wi} / \hat{\sigma}_r , \qquad (56)$$

to a given threshold, say 2.5. Here, the residuals are standardized by a robust estimate of their dispersion,

$$\hat{\sigma}_r = C \sqrt{(r_W^2)_{\nu:m}} , \qquad (57)$$

where C is a correction factor which has to be chosen in such a way that when the errors are Gaussian, $N(0,\sigma^2)$, $\hat{\sigma}_r$ must tend to σ for increasing m. This implies the limit of C is 1.4826. Additionally, few good measurements should be rejected by the test. Experience has shown that a good choice for C is

$$C = 1.4826 \left(1 + \frac{5}{m-n}\right). \qquad (58).$$

This scale estimate is a byproduct of the robust fit itself, from which it inherits a high breakdown point. In fact, it neither inflates nor deflates even when less than half the redundant measurements are bad.

In order to overcome the poor efficiency of the LMS at the Gaussian distribution, the identification of the outliers may be followed by a standard weighted least squares estimation computed from the good measurements.

C. APPLICATION TO POWER SYSTEMS

1. ALGORITHM

The procedure which has been applied to power systems by Mili *et al.* [40], proceeds by selecting samples of size n for which the system is observable, namely for which the Jacobian matrix $\mathbf{H}$ is of full rank. For each selected sample contained in the n–dimensional subvector $\mathbf{z}_n$, the nonlinear system

$$\mathbf{z}_n = \mathbf{h}_n(\mathbf{x}) \tag{59}$$

is solved through the Newton–Raphson algorithm,

$$\mathbf{H}_n(\mathbf{x}^{(k)}) \, \Delta\mathbf{x}^{(k)} = \mathbf{z}_n - \mathbf{h}_n(\mathbf{x}^{(k)}), \tag{60}$$

and the ν–th squared residual computed. Here $\mathbf{H}_n(\mathbf{x}^{(k)})$ is the (nxn) Jacobian sub–matrix associated with the n selected measurements, which have zero residuals. The LMS estimates are associated with the sample having the smallest value of the criterion (56).

Since the LMS criterion is not differentiable and possesses many local minima, combinatorial methods have to be used. The general method consists of considering all samples of size n among the m measurements, yielding a number of samples of $\binom{m}{n}$. However, this method is feasible only for very small systems. For larger systems, it is desirable to resort to the resampling technique proposed in [19]. The idea is to draw only a number k of random samples that guarantees a high probability P (typically 0.95) of drawing at least one non–contaminated sample. Now, given P and ϵ, the fraction of contamination against which we want to be protected, the minimum number of samples k that must be considered may be calculated through

$$P = 1 - (1 - (1 - \epsilon)^n)^k .$$
(61)

This formula clearly indicates that the larger the value of ϵ, the more samples we have to select; the largest number being the one associated with ϵ_{max}. Since in practice, it is desirable to have more than one good sample, a number of samples larger than k, say 2k, have to be selected. Table II reports some values of k for P = 0.95 and ϵ = .03.

Table II. Number of samples for P = 0.95 and ϵ = 0.03 .

n	m	f	k	2k
50	135	4	13	26
80	216	6	33	66
100	270	8	62	124

2. FAST DECOUPLED STATE ESTIMATION

Computing times may be dramatically decreased by representing the high voltage power system by decoupled models. Roughly speaking, these models account for the fact that small changes in the phase angle δ affect mainly the real powers whereas small changes in the voltage magnitude V affect mainly the reactive powers. Formally, we have

$$H_{PV} = \frac{\partial h_P}{\partial V} \simeq 0 , \qquad H_{Q\delta} = \frac{\partial h_Q}{\partial \delta} \simeq 0 .$$
(62)

Therefore, for decoupled models, Eq. (60) splits into two parts given by

$$\alpha\, \mathbf{H}_{P\theta}\, \Delta\theta = \mathbf{z}_P - \mathbf{h}_P(\theta, \mathbf{V}), \tag{63}$$

$$\beta\, \mathbf{H}_{QV}\, \Delta\mathbf{V} = \mathbf{z}_Q - \mathbf{h}_Q(\theta, \mathbf{V}), \tag{64}$$

where α and β are relaxation factors. Here, the first equation is associated with the $(N-1)$ selected measurements on real powers and the second one is associated with the N selected measurements on reactive powers and voltage magnitudes. For each drawn sample, the two equations are alternately solved by replacing the most recently updated value for δ or $\mathbf{V}$ in the subsequent equation.

Similarly to the fast decoupled load flow [66] and the fast decoupled WLS state estimator [42], further simplifications may be performed. They consist of (i) calculating the square sub–matrices $\mathbf{H}_{P\theta}$ and $\mathbf{H}_{QV}$ at the flat voltage profile ($V = 1$ p.u. and $\theta = 0$ rad.) and (ii) maintaining them constant during the iterative process. Since the same sample set and hence the same Jacobian sub–matrices may be used as long as the topology of the system remains unchanged, these sub–matrices can be factorized in off–line and stored in a compact form.

When the decoupled assumption is well satisfied, these two equations may be solved separately from each other. The idea is to compute the LMS estimate first for δ with all the voltage magnitudes set to 1 p.u., and then for $\mathbf{V}$ with δ set to the calculated LMS solution. This allows to draw two independent sets of real and reactive samples whose number is obtained from Eq. (61) by setting n to N–1 and N, respectively. This yields a substantial reduction of the total number of samples that have to be selected. It is this decoupled version that has been implemented and tested. Simulation results show that good convergence characteristics of the iterative equations (63) and (64) are obtained when appropriate values (between 0 and 1) for the relaxation factors are used. Note that this algorithm lends itself well to an implementation on parallel computer system (see Appendix A).

3. OBSERVABILITY ANALYSIS

In order to decrease the number of the selected samples which lead to an unobservable system, we draw randomly one measurement among each of the n fundamental sets defined as follows:

Definition. A fundamental set associated with a state variable is a set containing those measurements which have nonzero terms in the corresponding column of the Jacobian submatrices, $H_{P\theta}$ and H_{QV}.

This procedure is based on the following

Theorem. A necessary condition that the system is observable for a drawn sample is that each fundamental set has been selected, i.e. it contains at least one drawn measurement.

Proof. Let us assume that the system is observable for a drawn sample for which there exists one fundamental set that has not been selected. This means that there exists one state variable which is related to none of the selected measurements. It is clear that this variable cannot be estimated from the sample, which contradicts our assumption that the sample makes the system observable.   □

This theorem shows that there is a one–to–one correspondence between the fundamental sets and the state variables. It states that with each state variable of an observable system, we may associate a non–empty fundamental set containing all the measurements which are related to it and hence allow its estimation. From a topological point of view, we may associate with the phase angle (a voltage magnitude) of each bus, a fundamental set containing all the real (reactive) measurements situated on the incident lines as well as the adjacent nodes.

There exist other procedures which allow us to identify a large number of unobservable cases [40]. For instance, one has to reject any drawn sample containing real (reactive) power flow measurements situated on (i) both ends of a line, (ii) all the lines incident to a node whose real (reactive) power injection measurement was drawn, (iii) all the lines of a loop. Also one has to reject any reactive power sample which contains no voltage magnitude.

Concerning the remaining unobservable cases, they have to be identified by a test which ensures a necessary and sufficient condition for observability. Several methods can be used. The implemented technique consists of checking the absolute values of the pivot elements obtained in the course of the (re)factorization of the Jacobian submatrices [38].

V. SIMULATION RESULTS

The fast decoupled LMS estimator has been implemented and tested on several power systems. Here only two examples will be described and analyzed; the aim being to demonstrate the robustness of such a high breakdown point estimator in situation of multiple conforming bad data [9]. This implies that the wrong measurements are contaminated by gross errors that are not random, but are related and their effect is to corroborate each other. Two methods of obtaining such interacting bad data will be used.

The test data case is the IEEE 14–bus system whose one–line diagram is shown in Fig. 6. The line resistances of this system have been sufficiently decreased in order to comply with the decoupled assumption. Here, only the voltage angles will be estimated from a set of 34 measurements on 23 real power flows (FLP) and 11 real power injections (INP) ; the voltage magnitudes being set to 1 p.u.. This gives a global redundancy of $34/13 = 2.6$. Therefore, the maximum number f_{max} of bad data that can be identified is of $[(34–13)/2] = 10$, to which corresponds a maximum global breakdown point ϵ^*_{max} of $10/34 = 0.29$. If we assume that only 5 bad data may occur, yielding a fraction ϵ of contamination of $5/34 = 0.15$, then 48 samples have to be drawn when applying the resampling method. Among this measurement set, there are 5 obvious leverage points; these correspond to power flow 5–4, and the power injections at buses 2, 3, 6 and 12. The corresponding terms of the Jacobian matrix are large.

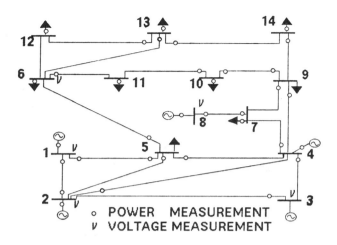

Figure 6. IEEE 14–Bus system.

Case 1: Multiple conforming bad data. Bus 5 was chosen since the associated fundamental set contains a large number of measurements, 9. In this test, 4 bad data were introduced in the real power flows 5–1, 5–2, 5–4 and 5–6, simulating a remote telemetry failure at bus 5. The values of the conforming gross errors were determined by taking the linearized equations corresponding to these 4 measurements from the Jacobian matrix ($\Delta z_P = H_{P\delta} \Delta \delta$), and substituting the true values of the angles for all buses except bus 5, which was arbitrarily given the value of 1 radian. The results for this case was reported in Table III. These show that the final retained sample includes only good measurements and LMS assigns them zero residuals. Moreover, all the bad data have been clearly identified as such by the residual test. The residuals were standardized by the σ of Gaussian noises fixed at 1 MW and the $\hat{\sigma}_r$ of the residuals estimated at 2.54 MW. However, it is worth noting that this standardization is not suitable for identifying the leverage points. For instance here, one good leverage point, the measurement on the power injection at bus 4 (INP 4), has been wrongly identified as bad. In fact, there is a need for a weighting factor which accounts for the distance of a point in the factor space.

TABLE III. LMS estimation results for case 1

Meas.	Measured Value	True Value	$|r_i|$	$|r_{Si}|$
FLP 2-1	43.91	46.38	0.039	0.015
FLP 3-2	-35.61	-36.58	0.533	0.210
FLP 5-1	449.8	-23.15	471.4	185.6
FLP 5-2	589.1	-13.34	600.9	236.6
FLP 5-4	2513.9	48.75	2461.8	969.2
FLP 5-6	396.8	-19.86	417.3	164.3
FLP 7-8	-10.77	-10.0	0.0	0.0
FLP 8-7	8.82	10.0	0.0	0.0
FLP 9-4	-4.25	-4.08	0.016	0.061
FLP 9-7	-13.36	-13.75	1.04	0.410
FLP 9-10	-11.26	-10.75	0.021	0.008
FLP 10-9	13.29	10.76	0.012	0.005
FLP 6-11	22.40	23.29	0.037	0.015
FLP 13-6	-25.22	-25.56	0.033	0.013
FLP 14-9	1.32	0.92	0.0	0.0
FLP 10-11	-19.81	-19.76	0.027	0.011
FLP 13-12	-3.27	-3.92	0.047	0.019
INP 1	69.97	69.55	0.041	0.016
INP 2	29.36	28.30	0.083	0.033
INP 4	-7.41	-7.80	41.83	16.47
INP 7	-0.33	0.0	1.035	0.408
FLP 2-1	-45.51	-46.37	0.00	0.00
FLP 2-4	25.57	24.65	0.00	0.00
FLP 1-5	24.24	23.17	0.00	0.00
FLP 4-3	18.32	17.65	0.00	0.00
FLP 4-7	2.11	3.75	0.00	0.00
FLP 13-14	16.09	15.99	0.00	0.00
INP 6	77.9	78.80	0.00	0.00
INP 8	9.38	10.0	0.00	0.00
INP 10	-9.79	-9.0	0.00	0.00
INP 11	-2.93	-3.5	0.00	0.00
INP 12	-6.63	-6.1	0.00	0.00
INP 13	-11.92	-13.5	0.00	0.00
INP 14	-15.39	-14.9	0.00	0.00

Case 2: Local breakdown. One extra wrong measurement at line 1–5 has been added to the 4 previously mentioned ones. Here, the conforming bad data were obtained by performing a load flow with the load at bus 5 increased by 200 MW. This creates a situation where out of the 9 measurements at bus 5, 5 are bad. The bad measurements are FLP 1–5, FLP 5–1, FLP 5–2, FLP 5–3, FLP 5–6, whereas the good measurements are INP 1, INP 2, INP 4 INP 6. Intuitively, it is expected that breakdown should occur since the majority of the measurements from which the phase angle at bus 5 is estimated are wrong; this issue is discussed in more detail in the next section. This is precisely what the simulation results showed. Indeed, the sample with the lowest value of the LMS criterion contained bad data. Note that it is the LMS estimator and not the resampling algorithm which breaks down since several selected samples are not contaminated, i.e. contained only good measurements. In fact at bus 5, the maximum number of bad data that any high breakdown point estimator can handle is only $[(9 - 1)/2] = 4$ and not 10 as given by the global breakdown point.

VI. THE LOCAL BREAKDOWN POINT CONCEPT

The previous example showed that the LMS estimator may breakdown locally in power systems. This can be explained by means of the local breakdown point concept. Let us introduce this important concept first in multiple linear regression and then in power systems. Note that similar problems were considered by Hampel in [3] and [67].

A. MULTIPLE LINEAR REGRESSION

Consider a linear model defined by

$$\mathbf{z} = \mathbf{H}\,\mathbf{x} + \mathbf{e}, \tag{65}$$

where $\mathbf{x}$ is the n–dimensional state vector which has to be estimated from the m–dimensional observation vector $\mathbf{z}$ to which the error vector $\mathbf{e}$ has been added, and H is the (m x n) observation matrix. The definition of the breakdown point which has been given so far implicitly assumes that the observation matrix H is full. This means that each measurement is function of all the state variables contained in x. However, in many applications and especially in large–scale systems, H is sparse. In that case, a measurement, say z_i, is often a linear combination of not all x_j ($j = 1, \cdots, n$) but only of some of them, those for which ℓ_{ij} are nonzero. On the other hand, for each state variable x_j we may associate a subset $Z_j = \{z_i\}$ of measurements z_i for which the i-th term ℓ_{ij} of the j-th column of H is nonzero. Only these measurement points have projections on the ℓ_j-axis different from the origin of the factor space. Such a subset has been called a *fundamental set* in Section IV.C.3. Let m_j be the number of measurements which determine the state variable x_j; m_j is the size of the fundamental set Z_j.

> **Theorem.** *Assume that all measurements are good except some of the fundamental set Z_j associated with the state variable x_j. The number of outliers among Z_j that any estimator can withstand is*
>
> $$f_{j,max} \leq [\frac{m_j - 1}{2}], \qquad (66)$$
>
> *which corresponds to a maximum local breakdown point of*
>
> $$\epsilon^*_{j,max} \leq [\frac{m_j - 1}{2}] / m_j . \qquad (67)$$

Proof. Without loss of generality, we may assume that the true state vector x is 0 and that all measurements are perfect with zero values except f_j measurements belonging to the Z_j fundamental set. The latter ones are supposed to be the first numbered measurements, $\{z_1, \cdots, z_{f_j}\}$. Now, suppose that these bad data are perfectly conforming, i.e. that they satisfy

$$z_i = \ell_{ij} \, \tilde{x}_j \qquad \text{for} \quad i = 1, \cdots, f_j . \qquad (68)$$

It is clear that if the wrong measurements are the majority ones in Z_j, i.e. if f_j is larger than $f_{j,max}$, then any high breakdown point estimator will give an estimate close to $\tilde{x}_j$, which may be arbitrarily far away from the true value, zero. The breakdown occurs regardless of the value of $f_{max}=[(m-n)/2]$, i.e. even if the local number $f_{j,max}$ is smaller than the global number f_{max}. □

B. THE POWER SYSTEM MODEL

If we linearize the power system nonlinear model given by Eq. (25) about the flat voltage profile ($V = 1$ p.u. and $\theta = 0$ rad.), we get two linear systems

$$z_P = H_{P\theta}\, \theta + e_P, \qquad (69)$$

$$z_Q = H_{QV}\, V + e_Q, \qquad (70)$$

where Eq. (69) relates the real power measurement vector z_P to the voltage phase angle vector θ and Eq. (70) relates the reactive power measurement vector z_Q to the bus voltage magnitude V. For large high voltage networks, the Jacobian submatrices $H_{P\theta}$ and H_{QV} are very sparse. Typically their fill–in ranges from 5% to 10%, with the percentage being smaller as the system size and voltage increases. This is because a power flow measurement contributes only 2 nonzero terms in the associated row of H whereas a voltage magnitude contributes only one nonzero term. The contribution due to a power injection measurement at a bus is equal to the number of the incident branches to this bus plus one, typically 4 nonzero terms. As a result, the number of measurements m_i which contribute to the estimation of a phase angle θ_i or to a voltage magnitude V_i are much smaller than the total number of available measurements m; m_i rarely exceeds ten whereas m is of the order of several hundred. Therefore the maximum number of bad data that can be identified locally is much smaller than the global one, which yields a value of the local breakdown points very different from the global one.

C. EXAMPLE OF THE IEEE 14–BUS SYSTEM

To illustrate this concept, let us consider the IEEE 14–bus system referred to previously. With 34 real power measurements and 13 unknowns θ, the total number of real bad data that can be identified is $[(34–13)/2] = 10$, giving a global breakdown of 0.29. Table IV shows the values of m_i and f_i for each of the unknowns θ_i associated with the i-th bus; note that the values for bus 1 has not been reported since it is the chosen slack bus. It can be seen that there are several buses (3, 8, 12, and 14) for which the maximum number $f_{i,max}$ of bad data that can be identified is just one. To illustrate this, we shall place 2 conforming bad data at bus 14 in order to demonstrate the local breakdown concept. The fundamental set associated with θ_{14} contains 4 real power measurements which are FLP 14–9, FLP 13–14, INP 13, INP 14, namely z_{20}, z_{23}, z_{33}, z_{34}, respectively. By setting all the θ_i to zero (the true values) except θ_{14}, we find that

$$z_{20} = 3.661\ \theta_{14}, \qquad z_{23} = -2.759\ \theta_{14},$$
$$z_{33} = -3.968\ \theta_{14}, \qquad z_{34} = 9.565\ \theta_{14}. \tag{71}$$

Table IV. Local measurement data and fundamental set sizes.

Bus #	Initial configuration		Improved configuration	
	Number of Meas.	$f_{i,max}$	Number of Meas.	$f_{i,max}$
2	8	3	8	3
3	4	1	6	2
4	8	3	9	4
5	9	4	7	3
6	7	3	10	4
7	7	3	7	3
8	4	1	3	1
9	9	4	9	4
10	5	2	7	3
11	5	2	7	3
12	4	1	5	2
13	7	3	7	3
14	4	1	5	2

Now, let two measurements, say z_{20} and z_{34} be wrong, corresponding to $\theta_{14} = 1.0$ rad., while z_{23} and z_{33} have good values, i.e. zero. The LMS in this case cannot determine which pair of data are bad and it chooses the wrong values. Table V gives the angles estimated by the LMS; clearly the estimated angle $\hat{\theta}_{14}$ has been corrupted by the 2 bad data. It may be driven to any arbitrary value by simply adjusting the values of the bad data.

Table V. Bus angles estimated by LMS for the IEEE 14–bus system.

Bus #	Angle (in degrees)
1	0.0000
2	-0.0092
3	0.0061
4	-0.0462
5	0.0024
6	-0.0634
7	-0.0748
8	-0.0710
9	-0.1100
10	-0.1050
11	-0.0941
12	57.2708
13	57.2067
14	57.1859

D. OPTIMAL METER PLACEMENT FOR BAD DATA IDENTIFICATION

The preceding analysis regarding the dominance of the local breakdown point over the global breakdown point has applications in the area of optimal meter placement. Until now, the meter placement methods proposed in the literature [68–69] have generally been confined to locating the measurements in such a way as to maximize the accuracy (i.e. minimize the trace of the covariance matrix of the error estimates of the state variables) of the WLS state estimator at the Gaussian distribution, with little or no regard to protection against bad data. The concept of local breakdown as defined can be used to provide a criterion for placing the measurements in a manner that provides such protection. This can be done by maximizing the minimum number of bad data that can be identified locally, which amounts to maximizing the size of the smallest fundamental set across the system. This is a typical combinatorial optimization problem. Algorithms for the solution of such problems are still a research topic ; the simulated annealing method [70] is one of the most promising methods for this purpose. Table IV shows the improvement that can be made in the distribution of measurements by a heuristic method.

VII. CONCLUSIONS

An algorithm for solving the fast decoupled LMS state estimation in power systems has been developed and tested. Resampling methods in conjunction with observability techniques make this algorithm applicable to large–scale systems. Concepts such as leverage point, fundamental set, and local breakdown point have been defined. Based on these concepts, meter placement methodologies which are optimal for bad data identification have been outlined.

Some of the issues which need further development are :

(i) Determination of relaxation factors which increase the convergence rate of the fast decoupled LMS iterative algorithm.

(ii) Development of bad data rejection rules based on LMS standardized residuals which are suitable for identifying bad leverage points. One possibility is to use the minimum volume ellipsoid estimator for the robustification of the the Mahalanobis distance as proposed by Rousseeuw and Van Zomeren in Ref. [71].

(iii) Practical implementation of the proposed optimal meter placement methodologies. Various algorithms for solving this class of combinatorial optimization are being investigated.

APPENDIX A. PARALLEL PROCESSING IMPLEMENTATION

This appendix compares the performance of the resampling algorithm on a single–processor and a 2–processor machine. The resampling algorithm when used in conjunction with the decoupled assumption, is perfectly suited for implementation on a multi–processor system. The Jacobian is pre–factorized and the LMS for each sample can then be computed independently, on separate processors if available. Both version of the algorithm were tested on a VAX 8800 computer with 2 processors, on the IEEE 30–bus system. The reduction in computing time due to the use of the 2 processors was 38 % ; ideally this would be 50 % but this is not achievable due to the increased computational overhead.

ACKNOWLEDGEMENTS

The authors would like to thank Professor A. G. Phadke for his support for this research as well as for the use of the computer facilities at the Power Systems Laboratory at Virginia Polytechnic Institute and State University.

REFERENCES

[1]   L. S. Van Slyck and J. Allemong, "Operating Experience with the AEP State Estimator", *IEEE Transactions on Power Systems*, Vol. 3, No. 2, May 1988, pp. 521–528.

[2]   M. M. Adibi and D. K. Thorne, "Remote Measurement Calibration, *IEEE Transactions on Power Systems*, Vol. PWRS–1, No. 2, May 1986, pp. 194–203.

[3]   F. R. Hampel, E.W. Ronchetti, P.J. Rousseeuw and W.A. Stahel, *Robust Statistics : the Approach based on Influence Functions*, John Wiley, 1986.

[4]   R. J. Beckman and R. D. Cook, "Outlier····s", *Technometrics*, Vol. 25, No. 2, May 1983, pp. 119–163.

[5]   V. Barnett and T. Lewis, *Outliers in Statistical Data*, John Wiley, 2nd edition, 1984.

[6]   D. M. Hawkins, *Identification of Outliers*, Chapman and Hall, 1980.

[7]   F. C. Schweppe, J. Wildes and D. B. Rom, "Power System Static State Estimation, Part I, II and III", *IEEE Transactions on Power Apparatus and Systems*, Vol. PAS–89, No. 1, January 1970, pp. 120–135.

[8]   E. Handschin, F.C. Schweppe, J. Kohlas and A. Fiechter, "Bad Data Analysis in Power System Estimation", *IEEE Transactions on Power Apparatus and Systems*, Vol. PAS–94, No.2, March/April 1975, pp. 329–337.

[9]   A. Monticelli, F. F. Wu and M. Yen, "Multiple Bad Data Identification for State Estimation by Combinatorial Optimization", *IEEE Transaction on Power Delivery*, Vol. PWRD–1, No. 3, July 1986, pp. 361–369.

[10]   K.A. Clements and P.W. Davis, "Multiple Bad Data Detectability and Identifiability: a Geometric Approach", *IEEE Transactions on Power Delivery*, Vol. PWRD–1, No.3, July 1986, pp. 355–360.

[11]   L. Mili, Th. Van Cutsem and M. Ribbens–Pavella, "Hypothesis Testing Identification: a New Method for Bad Data Analysis in Power System State Estimation", *IEEE Transactions on Power Apparatus and Systems*, Vol. PAS–103, Nov. 1984, pp. 3239–3252.

[12]   L. Mili, Th. Van Cutsem and M. Ribbens–Pavella, "Bad Data Identification Methods in Power System State Estimation – a Comparative Study", *IEEE Transactions on Power Apparatus and Systems*, Vol. PAS–104, Nov. 1985, pp. 3037–3049.

[13]   L. Mili, Th. Van Cutsem and M. Ribbens–Pavella, "Decision Theory for Fault Diagnosis in Electric Power Systems", *Automatica*, Vol. 23, No. 3, 1987, pp. 335–353.

[14]   L. Mili and Th. Van Cutsem, "Implementation of the Hypothesis Testing Identification in Power System State Estimation ", *IEEE Transactions on Power Systems*, Vol. 3, No. 3, Aug. 1988, pp. 887–893.

[15] N. Xiang , S. Wang. and E. Yu, "A New Approach for Detection and Identification of Multiple Bad Data in Power System State Estimation", *IEEE Transactions on Power Apparatus and Systems*, Vol. PAS–101, No. 2, Feb. 1982, pp. 454–462.

[16] H.M. Merrill and F.C. Schweppe, "Bad Data Suppression in Power System State Estimation", *IEEE Transactions on Power Apparatus and Systems*, Vol.PAS–90, No.6, Nov./Dec. 1971, pp. 2718–2725.

[17] M.R. Irving, R.C. Owen and M. Sterling, "Power System State Estimation using Linear Programming", *Proceedings of the IEE*, Vol. 125, No. 9, Sept 1978, pp 879–885.

[18] W. Kotiuga and M. Vidyasagar, "Bad Data Rejection Properties of Weighted Least Absolute Value Techniques Applied to Static State Estimation", *IEEE Transactions on Power Apparatus and Systems*, Vol. PAS–101, No. 4, April 1982, pp. 844–853.

[19] P.J. Rousseeuw and A.M. Leroy, *Robust Regression and Outlier Detection*, John Wiley, 1987.

[20] P.J. Huber, "Robust Estimation of a Location Parameter", *Annals of Mathematical Statistics*, Vol. 35, 1964, pp. 73–101.

[21] F. R. Hampel, *Contributions to the Theory of Robust Estimation*, Ph.D Thesis, University of California, Berkeley, 1968.

[22] J.W. Tukey, "A Survey of Sampling from Contaminated Distributions", in *Contributions to Probability and Statistics*, I. Olkin (ed.), Stanford University Press, 1960, pp. 448–485.

[23] P. J. Rousseeuw, "Least Median of Squares Regression", *Journal of the American Statistical Association*, Vol. 79, No. 388, 1984, pp. 871–880.

[24] F. R. Hampel, "Beyond Location Parameters : Robust Concepts and Methods", *Bulletin of the International Statistical Institute*, Book 1, Vol. 46, 1975, pp. 375–391.

[25] E. S. Pearson and C. Chandra Sekar, "The Efficiency of Statistical Tools and a Criterion for the Rejection of Outlying Observations", *Biometrika*, Vol. 28, 1936, pp. 308–320.

[26] G. E. P. Box, "Non–normality and Tests of Variance", *Biometrika*, Vol. 40, 1953, pp. 318–335.

[27] F. J. Anscombe, "Rejection of Outliers", *Technometrics*, Vol. 2, 1960, pp. 123–147.

[28] E. Lehmann, *Testing Statistical Hypotheses*, John Wiley, 1959.

[29] P. J. Huber, *Robust Statistics*, John Wiley, 1981.

[30] F. R. Hampel, "A General Qualitative Definition of Robustness", *Annals of Mathematical Statistics*, Vol. 42, 1971, pp. 1887–1896.

[31] F. R. Hampel, "The Influence Curve and its Role in Robust Estimation", *Journal of the American Statistical Association*, Vol. 69, 1974, pp. 383–393.

[32] P. Billingsley, *Probability and Measure*, John Wiley, 1986.

[33] D. L. Donoho and P. J. Huber, "The Notion of Breakdown Point", in *A Festschrift for Erich L. Lehmann*, P. J. Bickel, K. A. Doksum and J. L. Hodges (eds.), Wadsworth 1983.

[34] D. F. Andrews, P. J. Bickel, F. R. Hampel, P. J. Huber, W. H. Rogers and J. W. Tukey, *Robust Estimates of Location: Survey and Advances*, Princeton University Press, 1972.

[35] F. R. Hampel, "The Breakdown of the Mean Combined with some Rejection Rules", *Technometrics*, Vol. 27, No. 2, May 1985, pp. 95–107.

[36] D. C. Hoaglin, F. Mosteller and J. W. Tukey, *Understanding Robust and Exploratory Data Analysis*, John Wiley, 1983.

[37] G. R. Krumpholz, K. A. Clements and P. W. Davis, "Power System Observability: a Practical Algorithm Using Network Topology", *IEEE Transactions on Power Apparatus and Systems*, Vol. PAS–99, No. 4, July/Aug. 1980, pp. 1534–1542.

[38] A. Monticelli and F. F. Wu, "Network Observability", *IEEE Transaction on Power Apparatus and Systems*, Vol. PAS–104, No. 5, May 1985, pp. 1035–1048.

[39] F. R. Hampel, "Optimally Bounding the Gross–error–sensitivity and the Influence of Position in Factor Space", Proceedings of the ASA Statistical Computing Section, 1978, pp. 59–64.

[40] L. Mili, V. Phaniraj and P. J. Rousseeuw, "Least Median of Squares Estimation in Electric Power Systems", Submitted for Presentation at the IEEE Winter Meeting, Atlanta, Georgia, Feb. 1990.

[41] F. Broussole, "State Estimation in Power Systems: Detecting Bad Data Through the Sparse Inverse Matrix Method", *IEEE Transactions on Power Apparatus and Systems*, Vol. PAS–97, No. 3, May/June 1978, pp. 678–682.

[42] A. Garcia, A. Monticelli and P. Abreu, "Fast Decoupled State Estimation and Bad Data Processing", *IEEE Transactions on Power Apparatus and Systems*, Vol. PAS–98, No. 5, Sept./Oct. 1979, pp. 1645–1651.

[43] Th. Van Cutsem and M. Ribbens–Pavella, "Critical Survey of Hierarchical Methods for State Estimation of Electric Power Systems", *IEEE Transactions on Power Apparatus and Systems*, Vol. PAS–102, No. 10, 1983, pp. 3415–3424.

[44] A. Bose and K. A. Clements, "Real–Time Modeling of Power Networks", *Proceedings of the IEEE*, Vol. 75, No. 12, Dec. 1987, pp. 1607–1622.

[45] F. C. Schweppe and E. Handschin, "Static State Estimation in Electric Power Systems", *Proceedings of the IEEE*, Vol. 62, No. 7, July 1974, pp. 972–982.

[46] R. D. Cook and S. Weisberg, *Residuals and Influence in Regression*, Chapman and Hall, 1982.

[47] D. A. Belsley, E. Kuh and R. E. Welsch, *Regression Diagnostics: Identifying Influential Data and Sources of Collinearity*, John Wiley, 1980.

[48] W. W. Kotiuga, "Development of a Least Absolute Value Tracking State Estimator", *IEEE Transactions on Power Apparatus and Systems*, Vol. PAS–104, No. 5, May 1985, pp. 1160–1166.

[49] D. M. Falcão and S. M. de Assis, "Linear Programming State Estimation: Error Analysis and Gross Error Identification, *IEEE Transaction on Power Systems*, Vol. 3, No. 3, Aug. 1988, pp. 809–815.

[50] H. Muller, "An Approach to Suppression of Unexpected Large Measurement Errors in Power Systems", *Proceedings of the 5th PSCC Conference*, Cambridge, 1975, Paper no. 2.3/5.

[51] G. H. Couch, A. C. Sullivan and J. A. Dembecki, "Results from a Decoupled State Estimator with Transformer Ratio Estimation for a 5 GW Power System in the Presence of Bad Data", *Proceedings of the 5th PSCC Conference*, Cambridge, 1975, Paper no. 2.3/3.

[52] D. M. Falcão, S. M. Karaki and A. Brameller, "Nonquadratic State Estimation: a Comparison of Methods", *Proceedings of the 7th PSCC Conference*, Lausanne, 1981, pp. 1002–1006.

[53] D. M. Falcão, P. A. Cooke and A. Brameller, "Power System Tracking State Estimation and Bad Data processing", *IEEE Transactions on Power Apparatus and Systems*, Vol. PAS–101, No. 2, Feb. 1982, pp. 325–333.

[54] K. L. Lo, P. S. Ong, R. D. McColl, A. M. Moffatt and J. L. Sulley, "Development of a Static State Estimator, Part I, II and III", *IEEE Transactions on Power Apparatus and Systems*, Volume PAS–102, No. 8, Aug. 1983, pp. 2486–2500.

[55] F. Zhuang and K. Balasubramanian, "Bad Data Suppression in Power System State Estimation with a Variable Quadratic–constant Criterion", *IEEE Transactions on Power Apparatus and Systems*, Vol. PAS–104, No. 4, April 1985, pp. 1738–1744.

[56] P. J. Huber, "Robust Regression: Asymptotics, Conjectures and Monte Carlo", *Annals of Statistics*, Vol. 1 , No. 5., 1973, pp. 799–821.

[57] R. W. Hill, *Robust Regression when there are Outliers in the Carriers*, Ph.D thesis, Harvard University, 1977.

[58] C. L. Mallows, *On Some Topics in Robustness*, Unpublished Memorandum, Bell Telephone Laboratories, Murray Hill, N.J., 1975.

[59] W. S. Krasker and R. E. Welsch, "Efficient Bounded–influence Regression Estimation", *Journal of the American Statistical Association*, Vol. 77, No. 379, 1982, pp. 595–604.

[60] A. Marazzi, "On the Numerical Solution of Bounded Influence Regression Problems", *COMPSTAT 1986*, Physica–Verlag for IASC, 1986, pp. 114–119.

[61] A. Marazzi, "Solving Bounded Influence Regression Problems with ROBSYS", *Proceedings of the 1st International Conference on*

Statistical Data Analysis using the L_1 – norm and Related Methods, Neuchâtel Switzerland, 1987.

[62] R. A. Maronna, "Robust M–estimators of Multivariate Location and Scatter", *Annals of Statistics*, Vol. 4, No. 1, 1976, pp. 51–67.

[63] R. A. Maronna, O. H. Butos and V. J. Yohai, "Bias–and–efficiency Robustness of General M–estimators for Regression with Random Carriers", in *Smoothing Techniques for Curve Estimation*, T. Gasser and M. Rosenblass (eds.) , Lecture Notes in Mathematics 757, Springer–Verlag, 1979, pp. 91–116.

[64] A. F. Siegel, "Robust Regression using Repeated Medians", *Biometrika*, Vol. 69, No. 1 , 1982, pp. 242–244.

[65] P. J. Rousseeuw and V. J. Yohai, "Robust Regression by Means of S–estimators", in *Robust and Nonlinear Time Series Analysis*, J. Franke, W. Härdle, and R. D. Martin (eds.), Lecture Notes in Statistics, No. 26, Springer Verlag, 1984, pp. 256–272.

[66] B. Stott and O. Alsaç, "Fast Decoupled Load Flow", *IEEE Transactions on Power Apparatus and Systems*, Vol. PAS–93, No. 3, May/June 1974, pp. 859–867.

[67] F. Hampel, "Some Problems in Statistics", *Proceedings of the First World Congress of the Bernoulli Society for Mathematical Statistics and Probability*, Tashkent, USSR, Sept. 8–14, 1986.

[68] H. J. Koglin, "Optimal Measuring System for State Estimation", *Proceedings of the PSCC Conference*, Cambridge, U.K., Sept. 1975.

[69] S. Aam, L. Holten and O. Gjerde, "Design of the Measurement System for State Estimation in the Norwegian High–Voltage Transmission Network", *IEEE Transactions on Power Apparatus and Systems*, Vol. PAS–102, No. 12, Dec. 1983, pp. 3769–3777.

[70] E. Aarts and J. Korst, *Simulated Annealing and Boltzmann Machines: a Stochastic Approach to Combinatorial Optimization and Neural Computing*, John Wiley, 1989.

[71] P. J. Rousseeuw and B. C. Van Zomeren, "Unmasking Multivariate Outliers and Leverage Points", to appear in *Journal of the American Statistical Association*, Dec. 1989.

MATCHING ANALYTICAL MODELS
WITH EXPERIMENTAL MODAL DATA
IN MECHANICAL SYSTEMS

DANIEL J. INMAN
CONSTANTINOS MINAS*

Mechanical Systems Laboratory
Department of Mechanical and
Aerospace Engineering
University at Buffalo
Buffalo, NY 14260

I. INTRODUCTION

The discipline and practice of finite element modeling (FEM) of structures has become a sophisticated technology. The techniques of experimental modal analysis (EMA) have developed in the last ten years into a formal and well developed technology. Both these disciplines imply specific rules and structure to be used to develop a model. Both claim great successes in their ability to provide consistent and useful models. In principal, of course, the analytical model should be consistent with, or predict, the results obtained form vibration tests. However, this rarely happens. As a result, the finite element model must often be adjusted or modified, until it agrees with the test data. Until recently the adjustment of the finite element model has been accomplished on a ad hoc basis. Here several systematic approaches to modifying an FEM to agree with experimental modal data are presented.

Experimental modal analysis results in measured values of a structure's damping ratios, natural frequencies and mode shapes. These modal quantities can also be calculated by solving the eigenvalue/eigenvector problem for the FEM. The

* Currently with GE Corporate Research and Development, K-1, EP123, Schenectady, NY 12301

disagreement between the calculated (FEM) modal data and measured (EMA) modal data forms the topic of this chapter. It is desirable to have systematic analytical procedures for modifying the FEM until the calculated modal data agrees with that set of the experimental modal data which has been obtained with some degree of confidence.

The procedures suggested in this chapter recognize that the model correction problem in structural dynamics, is similar to the eigenstructure assignment problem common to control theory. Several methods of pole placement and eigenstructure assignment have been developed in the controls and systems theory communities. Two of these methods are exploited here and adapted to incorporate the special features and problems of matching analytical models to experimental data.

Depending on the complexity of a given structure, testing circumstances and the level of sophistication of test equipment, the quality of the modal data available for a given structure varies. In some circumstances only a few of the lower natural frequencies may be available. In some situations natural frequencies and modal damping ratios are measured. In other situations a larger number of natural frequencies (and damping ratios) as well as some mode shape information may be available. In many circumstances the measured mode shapes are complex and only some of the elements of a given mode shape are measured. In general structural tests can provide very accurate frequency measurements within an error of a few percent. However, it is not uncommon that the damping ratios are only measured to an accuracy of 10% to 20%. Mode shapes can also be very difficult to measure. Techniques are developed here which exploit these features in adjusting the FEM model.

The nature of the FEM almost always produces real mode shapes. This is true because the mechanism most commonly employed in modeling damping in FEM's is to use a damping model consisting of a linear combination of the mass and stiffness matrices. Proportional damping always results in real mode shapes. Hence, for those structures with measured complex mode shapes, the damping matrix of the analytical model must be incorrect. Therefore, a technique for modifying just the damping matrix is also of value.

II. PREVIOUS WORK

Early work in model correction focused on constructing a damping matrix assuming that the model of the mass and stiffness matrix were correct. A method which focused on measuring off diagonal elements of a damping matrix was proposed by Hasselman [1]. A perturbation technique was used to expand the damping matrix in terms of measured modal information. The complex mode shapes are treated as a perturbation of the real mode shapes provided by the corresponding undamped model. The diagonal elements of the damping matrix are those arising from the proportionally damped matrix usually assumed. The original work was analytical. Hasselman [2] extended his earlier results to consider experimental data and identified several problem areas. Difficulties are encountered when natural frequencies are repeated and/or clustered, when inaccurate test data is used and when the FEM is much larger than the experimental model. This later situation is sometimes referred to as *incomplete modal data*. Caravini and Thomson [3] presented a numerical technique for using the frequency domain (Fourier Transform) of the displacement response and input force vector calculated from measured data to construct damping coefficients for an n-degree of freedom system. The entries of the unknown damping matrix are used to form a vector. The method processes one frequency point at a time calculating the vector of damping coefficients that minimizes the difference between the estimated response and the measured response.

Beliveau [4] also used a method of eigenvalue/eigenvector perturbation to estimate the damping matrix from incomplete modal test data. In the Beliveau approach an initial estimate of the damping matrix is required and corrections are calculated by using the measured natural frequencies, damping ratios and mode shapes in a modified Newton-Rampson scheme.

Hanagud et al. [5] used an optimization method to minimize the difference between the analytical eigenvalue problem and the experimental eigenvalue problem. They also included a term consisting of the estimated damping matrix and its transpose in order to keep the estimated damping matrix symmetric. Hendrickson and Inman [6] used an optimization scheme to minimize a set of objective functions constructed from the eigenvalue problem for the damped system and measured modal data. The damping matrix is not necessarily real or symmetric. A similar formulation is used in Inman and Jha [7] except that a generalized inverse is used to compute the unknown damping matrix.

Several previous methods have focused only on systems without damping. These approximate methods correct the stiffness matrix of the FEM by using measured mode shapes and frequencies. Rodden [8], Berman and Flannelly [9], Baruch and Bar Itzbach [10], Kabe [11] and Kammer [12] used a variety of orthogonality conditions and cost functions in optimization schemes to update the stiffness matrix. Berman and Flannelly [9], Berman and Nagy [13] and Heylen [14] used orthogonality methods to update both the mass and the stiffness matrix of the undamped model.

Several previous methods have also addressed the problem of updating both the damping (non proportional) and stiffness matrix. Ibrahim [15] assumed that the unmeasured higher order modes coincide with the higher order FEM modes combined with the solution of the eigenvalue problem to provide updated damping and stiffness matrices. This method involves only simple matrix manipulations and inversions but does not necessarily produce symmetric corrected matrices. Hanagud et al.[5] attempts the same problem by adding symmetry while minimizing the error in the eigenvalue problem caused by the uncorrected model. Fuh, Chen and Berman [16] attacked this problem by emphasizing the smallest possible change is made in the analytical model based on orthogonality conditions.

All of the above methods are ad hoc in the sense that the criteria used to determine how the corrected FEM matrices are obtained based on a variety of interpretations of the source of error between measured and analytical modal data. Most use an optimization scheme and/or an orthogonality condition to formulate a corrected model. The author proposed in Minas and Inman [17] that the model correction problem is identical to the eigenstructure assignment problem developed in the control and systems theory literature. This provides a variety of methodologies and algorithms which can be readily adapted to the model correction problem. Since the pole placement and eigenstructure assignment problem form a fairly mature discipline, these algorithms are numerically sophisticated and well thought out. The requirement that the finite element model consist of matrix coefficients that are symmetric, however, requires that the algorithms from eigenstructure assignment and pole placement be modified.

Minas and Inman [18] first addressed using eigenstructure assignment coupled with a non linear programming technique to calculate symmetric corrections to the damping and stiffness matrix. Zimmerman and Widengren [19-20] used eigenstructure assignment combined with a generalized algebraic Riccatic equation to calculate symmetric corrections to the damping and stiffness matrices. The equivalence relations between various eigenstructure assignment methods used to produce symmetric correction matrices is discussed by Zimmerman and Widengren [19]. Minas and Inman [21] presented iterative methods of computing correction matrices and introduced a measure of the changes made in the finite element model by the correction algorithm. Minas and Inman [22] introduced the use of pole placement methods for those cases where the mode shapes are not known.

The methods of eigenstructure assignment and pole placement for correcting FEM's to match experimentally modal data are presented in this chapter. These methods were developed at the University at Buffalo's Mechanical System Laboratory by Minas [23-24] under the first author's supervision. The methods introduced by Zimmerman at the University of Florida are also briefly discussed. The eigenstructure assignment procedure used is that developed by Srinathkumer [25] and adapted to mechanical structures by Andy et al. [26]. The pole placement method used is discussed in Porter and Crossley [27].

III. BACKGROUND

A. The Nature of Finite Element Models

The analytical or finite element model discussed in this chapter is assumed to be of the form

$$M\ddot{x}(t) + D\dot{x}(t) + Kx(t) = B_0 u(t) \tag{1}$$

where M, D and K are nxn symmetric positive definite matrices with real valued entries reflecting the structure's mass, damping and stiffness properties respectively. The nx1 vector x(t) represents the displacement of the n degree of freedom system. Each coordinate of the vector x represents the displacement of one degree of freedom of the FEM. The overdots represent differentiation with respect to time so that $\dot{x}(t)$ is the structure's velocity vector and $\ddot{x}(t)$ is the structure's acceleration vector. The mx1 vector u(t) represents externally applied forces. The nxm real valued matrix B_0 represents the location where the forces are applied. Equation (1) and the vector x(t) are called the *physical coordinates* of the system because they correspond to the physical location of moving points on a given structure which are the *nodes* of the finite element model.

As mentioned in the introduction, the damping matrix D does not easily follow from the FEM. The matrices M and K are readily constructed in the finite element formulation in a systematic fashion. However the damping matrix D is often assumed to be proportional of the form

$$D = \alpha M + \beta K \tag{2}$$

in FEM procedure where α and β are constants. See Shames and Dym [28] for an introduction to FEM.

The analytical model presented by Eq.(1) is subject to initial conditions which are denoted

$$x(0) = x_0 \quad \text{and} \quad \dot{x}(0) = \dot{x}_0 \tag{3}$$

These are nx1 vectors of constants reflecting the initial state of the structure.

Equation (1) can also be represented in state space form by defining the 2nx1 state vector z to be

$$z(t) = \begin{bmatrix} \dot{x} \\ x \end{bmatrix} \tag{4}$$

This results in the standard linear state equation

$$\dot{z} = Az + Bu \tag{5}$$

where the augmented matrices A and B are defined by

$$A = \begin{bmatrix} -M^{-1}C & -M^{-1}K \\ I & 0 \end{bmatrix}, \quad B = \begin{bmatrix} B_0 \\ 0 \end{bmatrix} \tag{6}$$

The 2nx2n matrix A is called the state matrix, the 2nxn matrix B is called the input matrix and u(t) is now interpreted as a control vector. Here 0 and I denote the nxn matrix of zeros and the nxn identity matrix respectively. The matrix M^{-1} denotes the nxn matrix inverse of the non singular matrix M. Equation (5) and the vector z are called the *state variable coordinate system*.

A third coordinate system can be defined from Eq. (1) by calculating the eigenvectors of the undamped system (D = 0). Let u_i be the nx1 eigenvector of the matrix K normalized with respect to M such that

$$Ku_i = \omega_i^2 u_i \qquad (7)$$

$$u_i^T Mu_i = I \qquad (8)$$

where ω_i are the undamped natural frequencies of the structure. In this case the eigenvector u_i defines the mode shapes of the structure. If proportional damping is assumed, Eq. (1) can be transformed from physical coordinates to modal coordinates by substituting x = Uq(t) into Eq. (1) and postmultiplying by U^T. Here U is the nxn matrix of eigenvectors u_i defined by Eq. (8). This yields

$$U^TMU\ddot{q} + U^TDU\dot{q}(t) + U^TKUq(t) = U^TB_0u(t) \qquad (9)$$

Using Eqs. (7) and (8) and realizing that U^TDU is diagonal (see Inman [29] for instance), Eq. (9) becomes

$$\ddot{q}(t) + \text{diag}(2\zeta_i\omega_i)\dot{q}(t) + \text{diag}(\omega_i^2)q(t) = U^TB_0u \qquad (10)$$

where ζ_i are defined to be the ith modal damping ratios. Equation (10) defines the modal coordinate system and can be written as the n independent modal equations

$$\ddot{q}_i(t) + 2\zeta_i\omega_i\dot{q}_i(t) + \omega_i^2 q_i(t) = f_i(t) \qquad (11)$$

$$i = 1, 2, 3 \dots n$$

Here f_i denotes the ith element of the vector $U^TB_0u(t)$ and q_i is ith element of the vector q, i.e., the ith modal coordinate. Equation (11) is also the equation of a damped single degree of freedom model of an oscillating system and is used to provide nomenclature and intuition for the modal testing problem. Equations (10) and (11) and the vector q are called the *modal coordinate system*.

B. The Nature of Experimental Modal Analysis Models

EMA is based on measuring the forced response, or vibration, of a structure to known, or measured, signals. Usually an impulsive input force is provided.

Alternately a sinusoidal driving force is provided through a range of frequencies (called a swept sine test). In either case the strain, velocity, acceleration or displacement is measured at a variety of points. Transfer functions are calculated between the various inputs and outputs. The various methods of EMA are then applied to extract *measured* natural frequencies, $\hat{\omega}_i$, damping ratios $\hat{\zeta}_i$ and mode shapes $\hat{u}_i$ of the test structure. An introduction to EMA is provided by Inman [29] and a detailed account of EMA is given by Ewins [30]. The data produced by EMA, called *modal data*, varies in quality and quantity as mentioned in the introduction. The interpretation of the measured modal data is quite clear from Eq. (11) for those cases where the measured mode shapes are real. However, if the measured modes are complex, the interpretation is that the system is not proportionally damped and the meaning of modal damping ratios becomes generalized.

In the non proportionally damped case, the eigenvectors of the stiffness matrix do not diagonalize the damping matrix and hence do not become eigenvectors of the system, as they are in Eq. (9). Instead, the eigenvectors v_i are complex solutions to

$$(M\lambda_i^2 + D\lambda_i + K) \, v_i = 0 \tag{12}$$

where the complex numbers λ_i are the eigenvalues of the system. The interpretation of the measured complex mode shapes from EMA is that they are the complex eigenvectors satisfying Eq. (12). Likewise the natural frequencies and damping ratios measured by EMA are related to the eigenvalues of Eq. (12) by

$$\lambda_i = - \zeta_i \, \omega_i - \omega_i \sqrt{1 - \zeta_i^2} \; j \tag{13}$$

$$\lambda_{i+1} = - \zeta_i \, \omega_i + \omega_i \sqrt{1 - \zeta_i^2} \; j$$

where $j = \sqrt{-1}$ is the unit imaginary number. With this interpretation the results of an experimental vibration test consist of some of the eigenvalues $\{\lambda_i\}$ and eigenvectors $\{v_i\}$. A complete set of eigenvalues and eigenvectors is seldom available. Furthermore, not all of the elements of the measured mode shapes will be known. This is combined with the fact that the size of the FEM will be much larger than the number of measured frequencies and much larger than the number of measured elements of v_i.

C. The Control Model

While no control effort is actually used here, the feedback control formulation in both state space coordinates and in physical coordinates forms the language for the statement of the eigenstructure assignment problem and the pole placement problem. First consider the state space formulation of Eq. (5). The feedback control problem is to use the measurement of the response defined by the equation

$$y(t) = Cz(t) \tag{14}$$

where C is a $r \times 2n$ matrix of constants denoting which elements of the state vector $z(t)$ are measured and $y(t)$ is an $r \times 1$ vector of measurements. To calculate the control input $u(t)$, the form of output feedback control is used. Output feedback uses a control force $u(t)$ of the form

$$u(t) = Gy = GCz(t) \tag{15}$$

where the $m \times r$ matrix G is called a *gain matrix*. In this case the closed loop system is given by

$$\dot{z}(t) = Az(t) + BGC \, z(t) \tag{16}$$

$$\dot{z}(t) = (A + BGC) \, z(t) \tag{17}$$

The idea here is that the closed loop system defined by the new state matrix A+BGC will have more desirable properties than the open loop system defined by the state matrix A. These properties are adjusted by a judicious choice of the matrices B, G and C.

The pole placement problem for (17) is based on the eigenvalues of A which are the complex numbers λ_i satisfying the characteristic polynomial

$$\det (\lambda I - A) = 0 \tag{18}$$

Note here that λ_i are identical to those of the second order system of Eq. (12). The pole placement problem defines a desired set of eigenvalues $\{\lambda_i^d\}$ and poses the

problem: Calculate matrices B, G and C such that the eigenvalues of the closed loop system A+BGC contains the set $\{\lambda_i^d\}$. In the control problem, measurement and control actuator devices are constructed and electronic gains calculated to produce the matrices B,G and C.

Next consider the control formulation in physical coordinates. In this case the measurement equation becomes

$$y(t) = C_0 x(t) + C_1 \dot{x}(t) \tag{19}$$

where C_0 and C_1 are rxn matrices representing locations and gains of the measurements. Again the feedback control law is of the form u = Gy or

$$u = GC_0 x(t) + GC_1 \dot{x}(t) \tag{20}$$

Substitution of this form into Eq. (1) yields

$$M\ddot{x}(t) + (B_0 GC_1 + D)\dot{x}(t) + (B_0 GC_0 + K)x(t) = 0 \tag{21}$$

The eigenstructures assignment problem is to choose the matrices B_0, G, C_1 and C_0 such that the eigenvalues, λ_i and eigenvectors, v_i associated with the system of Eq. (22) contain the desired specified eigenvalues $\{\lambda_i^d\}$ and eigenvectors $\{v_i^d\}$.

Srinathkumar [25] showed that for controllable and observable systems (see Chen, [31] or Inman [29] for definitions) the elements of the feedback gain matrix G can be specified such that the max (m,r) closed loop eigenvalues can be assigned, max (m,r) eigenvectors can be particially assigned and the min (m,r) entries in each eigenvector can be partially assigned. In the control problem m is the number of actuators and r is the number of sensors. Since there are no sensors and actuators, only matrices, in the model correction problem, the integers m and r are specified by choosing the rank of the matrices B and C respectively.

D. A Simple Example

To solidify the eigenvalue assignment, or pole placement problem, and how it can be used to solve the model correction problem, consider the single degree of freedom damped oscillatior described by

$$m\ddot{x}(t) + c\dot{x}(t) + kx(t) = u(t) \tag{22}$$

Where m, c and k are the scalar mass, damping and stiffness coefficients respectively, and the time dependent scalars x(t) and u(t) represent the displacement and applied force respectively. The eigenvalues of (22) are, of course, the solution to

$$m\lambda^2 + c\lambda + k = 0 \tag{23}$$

which describes the analytical eigenvalues. The measured eigenvalues have the form

$$\hat{\lambda}_1 = -\hat{\zeta}\hat{\omega} - \hat{\omega}\sqrt{1-\hat{\zeta}^2}\, j$$

and   (24)

$$\hat{\lambda}_2 = -\hat{\zeta}\hat{\omega} + \hat{\omega}\sqrt{1-\hat{\zeta}^2}\, j$$

where $\hat{\omega}$ is the measured undamped natural frequency and $\hat{\zeta}$ is the measured damping ratio as determined from a modal test. The model correction problem is to calculate corrections Δc and Δk such that

$$m\ddot{x} + (c+\Delta c)\dot{x} + (k+\Delta k)\, x = 0$$ (25)

has $\hat{\lambda}_1$ and $\hat{\lambda}_2$ as its eigenvalues. On the other hand the pole placement problem is to calculate a control, u(t) such that Eq. (22) has the specified complex numbers λ_1^d

and λ_2^d as its eigenvalues.

To adapt the pole placement closed loop control algorithm to solve the model correction problem, it is required to set $\lambda_1^d = \hat{\lambda}_1$ and $\lambda_2^d = \hat{\lambda}_2$ and to choose u(t) to

be of the form

$$u(t) = -\,g_1 x(t) - g_2 \dot{x}(t)$$ (26)

which is called full state feedback. Then Eq. (22) becomes

$$m\ddot{x} + (g_2+c)\,\dot{x} + (g_1+k)\, x = 0$$ (27)

which has characteristic equation

$$m\lambda^2 + (g_2+c)\lambda + (g_1+k) = 0$$ (28)

The characteristic equation for the measured eigenvalues is

$$(\lambda - \hat{\lambda}_1)\,(\lambda - \hat{\lambda}_2) = \lambda^2 - (\hat{\lambda}_1 + \hat{\lambda}_2)\,\lambda + \hat{\lambda}_1\hat{\lambda}_2 = 0$$ (29)

Comparing coefficients of (29) and (28) yields that

$$g_1 = m\,\hat{\lambda}_1\,\hat{\lambda}_2 - k$$
$$g_2 = -\,m\,(\hat{\lambda}_1 + \hat{\lambda}_2) - c$$ (30)

which solves the pole placement problem. That is, with g_1 and g_2 given the values specified by (30) the closed loop system defined by (22) and (26) will have the desired eigenvalues.

Recognizing g_1 as Δk and g_2 as Δc and using the measured modal data as given by Eq. (24) yields that the solution to the model correction problem is

$$\Delta k = m\hat{\omega}^2 - k$$
$$\Delta c = 2m\hat{\zeta}\hat{\omega} - c$$ (31)

This last set of expressions indicates how to adjust the analytical model (m,c,k) to agree with measured modal data for the simple single degree of freedom case.

E. The Problems

The work of Srinathkumar [25] and Andy et al. [26] extend this simple idea to multiple degree of freedom structural models. The work of Minas [23-24] and Zimmerman [19] extends this contribution to the model correction problem. The four basic problems considered here are:

Problem 1. Develop a method for correcting the stiffness matrix of an undamped FEM of a structure based on measured mode shapes and natural frequencies.

Problem 2. Develop a method for constructing a damping matrix for an undamped FEM of a structure based on measured damping ratios, natural frequencies and (complex) mode shapes.

Problem 3. Develop a method for correcting the damping and stiffness matrices of a FEM of a structure based on measured damping ratios, natural frequencies and (complex) mode shapes.

Problem 4. Develop a method for correcting the state space formulation, or response model, of a FEM of a structure based on a limited amount of measured modal data and no mode shape information with a minimum change in the original model.

F. Motivation

Before proceeding with the solution of the four problems outlined above it is important to motivate the need for a model improvement method by examining several different models of a simple structure and the data obtained from EMA for the same structure. A FEM of a 15 node truss was obtained using a standard FEM software package (MSC/PAL). The same truss was measured experimentally using standard EMA procedures (SDRC Modal Plus) obtained by processing the frequency response functions in the frequency range of 0-256 Hz. The results are presented in Table III-1 which show the natural frequencies computed from the FEM versus those obtained form EMA.

A Comparison of Experimental and Analytical Natural Frequencies for a Space Truss.

	Analytical (FEM)	Experimental (EMA)
	ω in Hz	
1	1.38	1.07
2	4.56	3.54
3	10.88	7.94
4	26.98	22.54
5	29.68	32.61
6	30.94	-
7	42.63	40.35
8	53.79	52.51
9	68.46	61.41
10	72.61	65.62
11	82.93	78.24
12	101.93	84.43
13	102.88	-
14	116.52	91.74
15	236.64	187.13

Table III-1

Table III-1 shows clearly that there are errors of up to 30% in the lower frequencies and that for some nodes there is no clear correspondence between the analytical natural frequencies and experimental natural frequencies. In this particular case, the difficulty may be due to the fact that the truss is composed of a variety of materials (joints, tubes and bolts) which are not easily accounted for in the FEM. This table clearly indicates a need to adjust or correct the analytical model to coincide with the measured frequencies. While, one can certainly obtain better agreement by improving both the FEM and EMA effort, this example does illustrate that using relatively sophisticated commercial packages can lead to a measurable disagreement between the analytical and theoretical models

IV. PROBLEM 1 MODEL CORRECTION FOR THE UNDAMPED STRUCTURE

While the majority of interest in model correction methods is based on the presence of damping in the model and in the measurements, this undamped case is treated to provide an introduction to the problem and to introduce the method proposed by Andry et al. for eigenstructure assignment. For simplicity let the measured modal data be complete so that m=r=n. Let $\hat{U}$ denote the nxn matrix of measured mode shapes which in this case is assumed to be real valued. Let $\hat{\omega}_i$ denote the measured natural frequencies so that

$$\hat{U}^T K_0 \hat{U} = \text{diag}(\omega_i^2) \tag{32}$$

where K_0 is the stiffness matrix that the FEM model should yield (but doe not). Here $\hat{U}$ is normalized such that

$$\hat{U}^T M U = I$$

where M is the analytically determined mass matrix. Hence, the experimental model satisfies the equation

$$M\ddot{x} + K_0 x = 0 \tag{33}$$

where

$$K_0 = \hat{U}^{-T} \text{diag}(\hat{\omega}_i) \hat{U}^{-1} \tag{34}$$

Multiplying Eq. (33) by M^{-1} yields that the measured model must satisfy

$$\ddot{x}(t) + M^{-1} K_0 x(t) = 0 \tag{35}$$

On the other hand, the eigenstructure assignment problem given by Eq. (21) with $D=C_1=0$ yields that x(t) must satisfy

$$\ddot{x}(t) + M^{-1}(B_0 G C_0 + K) x(t) = 0 \tag{36}$$

The model correction problem becomes: choose B_0, G and C_0 such that (36) has the same eigenvalues and eigenvectors as Eq. (35). This will happen if the coefficients of x(t) are the same or if

$$B_0 G C_0 + K = K_0 = \hat{U}^{-T} \text{diag} \hat{\omega}_i \hat{U}^{-1} \tag{37}$$

Hence, if B_0, G and C_0 are chosen such that their product

$$B_0GC_0 = \hat{U}^{-T} \text{diag} \hat{\omega}_i \ \hat{U}^{-1} \tag{38}$$

then the corrected model will have the same natural frequencies and mode shapes as the analytical model.

Note that a major problem here, is that the product B_0GC_0 may not be symmetric and positive definite. Hence, an added difficulty is to calculate a symmetric value. Another difficulty occurs if all of U and ω_i are not known. This is frequently the case. Then, C_0 and B_0 are not square matrices, which further complicates the calculation of the corrected matrix G.

V. PROBLEM 2 MODEL CORRECTION FOR ADDING DAMPING

In this problem it is assumed that the FEM yields a satisfactory mass and stiffness matrix but that no damping matrix is available. Test data in the form of complex mode shapes and eigenvalues are assumed to be available for use in constructing the damping matrix. The modal data available is assumed to be incomplete.

A rearrangement of the eigenvalue problem given in Eq. (12) yields that

$$D\hat{v}_i = -\frac{1}{\hat{\lambda}_i} (\hat{\lambda}_i^2 M + K) \hat{v}_i \tag{39}$$

Following Inman and Jha [7] the right side of Eq. (39) is defined to be the vector f_i which is known since M and K are given by the FEM and $\hat{\lambda}_i$ and $\hat{v}_i$ are measured. Since $\hat{v}_i$ and λ_i are complex this results in two equations.

$$D\hat{v}_i = f_i \tag{40}$$
$$\hat{v}_i^* D = f_i^* \tag{41}$$

where the * denotes complex conjugate transpose. Equation (40) and (41) are relationships in the unknown elements of the matrix D. Rearranging these elements into a vector d results in

$$R_i d = b_i \tag{42}$$

where the entries of R_i are the real and imaginary parts of $\hat{v}_i$ and the vector b_i is defined by

$$b_i = \begin{bmatrix} Re \ f_i \\ -Im \ f_i \end{bmatrix} \tag{43}$$

Here d is a $(n^2+n)/2 \times 1$ vector and each R_i is a $2n \times \dfrac{n^2+n}{2}$ matrix.

Next define R to be the augmented matrix formed from each R_i by

$$R = \begin{bmatrix} R_1 \\ R_2 \\ . \\ . \\ . \\ R_m \end{bmatrix} \tag{44}$$

Also let b denote the vector formed from each b_i, so that

$$b = \begin{bmatrix} b_i \\ . \\ . \\ . \\ b_m \end{bmatrix} \tag{45}$$

then Eq. (40) becomes the system of linear equations

$$Rd = b \tag{46}$$

which may be solved for the damping vector d. Here as before m denotes the number of measured mode shapes and eigenvalues.

Since (46) is a linear system of equations to be solved three cases are possible. The system can be overdetermined, underdetermined or determined depending on the size and rank of the matrix R. In the case where R is of full rank, R^TR has a left generalized inverse and the solution for d becomes

$$d = (R^TR)^{-1}R^Tb \tag{47}$$

This least squares solution can be computed in a variety of ways (Golub and Van Loan, [32]). The unknown damping matrix D is then calculated from d. Here symmetry is assumed for the calculated damping matrix D by first casting the elements of the symmetric matrix D into the elements of the vector d.

It is shown in Minas [24] that this procedure is also successful when incomplete modal data is used. Several examples are also given to both verify and illustrate the method. This method is introduced briefly here, as the following two methods require that a damping matrix exist for the FEM before the pole placement and eigenstructure assignment method can be used. This is required because these control methods assume that the original systems be asymptotically stable.

VI. PROBLEM 3 MODEL CORRECTION BY EIGENSTRUCTURE ASSIGNMENT

This method assumes an asymptotically stable FEM model of a structure is available. It is further assumed that a modal test has been performed on the same structure using force impulses and response measurements on a grid which coincides with the analytical FEM grid. Not all nodes need to be measured and the EMA data is allowed to be incomplete. The original FEM model is of the form given in Eq. (12). The model correction scheme starts with the closed loop state feedback form given by Eq. (21).

Again let the measured eigenvalues be denoted by $\hat{\lambda}_i$ and the measured eigenvectors by $\hat{v}_i$. Then from Eq. (21) these measured values should satisfy

$$(I\hat{\lambda}_i^2 + M^{-1}D\hat{\lambda}_i + M^{-1}K)\,\hat{v}_i = (-M^{-1}B_0GC_1\hat{\lambda}_i - M^{-1}B_0GC_0)\,\hat{v}_i \qquad (48)$$

The problem now is to compute B_0, G, C_1 and C_2 such that (48) holds. Then the structural model (FEM), corrected by B_0GC_1 and B_0GC_0, will have the measured set $\hat{\lambda}_i$ and $\hat{v}_i$ as its eigenvalues and eigenvectors.

Since it is assumed here that none of the measured eigenvalues is an eigenvalue of the original FEM (otherwise they would be no need to correct the model) the inverse of the matrix $(\lambda^2 + M^{-1}D\lambda + M^{-1}K)$ exists. Hence, Eq. (48) can be manipulated to yield

$$\hat{v}_i = (\hat{\lambda}_i^2 + M^{-1}D\hat{\lambda}_i + M^{-1}K)^{-1}M^{-1}B_0G\,(\hat{\lambda}_iC_1 + C_0)\,\hat{v}_i \qquad (49)$$

Equation (49) places a restriction on the measured eigenvectors. It implies a connection between the measured eigenvector and the FEM. Specifically it states that the measured eigenvector must lie in the subspace defined by the right hand side of equation (49). This requires that the analytical model (FEM) be a very good representation of the test structure. Otherwise, the measured eigenvector will be skew to the space defined by Eq. (49). This is discussed in more detail by Andry et al. [26].

Next recall the statement of the eigenstructure assignment theorem [25] that indicates that only m elements of the eigenvectors may be assigned by the algorithm. This states that only m elements of the measured mode shape can be used in the model correction algorithm. That is, the corrected model will have the m elements of the measured mode shape as part of its eigenvector. However, in this application rather then being a restriction, this coincides very well with the practical aspect of EMA. Namely, only partial modal data is in general available. The order of the FEM is always larger than the order of the EMA model. Hence, not all elements of the measured eigenvectors are known. Thus, $\hat{v}_i$ are partitioned as

$$\hat{v}_i = \begin{bmatrix} z_i \\ d_i \end{bmatrix} \tag{50}$$

where z_i is mx1 and d_i is an (n-m) x 1 vector of unmeasured components.

The subvector d_i is chosen such that $\hat{v}_i$ satisfies the restriction implied by Eq. (49). To implement this the matrix R_i is defined to be

$$R_i = (\lambda_i^2 + M^{-1}D\lambda_i + M^{-1}K)^{-1}M^{-1}B_0G(C_1\lambda_i + C_0) \tag{51}$$

and partitioned as

$$R_i = \begin{bmatrix} U_i \\ Q_i \end{bmatrix} \tag{52}$$

Where U_i is an mxm matrix with columns $g_1, g_2...g_m$ and Q_i is an (n-m) x m matrix with columns $q_1, q_2...q_m$. Each of the vectors q_i and g_i are complex. If G_i is non singular its columns are linearly independent and the vector z_i can be written as a linear combination of the g_i so that

$$z_i = a_1g_1 + a_2g_2 + a_mg_m \tag{53}$$

Defining the vector a by

$$a = [a_1 \quad a_2 \quad a_3... a_m]$$

yields that

$$z_i = U_ia \tag{54}$$

Similarly the vector d_i becomes

$$d_i = Q_ia \tag{55}$$

Since U is nonsingular (54) and (55) can be solved for the vector d_i in terms of the vector z_i. This yields

$$d_i = Q_iU_i^{-1}z_i \tag{56}$$

Therefore the ith measured eigenvector will be of the form

$$v_i = \begin{bmatrix} z_i \\ Q_i U_i^{-1} z_i \end{bmatrix} \tag{57}$$

Equation (57) states how this mathematical restriction relating the measured mode shapes to the analytical model forces the unmeasured elements of the mode shape (d_i) to be determined. Hence the new analytical mode shape, or corrected mode shape becomes

$$\hat{v}_i = \begin{bmatrix} z_i \\ Q_i U_i^{-1} z_i \end{bmatrix} \tag{58}$$

This is equivalent to the relationship derived by Berman (1983) using other methods.

Equation (48) can be written down m times (one for each measured pair of eigenvalues and eigenvectors). These m equations can be gathered together and written in matrix form as

$$V\Lambda^2 + M^{-1}DV\Lambda + M^{-1}KV = - M^{-1}B_0GC_1V\Lambda - M^{-1}B_0GC_0V \tag{59}$$

Here Λ is the diagonal matrix consisting of the m measured eigenvalues $\hat{\lambda}_i$ and V is the nxm matrix with columns consisting of the mx1 measured eigenvectors $\hat{v}_i$. Equation (59) can now be solved for the matrix G to yield

$$G = (B_0^T B_0)^{-1}B_0^T[MV\Lambda^2 + DV\Lambda + KV][C_1V\Lambda + C_0V]^{-1} \tag{60}$$

Here the matrix B_0, C_1, and C_0 can be arbitrarily chosen. In particular however, C_1 and C_0 must be chosen such that $C_1V\Lambda + C_0\Lambda$ is non singular. Once these matrices are fixed, the matrix in G is calculated by using Eq. (60). Equation (21) then gives the correction in the stiffness matrix as B_0GC_0 and the correction in the damping matrix as B_0GC_1. Thus the new or corrected analytical model will be

$$M\ddot{x} + (D + B_0GC_1)\dot{x} + (K + B_0GC_1) x = 0 \tag{61}$$

This analytical model will now have exactly the measured modal data: ζ_i, ω_i and $\hat{v}_i$

In general, the correction matrices B_0GC_1 and B_0GC_0 will not be symmetric. Hence, if the nature of the analytical model requires them to be symmetric, i.e., no gyroscopic forces or circulatory forces are present, then further correction is required. This is accomplished by iterating on the choice of C_0 and C_1 until symmetric correction matrices result.

One method of forcing the correction matrices to be symmetric is to solve the problem analytically as given by Eq. (60) for a starting choice of C_0, C_1 and B_0. Then solve for a new C_0, C_1 and G by minimizing a cost function consisting of the skew symmetric part of the correction matrices B_0GC_0 and B_0GC_1 using the original corrected damping and stiffness. A standard unconstrained optimization method is used. This optimization produces symmetric correction matrices, but causes the eigenvalues and eigenvectors to deviate slightly from their measured values. This procedure is described in detail in the following.

Recall that a non symmetric matrix can always be written as the sum of a symmetric matrix and a skew symmetric matrix. The skew symmetric part of a matrix A is given by

$$A_{ss} = \frac{A - A^T}{2} \tag{62}$$

Thus the skew symmetric parts of the correction matrices are given by

$$\Delta D_{ss} = B_0GC_1 - C_1^TG^TB_0^T$$

$$\Delta K_{ss} = B_0GC_0 - C_0^TG^TB_0^T \tag{63}$$

where ΔD_{ss} denotes the skew symmetric part of the correction in the damping matrix and ΔK_{ss} denotes the skew symmetric part of the correction in the stiffness matrix.

The expression in (63) and (64) for B_0G and $G^TB_0^T$ is taken from Eq. (60) to be

$$B_0G = -(MV\Lambda^2 + DV\Lambda + KV)(C_1V\Lambda + C_0V)^{-1} \tag{65}$$

and

$$G^TB_0 = -(C_1V\Lambda + C_0V)^{-T}(MV\Lambda^2 + DV\Lambda + KV)^T \tag{66}$$

The objective function J, chosen for the optimization is

$$J = \|\Delta K_{ss}\| + \|\Delta D_{ss}\| \tag{67}$$

where ΔK_{ss} and ΔD_{ss} are expressed in terms of Eq. (65) and (66) where the vertical bars denote the standard Euclidian matrix norm.

This objective function is minimized over the elements of C_1 and C_0 for a given choice of B_0. The unconstrained optimization routine CODIR from IMSL was used which employes the conjugate directions technique to minimize J.   If the optimization routine reduces J to zero, the procedure is finished. If J is not zero then the correction coefficients are not perfectly symmetric. Therefore only the symmetric parts of the calculated correction are used. This causes the modal data of the corrected model to deviate from the measured modal data.   An iteration procedure is then applied to the corrected model until the elements of the matrix G become negligibly small, yielding an almost symmetric corrected model which faithfully reproduces the measured modal data.

This procedure is summarized by the following 6 steps

1. Choose B_0

2. Assign initial values of C_0 and C_1 and calculate J

3. Minimize J to get new values of C_0, C_1 and G

4. Calculate the corrected model

$D_{new} = D + B_0 G C_1$

$K_{new} = K + B_0 G C_0$

5. If $J = 0$, stop. If not, D_{new} and K_{new} are not symmetric so set

$$D_{new} = \frac{D_{new} + D_{new}^T}{2}$$

$$K_{new} = \frac{K_{new} + K_{new}^T}{2}$$

6. Iterate until the desired accuracy in the corrected model's modal data is achieved.

This procedure results in a corrected FEM which will reproduce the measured modal data.

As an example of this procedure, consider a (fictitious) 4 degree of freedom non proportionally damped system with known mass, damping and stiffness matrix. This will constitute the FEM of a structure. It is assumed that modal test data for the same structure is available in the form of 2 eigenvalues and 2 mode shapes. The suggested procedure is then used to correct the FEM until the new analytical model yields eigenvalues and eigenvectors agreeing with the measured values. Here it is assumed that the damping matrix in the FEM has already been adjusted by the methods of Section V and hence is non proportional.

The analytical model is given by

$$\begin{bmatrix} 2 & 0 & 0 & 0 \\ 0 & 3 & 0 & 0 \\ 0 & 0 & 4 & 0 \\ 0 & 0 & 0 & 2 \end{bmatrix} \ddot{x} + \begin{bmatrix} 6 & -2 & -4 & 0 \\ -2 & 2 & 0 & 0 \\ -4 & 0 & 6 & -2 \\ 0 & 0 & -2 & 2 \end{bmatrix} \dot{x} + \begin{bmatrix} 4 & -1 & -2 & 0 \\ -1 & 2 & -1 & 0 \\ -2 & -1 & 5 & -2 \\ 0 & 0 & -2 & 2 \end{bmatrix} x = 0$$

The eigenvalues are calculated to be $\{\lambda_i\} - \{-3.2813, -0.7269, -0.7681 \pm 1.0158j,$
$-0.0126 \pm 0.2749j, -0.2986 \pm 0.7887j\}$. The matrix B_0 is chosen to be

$$B_0 = \begin{bmatrix} 1 & 0 \\ 0 & 0 \\ 0 & 0 \\ 0 & 1 \end{bmatrix}$$

The measured modal data is assumed to yield

$$\hat{\lambda}_{1,2} = 1 \pm j$$

$$\hat{v}_1 = \begin{bmatrix} 1 \\ -1-.05j \end{bmatrix}, \ \hat{v}_2 = \begin{bmatrix} 1 \\ -1+.05j \end{bmatrix}$$

Using these values, the optimization procedure produces a corrected model defined by (here the mass matrix remains unchanged)

$$D_{new} = \begin{bmatrix} 5.8095 & -2.0246 & -3.9628 & -.1777 \\ -2.0246 & 2.000 & 0 & .0174 \\ -3.9628 & 0 & 6.0 & -2.0287 \\ -.777 & .0174 & -2.0287 & 2.9560 \end{bmatrix}$$

and

$$K_{new} = \begin{bmatrix} 4.4298 & -.9928 & -2.0011 & -1.5036 \\ -.9928 & 2.0 & -1 & .-.0098 \\ -2.0011 & -1 & 5 & -2.0009 \\ -.1.5036 & .0098 & -2.0009 & 3.6639 \end{bmatrix}$$

This corrected analytical model produces eigenvalues and eigenvectors given by

$$\lambda_{1,2} = -1 \pm .9999j$$

$$v_1 = \begin{bmatrix} 1 \\ -.9908 - 0.0868j \\ xxx \\ xxx \end{bmatrix}, \ v_2 = \begin{bmatrix} 1 \\ -.9908 + 0.0868j \\ xxx \\ xxx \end{bmatrix}$$

Where the xxx denotes that these elements are not measured or altered. The eigenvalues and the eigenvectors and hence the modal data of the corrected model is much closer to the "experimentally" obtained data.

This analytical model can be moved closer to the experimental data, if so desired, by iterating. After ten iterations, the corrected model becomes

$$D_{new} = \begin{bmatrix} 5.8095 & -2.0246 & -3.9582 & -.1991 \\ -2.0246 & 2 & .0126 & .0013 \\ -3.9582 & .0126 & 6.0034 & -2.0312 \\ -.1991 & 0013 & -2.0312 & 2.9528 \end{bmatrix}$$

$$K_{new} = \begin{bmatrix} 4.4391 & -1.0016 & -1.994 & -1.5048 \\ -1.0016 & 1.9679 & -1.0037 & .-.0114 \\ -1.9994 & -1.0037 & 5 & -2.0009 \\ -.1.5048 & .0114 & -2.0009 & 3.6639 \end{bmatrix}$$

This corrected analytical model has eigenvalues an eigenvectors containing exactly the measured values given above.

This low order example with fictitious data is intended to illustrate and verify the procedure. Larger order examples and applications of this method to actual measured data are given in Minas [24]. The method presented here provides a systematic method of updating analytical models with experimentally observed data. This proposed method is limited to updating and improving the stiffness and damping matrix. It does not improve the mass matrix.

Disadvantages of this method include the fact that the method does not result in corrected damping and stiffness matrices having the same element structure as the original model. Hence, any physical interpretation of the elements of the original model is not retained. Additional coupling is often introduced by the above procedure. The procedure is iterative and uses an optimization method, both of which are not desirable.

Zimmerman and Widengren [19-20] approach the problem in a non iterative mode by solving directly for symmetric correction factors. This is done by first using the restriction on the measured eigenvectors given by Eq. (49). In Zimmerman's method, this restriction is used first to transform the measured eigenvectors into the best achievable eigenvector, denoted $\hat{v}_{ia}$, by calculating the least squares projection of the measured eigenvectors onto the space defined by Eq. (49). Thus

$$\hat{v}_{ia} = L_i (L_i^* L_i)^{-1} \hat{v}_i \tag{68}$$

where

$$L_i = (M\lambda_i^2 + D\lambda_i + K)^{-1} B_0 \tag{69}$$

Next, the matrix G of Eq. (61) is calculated based on forcing the correction matrices B_0GC_0 and BGC_1 to be symmetric. This leads to the solution of a generalized algebraic Riccati equation. This approach provides an alternative to the iterative optimization approach proposed here. However, it does require the solution of a Riccati type equation which presents other computational considerations.

It should be noted that if the EMA and FEM have eigenvalues that are the same or nearly the same, then the inverse used in Eq. (69) and (49) does not exist. This difficulty is avoided by not correcting for the measured modal data that is already in agreement with the FEM modal data. The user is free to choose which measured modal data to use. Hence, modal data with a low confidence factor need not be used in the correction procedure. These methods also do not guarantee that the corrected model will not have completely different modal data for those modes not corrected for. This results because changes in the matrix G effects all eigenvalues and eigenvectors, not just those used in the algorithm. In this case the procedure must be repeated until all of the modal data of the corrected FEM is satisfactory.

VII. PROBLEM 4 MODE CORRECTION BY POLE PLACEMENT

As mentioned above, the correction of the FEM to accommodate some measured modes often causes other modal data to shift to undesirable values. In addition, many modal tests do not produce reliable measurements of mode shapes. In fact, many EMA techniques do not attempt mode shape measurement. In these cases, it is appropriate to consider using pole placement methods rather than eigenstructure assignment techniques. Such methods do not use eigenvector information. Such a method is developed here.

Consider the state space formulation of the structural dynamics problem as given in Eqs. (4), (5) and (6). In this case a correction to the state matrix A denoted δA, is sought of the form

$$\delta A = \begin{bmatrix} \delta A_{11} & \delta A_{22} \\ 0 & 0 \end{bmatrix} \tag{70}$$

where δA_{11} is an nxn matrix of corrections to $M^{-1}D$ and δA_{22} is an nxn matrix of corrections to $M^{-1}K$. The original analytical model is represented by the state matrix A and the corrected or updated model is given by

$$A_{new} = A + \delta A \tag{71}$$

As before the correction matrix is calculated by determining a control gain matrix G, measurement matrix C and input matrix B such that

$$\dot{z} = Az + Bu = Az + BGCz = (A + BGC)z \tag{72}$$

so that

$$\delta A = BGC \tag{73}$$

In the pole placement problem B, G and C are all real matrices. In this application, however, the matrix C is allowed to be complex valued. Let u_i denote the eigenvector of the matrix A and let v_i denote the eigenvector of the matrix A^T. Both these sets of vectors are complex and they are orthogonal to each other, i.e.,

$$v_j^* u_i = \delta_{ij} \tag{74}$$

Here the * denotes the complex conjugate transpose. The vectors v_j are easily calculated form knowledge of the initial analytical model.

The matrix C of Eq. (73) is formed from the eigenvector of A^T by

$$C = \begin{bmatrix} v_1^* \\ v_2^* \\ \cdot \\ \cdot \\ \cdot \\ v_{2m}^* \end{bmatrix} \tag{75}$$

where m is again the number of measured modes. The matrix B in the correction matrix of Eq. (73) is chosen to be

$$B = \begin{bmatrix} I_n \\ 0 \end{bmatrix} \tag{76}$$

where I_n is the nxn identity matrix. Note that this choice of input matrix B renders the analytical model fully controllable so that any or all of the eigenvalues of the original analytical model can be modified.

In order to calculate the matrix G and hence the correction δA, consider the eigenvalue problem A i.e.,

$$Au_k = \lambda_k u_k \tag{77}$$

Next consider multiplying A_{new} times the analytical eigenvector u_k for all values of the index $k > 2m$ corresponding to unmeasured modes:

$$A_{new}u_k = Au_k + BGCu_k \tag{78}$$

Note that $BGCu_k = 0$ for all $k > 2m$. Hence

$$A_{new}\, u_k = Au_k = \lambda_k u_k \tag{79}$$

or

$$A_{new}u_k = \lambda_k u_k \quad k > 2m \tag{80}$$

and the eigenvectors of the modified or corrected model and the analytical model are the same for all unmeasured modes as desired.

On the other hand, the eigenvalue problem for those indices corresponding to measured modes becomes

$$A_{new}\, u_k = Au_k = BGCu_k$$
$$= \hat{\lambda} u_k + \sum_{i=1}^{n} G_{ik}\, b_k \quad k \le 2m \tag{81}$$

where b_k are vectors made from the columns of the matrix B, G_{ik} are the elements of the matrix G and $\hat{\lambda}_k$, $k < 2m$, are the measured eigenvalues.

Denote the eigenvectors of A_{new} by u_{Nk}, $k \le 2m$. These are not available from measurement nor can they be calculated. However, they can be written as a linear combination of the known eigenvectors of A, i.e.,

$$u_{Nj} = \sum_{k=1}^{2n} Q_{jk}\, u_k \quad j = 1, 2 \ldots 2m \tag{82}$$

where the Q_{jk} are expansion coefficients. Likewise the matrix B can be factored as a linear combination of the eigenvectors of A so that

$$B = Pu \tag{83}$$

where

$$u = [u_1\ u_2 \ldots u_{2n}] \tag{84}$$

and P is defined by

$$P = Bu^{-1} \tag{85}$$

Such a factorization is possible because the set of eigenvectors of A is complete.

Next consider the eigenvalue problem for the measured eigenvalues $\hat{\lambda}_i$, $i = 1, 2 \ldots 2m$. It is desired to have this data satisfy

$$A_{new}\, u_{Nj} = \hat{\lambda}_i u_{Nj} \quad j = 1, 2 \ldots 2m \tag{86}$$

Substitution of this equation into the expansion for the unknown (and unmeasured) eigenvectors given by Eq. (82) and the statement of the model correction formula given by Eq. (81) yields

$$(A + BGC) \sum_{k=1}^{2n} Q_{jk}u_k = \hat{\lambda}_j \sum_{k=1}^{2n} Q_{jk}u_k \ , \ j = 1, 2 \ ... \ 2m \tag{87}$$

Using Eq. (83) this becomes, after some manipulation

$$\sum_{k=1}^{2n} Q_{jk}\lambda_k u_k + \sum_{i=1}^{n} (\sum_{k=1}^{2n} P_{ki}u_k \sum_{l=1}^{2m} G_{il} Q_{jl}) = \hat{\lambda}_j \sum_{k=1}^{2n} Q_{jk}u_k \ , \ j = 1, 2 \ ... \ 2m \tag{88}$$

This vector equation can be re-written after some manipulation, as the 2(nxm) scalar equations

$$(\hat{\lambda}_j - \lambda_k)Q_{jk} - \sum_{i=1}^{n} P_{ki} \sum_{l=1}^{m} G_{il}Q_{jl} = 0 \ , \ \text{for } k = 1, 2 \ ... \ 2m, j = 1, 2 \ ... \ 2n \tag{89}$$

Next rearrange the above set of scalar equations into the matrix equation

$$\{(\hat{\lambda}_j - \lambda_k)\delta_{jk} - \sum_{i=1}^{n} P_{ki} \sum_{l=1}^{m} G_{il}\} \ q_j = 0 \ , \ j = 1, 2 \ ... \ 2m \tag{90}$$

where

$$q_j = [Q_{j1} \ Q_{j2} \ \ Q_{j2n}]^T \tag{91}$$

which is never zero. Since this is not zero, the matrix coefficient contained in the brackets of Eq. (90) must be singular. Therefore its determinant must vanish so that

$$\det \{(\hat{\lambda}_j - \lambda_k)\delta_{jk} - \sum_{i=1}^{n} P_{ki} \sum_{l=1}^{2m} G_{il}\} = 0 \text{ for } j = 1, 2 \ ... \ 2m \tag{92}$$

If B is chosen such that $P_{ij} \neq 0$, then this yields

$$\hat{\lambda}_i - \lambda_i = \sum_{j=1}^{n} P_{ij} \ G_{ji} \tag{93}$$

which provides a means of calculating the correction matrix by solving for G_{ji}.

It is desirable that the analytical model be changed as little as possible. Hence a cost function penalizing large values of G_{ij} is constructed that is subject to the equality constraint given by Eq. (93). This suggests the set of cost functions S_i defined by

$$S_i = \sum_{j=1}^{n} G_{ji}^2 + L_i (\hat{\lambda}_i - \lambda_i - \sum_{j=1}^{n} P_{ij} \ G_{ji}) \tag{94}$$

where L_i are Lagrange multipliers used to attached the equality constraint. Minimizing the function S_i yields that

$$G_{ji} = \frac{(\hat{\lambda}_i - \lambda_i) P_{ij}}{\sum\limits_{i=1}^{n} P_{ij}^2} \tag{95}$$

This solution for the matrix G is then combined with B and C to yield the correction matrix δA from Eq. (73) and hence the corrected analytical model given by

$$A_{new} = A + \delta A \tag{96}$$

The new analytical model has the following desired properties:

1)   The matrix A_{new} has the measured eigenvalues (modes) as its eigenvalues (i.e., $-\zeta_1 \omega_i \pm \omega_i \sqrt{1-\zeta^2}$ j).

2)   The eigenvalues of the analytical model that are not measured remain unchanged.

3)   The matrix δA represents the "smallest" change from the original analytical model which reproduces the measured eigenvalues.

Next consider a simple numerical example consisting of a four degree of freedom system. No experiment is performed here but rather the analytical system is given then perturbed to simulate the correct model from which "measurements" are obtained. The FEM which is to be corrected by the pole placement method suggested above is taken to be the perturbed model.

The true or correct model is given by Eq. (1) with coefficient matrices

$$M = \text{diag} (1.0, 2.5, 2.0, 1.5) \tag{97}$$

$$D = \begin{bmatrix} 3.5 & -.25 & 0 & 0 \\ -.25 & .30 & -.05 & 0 \\ 0 & -.05 & .20 & -.15 \\ 0 & 0 & -.15 & .15 \end{bmatrix}, \quad K = \begin{bmatrix} 3.0 & -2.0 & 0 & 0 \\ -2.0 & 3.5 & -1.5 & 0 \\ 0 & -1.5 & 2.5 & -1.0 \\ 0 & 0 & -1.0 & 1.0 \end{bmatrix}$$

This model yields "measured" natural frequencies and modal damping ratios of:

$$\begin{aligned} \omega_1 &= .2924 , & \zeta_1 &= .0146 \\ \omega_2 &= .8288 , & \zeta_2 &= .0433 \\ \omega_3 &= 1.3410 , & \zeta_3 &= .0609 \\ \omega_4 &= 1.9340 , & \zeta_4 &= .1102 \end{aligned} \tag{98}$$

This data yields measured eigenvalues of

$$\hat{\lambda}_{1,2} = -.0043 \pm .2942j \quad , \quad \hat{\lambda}_{3,4} = -.0359 \pm .8280j$$

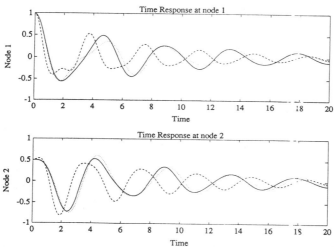

Figure 1. The Response of nodes 1 and 2 for each of the three models. The solid line represents the simulated experimental response, the dashed line represents the response of the uncorrected FEM and the dotted line represents the response of the FEM corrected by the pole placement procedures suggested here.

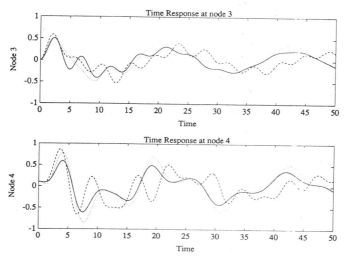

Figure 2. The Response of nodes 3 and 4 for each of the three models. The solid line represents the simulated experimental response, the dashed line represents the response of the uncorrected FEM and the dotted line represents the response of the FEM corrected by the pole placement procedures suggested here.

$$\hat{\lambda}_{5,6} = -.0817 \pm 1.3385j \ , \quad \hat{\lambda}_{7,8} = -.2131 \pm 1.9222j \qquad (99)$$

It is assumed that the analytical FEM is a perturbation of the correct model given by Eq. (1) with the following coefficients

$$M = \text{diag } (1.25, 2.0, 1.5, 1.0)$$

$$K = \begin{bmatrix} 2.0 & -1.5 & 0 & 0 \\ -1.5 & 3.5 & -2.0 & 0 \\ 0 & -2.0 & 3.25 & -1.25 \\ 0 & 0 & -1.25 & 1.25 \end{bmatrix} \qquad (100)$$

Note that every element of the FEM deviates at least 30% from the corresponding element of the correct model given by Eq. (97). In addition a damping matrix must be contrived since the FEM does not normally provide such information.

A proportional damping matrix is constructed from the FEM by using the measured modal data given in (98) by using the simple formula

$$D = MU[2 \text{ diag } (\zeta_1\omega_1, \zeta_2\omega_2, \zeta_3\omega_3, \zeta_4\omega_4)]U^{-1} \qquad (101)$$

where U is the matrix of eigenvectors of the matrix $M^{-1}K$. This procedure for constructing a damping matrix is discussed in Inman [29] and examples are given in Minas [24]. Using the data proposed here, the FEM matrix is

$$D = \begin{bmatrix} .1057 & -.0856 & .0904 & -.0563 \\ -.0856 & .3471 & -.2762 & .0742 \\ .0904 & -.2762 & .3348 & -.0942 \\ -.0563 & .0742 & -.0942 & .0948 \end{bmatrix} \qquad (102)$$

Alternately the methods of Section V could be used to obtain a more precise estimate of the damping matrix D. However, the methods of Section V are computational expensive compared to Eq. (101) and all that is needed in this example is a starting point, as the model correction method will continue to adjust the value of the matrix D. Note that the estimate is proportional. This also represents a typical EMA/FEM incompatibility.

The eigenvalues of the FEM can now be computed using Eqs. (100) and (102) to be

$$\lambda_i = \{-.2085 \pm 1.8810j, -.0214 \pm 1.4717j, -.0159 \pm 2598j,$$
$$- 0422 \pm .9743j\} \qquad (103)$$

Note that the FEM eigenvalues do not agree with the measured eigenvalues, $\hat{\lambda}_j$, given in Eq. (99). Thus a correction procedure is required to change the FEM to match the EMA data. Using the correction method of this chapter yields the state matrix A_{new}. The eigenvalues of A_{new}, λ_i^{new}, calculated for comparison with the measured modal data are

$$\lambda_{1,2}^{new} = .0040 \pm .2926j \qquad \lambda_{3,4}^{new} = -.0360 \pm .8251j$$

$$\lambda_{5,6}^{new} = -.0818 \pm 1.3306j \qquad \lambda_{7,8}^{new} = -.2132 \pm 1.9215j \tag{104}$$

These compare extremely well with the measured modal data, $\hat{\lambda}_i$, given in Eq. (99).

The purpose of having an analytical model however, is to be able to predict the response of the test structure to other disturbances. In EMA such analytical models are often called *response models*. Plots of the response of each of the four degrees of freedom are made for an arbitrary excitation, to illustrate this ability (and inability) to predict a response. In order to make the comparison the response of 3 systems for each degree of freedom are illustrated. The 3 responses are:
- the correct model, or simulated experimental structure which is indicated by a solid line and consists of the response generated by the system of Eq. (97).
- the FEM model which is indicated by a dashed line and consists of the response generated by the system of Eqs. (100) and (102).
- the improved or corrected model which is indicated by the dotted line and consists of the response generated by Eq. (96) as calculated using Eq. (95) and the measured data of Eq.(99).

Each of these responses is calculated by numerically simulating the transient response to an initial displacement of $x(0) = [1 \quad .5 \quad 0 \quad .1]^T$ and $\dot{x}(0) = 0$. The displacement responses are plotted in Figure 1-2, for each node as indicated.

It is clear that the corrected model is much closer to the actual (correct) response than the uncorrected FEM response is. Careful examination indicates that the frequency (and damping) content of both the simulated experimental structure's response and the corrected model's (A^{new}) response are the same while that of the FEM is response different. The observation that the corrected model does not exactly correspond with the response of the actual or true model should not be too surprising because no eigenvector or mode shape information was used in constructing the improved model. Hence, a perfect reproduction of the response should not be expected for this approach. This case represents the best that can be done in EMA model improvement without measuring mode shapes.

In terms of producing a corrected model which faithfully reproduces the structures measured natural frequencies, this proposed method provides a substantial improvement over existing methods. The method presented in this section involves only matrix multiplications to correct the analytical model whereas other methods require more complicated computational schemes.

The method also offers relatively accurate corrections as can be seen by comparing it to the results presented by Kammer [12] for a six degree of freedom system. Kammer's frequencies are presented in Table VII-1 for a variety of noise levels

A Comparison of Natural Frequencies of Kammer's Example for the Method Prepared Here for Various Levels of Noise

FEM Frequency	0% Noise		10% Noise		20% Noise		Measured Frequency
	Kammer's Method	This Method	Kammer's Method	This Method	Kammer's Method	This Method	
5.216	5.258	5.257	5.173	5.173	5.456	5.468	5.258
17.204	17.626	17.623	17.629	17.625	17.589	17.589	17.626
23.514	23.171	23.168	23.154	23.150	23.195	23.192	23.171
25.495	26.060	-	24.764	-	24.576	-	26.060
36.798	40.169	-	31.159	-	31.159	-	40.169
41.256	47.464	-	39.920	-	63.858	-	47.464

Table VII-1

in the data compared with the pole placement method suggested here, the original model's frequencies and the measured frequencies. The method presented here used only the first 3 measured natural frequencies whereas the Kammer method used both the first 3 natural frequencies and the first 3 mode shapes. The last 3 natural frequencies are not listed for the method introduced here because they are identical to the analytical frequencies. Recall that this method changes only those frequencies that are used from the measured set of data, and leaves the other frequencies unchanged.

VIII. SUMMARY AND DISCUSSION

For problems concerned with correcting analytical FEM models to fit EMA data have been presented and discussed. The methods presented represent an improvement over methods usually employed in model correction by connecting the problem to the discipline of eigenstructure assignment and pole placement common to control theory. The methods presented provide a choice of approaches to take depending on what kind of experimental data is available and what type of analytical model is desired.

If only a response model is required and mode shapes cannot be measured or are not measured very accurately (the common case) then the method of Section VII is appropriate. If damping is not important the method of Section IV may be useful. On the other hand if the mass and stiffness models are acceptable the methods of Section V may be used to construct a damping matrix.

If both mode shapes as well as natural frequencies and modal damping ratios are measured and available then the methods of Section VI are useful. This method also produces corrected models with symmetric positive definite matrix coefficients.

All the methods presented here recognize that the measure modal data is almost always incomplete. That is, the methods accommodate the fact that the measured modal data consists of fewer natural frequencies and damping ratios than the analytical model is likely to have. In addition the methods recognize that the measured mode shapes seldom have as many entries as their analytical counter parts.

The methods presented here consist mainly of adapting known and previously developed pole placement and eigenstructure assignment methods to the model correction problem. The model correction problem addressed here consists of a method of adjusting the FEM model of a structure so that its modal data (i.e., mode shapes, natural frequencies and damping ratios) agrees with the modal data obtained from EMA test results. The approaches suggested here provide a systematic solution to the model correction problem which has previously been addressed in a less systematic fashion.

Acknowledgements

The author gratefully acknowledges the support of NSF Grant No. MSM-8351807 and AFOSR Grant # AFOSR-F49620-86C-0011. The original motivation for this work and initial support was provided by contracts with Firestone Tire and Rubber Co. and Harrison Radiator Division of General Motors Corporation who provided matching funds for the NSF award.

References

1.   T.K. Hasselman, "A Method of Constructing a Full Modal Damping Matrix from Experimental Measurements," AIAA Journal, Vol. 10, 526-527   (1972).

2.   T.K. Hasselman, "Model Coupling in Lightly Damped Structures," AIAA
Journal, Vol. 14, 1627-1628 (1976).

3.   P. Caravini and W.T. Thomson, "Identification of Damping Coefficients in Multidimensional Linear Systems," ASME Journal of Applied Mechanics, Vol. 41, 379-382 (1974).

4.   J. Beliveau, "Identification of Viscous Damping in Structures from Modal Information," ASME Journal of Applied Mechanics, Vol. 43, 335-338 (1976).

5.   S. Hanagud, M. Meyyappa, Y.P. Cheng and J.I. Craig, "Identification of Structural Dynamics Systems with Nonproportional Damping," Proceedings of the 25th SDM Conference, Palm Springs, California, May, 283-291 (1984).

6.   D.J. Inman and W.L. Hendrickson, "Identification of a Damping Matrix from Experimental Measurements," Proceedings of the 5th VPI&SU Symposium on Dynamics and Control of Structures, 19-26 (1985).

7. D.J. Inman and S.K.Jha, "Identification of a Damping Matrix for Tires," Proceedings of the 4th International Modal Analysis Conference, Vol. II, 1078-1080 (1986).

8. W.P. Rodden, "A Method for Deriving Structural Influence Coefficients from Ground Vibration Tests," AIAA Journal Vol. 5, No. 8, 991-1000 (1967).

9. A. Berman and W.G. Flannely, "Theory of Incomplete Models of Dynamic Structures," AIAA Journal, Vol. 9, Aug., 1481-1487 (1971).

10. M. Baruch, M. and I.Y. Bar Itzhack "Optimal Weighted Orthogonalization of Measured Modes," AIAA Journal, Vol. 16, April 346-351 (1978).

11. A.Kabe, "Stiffness Matrix Adjustment Using Mode Data," AIAA Journal, Vol. 23, No. 9, Sept., 1413-1436 (1985).

12. D.C. Kammer, "An Optimum Approximation for Residual Stiffness in Linear System Identification," Proceedings of the 28th SDM Conference, Monterey, California, 277-287 (1987).

13. A. Berman and E.Y. Nagy, "Improvement of a Large Analytical Model Using Test Data," AIAA Journal, Vol. 21, Aug., 1168-1173 (1983).

14. W. Heylen, "Optimization of Model Matrices by Means of Experimentally Obtained Dynamic Data," Proceedings of the 1st International Modal   Analysis Conference, Orlando, Florida, Nov., 32-38 (1982).

15. S.R. Ibrahim, "Dynamic Modeling of Structures from Measured Complex Modes," AIAA Journal, Vol. 21, No. 6, 898-891 (1983).

16. J. Fuh, S. Chen and A. Berman, "System Identification of Analytical Models of Damped Structures," Proceedings of the 25th SDM Conference, Palm Springs, California, May, 112-116 (1984).

17. C. Minas and D.J. Inman, "Correcting Finite Element Models with Measured Modal Data Using Eigenstructure Assignment Methods," Proceedings of the 4th International Modal Analysis Conference, 583-587 (1987).

18. C. Minas and D.J. Inman, "Correcting Finite Element Models with Measured Modal Results Using Eigenstructure Assignment Methods,

Proceedings of the 6th International Modal Analysis Conference, Orlando, Florida, Feb. 583-587 (1988).

19. D.C. Zimmerman and M. Widengren, "Model Correction Using a Symmetric Eigenstructure Assignment Technique," Proceedings of the 30th Structures, Dynamics and Material Conference 1947-1954 (1989) also to appear in the AIAA Journal.

20. D.C. Zimmerman and M. Widengren, "Equivalence Relations for Model Correction of Non Proportionally Damped Linear Systems," Proceedings of the 7th VPI&SU/AIAA Symposium on the Dynamics and Control of Large Structures (1990).

21. C. Minas and D.J. Inman, "Matching Finite Element Models to Modal Data,"ASME Journal of Vibration and Acoustics (accepted for publication, paper No. 88-487) (1990).

22. C. Minas and D.J. Inman, "Model Improvement by Pole Placement Methods," Vibration Analysis-Techniques and Applications, T.S. Sankar, ed, ASME, pp. 179-185 (1989).

23. C. Minas, "Correcting Finite Element Models with Measured Modal Results Using Eigenstructure Assignment Methods," M.S. Thesis State University of New York at Buffalo (1988).

24. C. Minas, "Modeling and Active Control of Flexible Structures," Ph.D. dissertation, State University of New York at Buffalo (1989).

25. S. Srinathkumar, "Eigenvalue/eigenvector Assignment Using Output Feedback," IEEE Transactions on Automatic Control, AC-23, 79-81 (1978).

26. A.N. Andry, Jr., E.Y. Shapiro and J.C. Chung, "Eigenstructure Assignment for Linear Systems," IEEE Transactions on Aerospace and Electronic Systems, Vol. AES-19, No. 5, 711-729 (1983).

27. B. Porter and R. Crossley, Modal Control Theory and Applications, Taylor and Francis, London, England, (1972).

28. I.H. Shames and D.L. Dym, Energy and Finite Element Methods in Structural Mechanics, Hemisphere Publishing Corp. Chapter 16, (1985).

29. D.J. Inman, Vibration with Control, Measurement and Stability, Prentice-Hall, Inc.Englewood Cliffs, NJ, (1989).

30. D.J. Ewins, <u>Modal Testing: Theory and Practice</u>, Research Studies Press Ltd., England, (1986).

31. C.T. Chen, <u>Linear System Theory and Design</u>, CBS College Publishing, Holt, Rinehart and Winston, (1984).

32. G.H. Goloub and C.F. Van Loan, "Matrix Computations," John Hopkins Press (1985).

TECHNIQUES IN INDUSTRIAL CHEMICAL SYSTEMS OPTIMIZATION

DAVID M. HIMMELBLAU

The University of Texas at Austin
Austin, Texas 78712

I. INTRODUCTION

The oil, chemical, and petrochemical industries are presently operating under conditions quite different from those of the last decade. Changes have occurred because of several factors. The most important influences have been concern for energy conservation, environmental constraints, and the evolution of computer capabilities. The motivation for optimization comes from the fact that even a small increase in the operating efficiency of a process plant can be transmitted into a significant change in profits. In operations, one distinguishing feature of process plants is large daily throughputs so that engineers are always seeking fractional improvements. If a plant can be modeled in a reasonably representative fashion, i.e. in such a way that inputs to the plant give outputs that can be imitated by simulation programs, then a better set of inputs can be devised using the model leading to a better set of outputs from the plant. The question is: how can the simulation program determine a 'better set of inputs'? That is where optimization programs come in. Just as there are well-established programs for simulation, there are equally well-tested codes for optimization . What must be done is to mesh these mathematical optimization programs with the simulation codes effectively.

In the broadest sense, the general optimization problem is to find an extremum of an objective function subject to equality and/or inequality constraints. The objective function and constraints may be linear and/or nonlinear. Usually problems involving integer valued variables and those with constraints comprised of differential equations are treated as two separate categories or special cases of the general nonlinear programming problem. We will let the continuous functions $f(\mathbf{x})$ denote the objective function, $h_1(\mathbf{x}), \ldots, h_m(\mathbf{x})$ denote the equality constraints, and $g_{m+1}(\mathbf{x}), \ldots, g_p(\mathbf{x})$ denote the inequality constraints, where $\mathbf{x} = [x_1, \ldots, x_n]^T$ is a column vector of variables $x_1, \ldots, x_n$, in n-dimensional Euclidean space. The variables $x_1, x_2, \ldots, x_n$ may be design parameters, controller adjustments, instrument readings, etc.

The nonlinear programming problem can be formally stated as

$$\text{Minimize: } f(\mathbf{x}) \quad \mathbf{x} \varepsilon E^n \tag{1}$$

subject to m linear and/or nonlinear equality constraints

$$h_j(\mathbf{x}) = 0 \quad j = 1, \ldots, m \tag{2}$$

and $(p - m)$ linear and/or nonlinear inequality constraints

$$g_j(\mathbf{x}) \geq 0 \quad j = m + 1, \ldots, p \tag{3}$$

Although in some codes the linear algebraic equality constraints can be explicitly solved for selected variables and those variables eliminated from the problem, reducing the problem to one of smaller dimensions, often the equality constraints can be solved only implicitly and must be retained.

Each independent equality constraint absorbs one degree of freedom in the process model and results in one dependent variable being generated. It is usually assumed that the analyst prepares the process-model statements carefully enough so that the equalities are independent, for if by omission or error he or she includes two redundant or otherwise dependent equations, then the apparent number of degrees of freedom will be different from the actual number. The number of residual degrees of freedom is an important concept in any type of optimization subject to equality constraints, because if the number of variables equals the number of independent equality constraints, no optimization need take place–the values of all the variables can be determined directly from the simultaneous solution of the system of equality constraints, $h_j(\mathbf{x}) = 0, j = 1, \ldots, m$. If the number of variables whose values are unknown exceeds the number of independent equality constraints, m, an optimal solution for a process is sought by adjusting the $(n - m)$ decision variable until the objective function attains its optimum value. If the number of independent equality constraints exceeds the number of unknown variables, optimization is also required, except that in this case the objective function must consist of some different type of criterion such as least squares.

II. FEATURES OF SOLVING OPTIMIZATION PROBLEMS IN THE CHEMICAL INDUSTRY

Fig. 1 illustrates the relation between the actual process, the analysis of the process, and its mathematical representation for a simplified smeltor. The following steps outline the general technique used for analysis and solution of optimization problems in the chemical process industries.

A. DEFINITION OF THE OBJECTIVE FUNCTION (COST MODEL, REVENUE FUNCTION, CRITERION)

The first step in any analysis is to define a suitable objective or revenue function for the problem. We would like to find the best answer out of all possible answers, but sometimes it is rather difficult to quantify a verbal statement or concept of the problem into a meaningful objective function. For example, it is clear that a reasonable goal in a plant is to maximize the energy efficiency of the plant. If coal were used as a raw material, then a superficial analysis might indicate the most efficient way to product electricity for power would be to burn the coal in a boiler. However, it is clear that this procedure may not be entirely feasible since the coal may have a high sulfur content. Therefore one could adopt a second strategy, which is to minimize the amount of pollution occurring in the power plant. This goal alone, of course, is not reasonable either since this arrangement could result in zero electricity production. Thus noncompatible criteria exist, some of which can be accommodated by constraints, and some of which may have to be treated by multiobjective function optimization techniques (which we do not discuss here).

A more poignant example can be pointed out with respect to the internal combustion engine. A number of years ago it was suggested that engine modifications be introduced in the air-fuel ratio to minimize the amount of nitrogen oxide produced in internal combustion. However, the operating conditions to do this led to an increase in pollution due to hydrocarbons emissions. As a result, catalytic convertors have had to be added to engines. We are now aware of the fact that the catalytic convertor, while minimizing hydrocarbon pollution, yields sulfuric acid in the atmosphere. From these two examples, it should be clear that not only can the formulation of objectives be a very difficult process, but we must write them in mathematical terms, a nontrivial task.

What is required in Step 1 is to express the objective function, which may be cost or profit, or energy use and material use, or combinations thereof in terms of the measurable inputs, states, and outputs of the system properly meshed with costs (if money is involved rather than efficiency or yield or throughout). Often the formulation of the objective function requires the association of various variables in a process with internal costs and prices. As an example, an important process in a refinery is a fluid catalytic cracker, which takes a heavy crude oil and breaks the large oil molecules into smaller ones, such as gasoline and aviation fuel. There are many operating variables which affect the product distribution from the cracker, e.g., feed composition and flow rates, temperature, pressure, steam consumption, catalytic rate, etc. How to associate valid specific costs and prices with these variables many of which are not salable products is beyond our scope here, but certainly influences the optimal solution.

Another facet of formulating objective functions is to determine the appropriate interest or discount factor for capital expenditures, operating costs, and future income stream. The amount of money and the point in time at which it is expended are very critical for any cost or revenue analysis. Money expended at the beginning of a project must be weighted differently than money expended five years later.

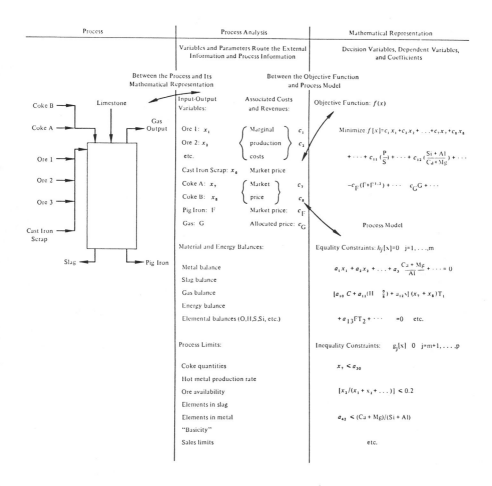

Fig. 1 Relation between the process and its mathematical representation.

B. DEVELOPMENT OF PROCESS MODELS, VARIABLES, AND THE
   PROCESS CONSTRAINTS

By model of the process we mean the equations that represent the input-output relations for the process or the equivalent computer codes. Such an input-output model is often based upon the physics and chemistry of the process (or approximations thereof), or based on empirical relations whose coefficients are estimated from the data available for the process. Usually it is necessary to take the physically based or empirical model and compress it or simplify it into functional forms that can be readily handled during the optimization procedure.

Two general formulations of process models exist. One formulation consists of equations and inequalities [1,2]; the other is comprised of computer subroutines [3,4]. The latter is by far the more common in the chemical industry, but less flexible because although an output can be obtained from an input to a subroutine, the reverse is not feasible, and furthermore codes cannot be analytically differentiated. Instead, the inputs must be perturbed to get finite difference substitutes for derivatives. However, the subroutines are portable and can provide nonlinear models of different degrees of complexity for essentially the same processing functions.

Thus, the first formulation, that using so called equation based models, literally consist of equality constraints (equations) and inequality constraints. For a single process unit, development of a model is reasonably straight forward because much is known about such equipment. However, for a larger system, which may involve the combination of many process units, such as a complete plant, the identification of equality constraints for a process model is much more difficult. In Fig. 2 we see a process flow diagram for a small scale chemical plant. Material and energy balances provide some of the relationships between the system variables,. If the material and energy balances were written for each individual unit, and the interconnection equations listed, we would have a large set of equality constraints which are interrelated. It would then be helpful to combine and eliminate superfluous variables from these equations in order to make the problem more mathematically tractable. With 5,000 equations, some nonlinear, in a plant model, reduction of dimensionality becomes essential.

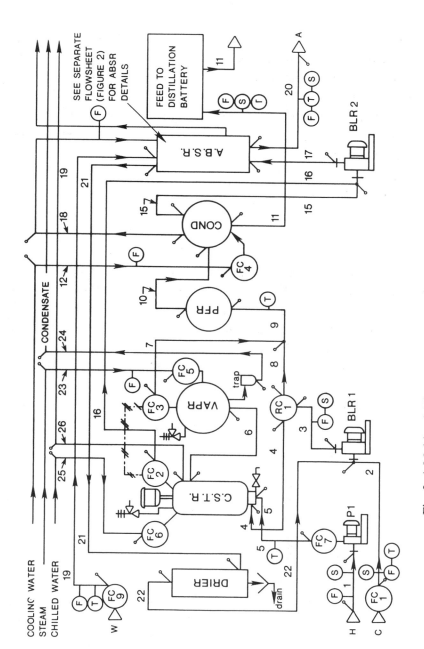

Fig. 2 A Multi-unit Plant

One comment should be made about the variables in a plant model. From the viewpoint of carrying out an optimization, three types of variables occur. First are the controllable variables which may or may not become the independent variables during the optimization calculations.

Second are the uncontrollable external inputs. These might include, for example environmental effects, such as the weather, on heat transfer and the temperature of cooling water; feedstock composition; or limitations on the amount of material that can be made at a given plant. Third would be the state or dependent variables. In a large plant with 5,000 equations and variables, only a few, say 10 or 20, would be in the controllable set, so that the degrees of freedom are really quite small in practice.

Inequality constraints are the second type of constraint mentioned above. Almost all variables have upper and lower bounds. In addition, we often encounter temperature or pressure limitations in design of vessels, governmental restrictions on transportation and pollution, quality specifications set by the customer, and the effects of competition on production rates and prices.

If the process is represented by a collection of modules each containing code equivalent to specific equipment or tasks, and inequalities related to each subsystem or piece of equipment, then the modules must be collected together and an executive program coded so that any module may be used in conjunction with the others. Examine Fig. 3. Each module contains the equipment

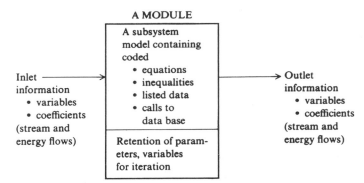

Fig. 3   A typical process module showing the necessary interconnections of formation. [Taken from T. F. Edgar and D. M. Himmelblau, "Optimization of Chemical Processes", McGraw-Hill Book Co., New York, 1988 by permission]

sizes, the material and energy balance relations, the component flowrates, and the temperatures, pressures, and phase conditions of each stream that enters and leaves the physical equipment represented by the module. Values of certain of theses parameters and variables determine the capital and operating costs for the units. Of course, the interconnections set up for the modules must be such that information can be transferred from module to module concerning the streams, compositions, flow rates, coefficients, and so on. In this way the modules comprise a set of portable building blocks that can be arranged in general ways to represent any process. An executive routine calls the modules in the proper order, transmits information from a library of calculational subroutines, and picks out information on physical properties from an associated data base. Both sequential and simultaneous calculational sequences have been proposed for solving optimization problems involving modules as well as for the equation-oriented approach (and intermediate mixtures of the two are possible as well). Either the program and/or the user must select the decision variables for recycle and provide estimates of certain stream values to make sure that convergence of the calculations occurs, especially in a process with many recycle streams. Reviews by Evans [5] and Rosen [6] point out many of the problems and practices pertaining to optimization using process models that we do not have the space to discuss here.

C.   APPLY A SUITABLE OPTIMIZATION CODE MESHED WITH THE PROCESS MODEL

Once the plant model is formulated either as equations or combinations of modules, the next step is the computation of the optimum. Quite a few techniques exist to obtain the optimal solution for a problem. We will describe several methods in detail later on. In general, the solution of most optimization problems involves the use of a digital computer to obtain numerical answers. It is fair to state that over the past 15 years, substantial progress has been made in developing efficient and robust digital methods for optimization calculations. Much is known about which methods are most successful, although comparisons of candidate methods often are of an ad hoc nature based on test cases of simple problems. Virtually all numerical optimization methods involve iteration, and the effectiveness of a given technique often depends on a good first guess as to the values of the variables at the optimal solution.

How can suitable initial guesses for the independent variables be obtained? Because the problem contains nonlinear functions, more than one extremum may exist, a feature absent from linear analysis. Consequently, if the initial guesses for the variables are too far away from the extremum sought, the optimization may not terminate at the global or desired extremum, but at some other extremum. Often approximate optimal values of the independent variables will be known from earlier studies or from physical reasoning. The ultimate resort is to try several starting vectors in the feasible range and ascertain whether or not they all yield the same value of the criterion at the extremum, but there are hazards in this approach, as can be seen from the comments made in connection with the difficulties in forming a suitable model. If all of the functions can be differentiated, homotopy procedures exist to get all of the solutions of the resulting set of nonlinear equations, but these have not yet been meshed with optimization codes [7,8].

How can the numerical computational errors be reduced during the solution? Errors resulting from truncating functions reduce the effectiveness of many algorithms. Stability is concerned with whether the intermediate solutions of approximating nonlinear programming problems converge in the limit to the solution of the original problem. Round off error also can play a troublesome role in optimization, particularly when derivatives are approximated by difference schemes.

D.   CHECK ANSWERS AND EXAMINE THE SENSITIVITY OF THE RESULTS

The last step involves checking the candidate solution to determine that it is indeed optimal. In some problems you can check that the sufficient conditions for an optimum are satisfied. More often an optimal solution may exist, yet you cannot demonstrate analytically that the sufficient conditions for optimization are satisfied. All you can do is show by repetitive numerical calculations that the value of the objective function is superior to all known alternatives.

A second consideration is how sensitive is the optimum to changes in parameters in the problem statement. A sensitivity analysis for the objective function value is always important because the value of the objective function may be quite insensitive to certain variables or the values of the constraints may be quite sensitive, so that in either case a clear cut extremum cannot be located. A local extremum which is sensitive to small variations in some of the variables may be less desirable than one which has a poorer value for the objective function but is less sensitive. In any case, an analyst wants to know something about the region immediately surrounding the extremum. The solution will seldom be exact, and he or she would like to know the effect of the approximation involved on the conclusions inferred from the solution to the problem. In addition, you often want to know what the effect would be of tightening or loosening the constraints by small amounts. It may be possible to compute the eigenvalues and eigenvectors of the Hessian matrix (matrix of the second partial derivatives of the objective function with respect to the variables) of the objective function. Then the eigenvectors can be interpreted in terms of the principal axes of the objective function, while the eigenvalues give a measure of its curvature along these directions, so that the effect of various perturbations of the extremal solution can be evaluated. In the case of constrained problems, this analysis is carried out in the subspace parallel to the tangent planes of the set of active constraints. If the Lagrange multipliers are computed, they provide an estimate of the effect of changes in the constraint conditions. However, such an analytical treatment is very difficult and time consuming to carry out for large scale processes.

E.   DIFFICULTIES IN FORMULATING THE OBJECTIVE FUNCTION AND CONSTRAINTS

Certain features of the functions involved in the objective function and constraints can lead to difficulties in applying optimization codes to solve optimization problems. Some examples are as follows:

(1)   The criterion to be optimized or one or more of the constraints may become *unbounded* in the range of the search for the extremum, or the partial derivatives of the functions in the model may become unbounded. Models with polynomials in the denominator are particularly subject to this hazard, as for example

$$y = \frac{b_0 + b_1 x_1}{b_2 x_1 + b_3 x_2}$$

in which the function and the first partial derivation of y with respect to x_1, become unbounded when $b_2 x_1 = -b_3 x_2$. The way to overcome this difficulty is to suitable restrict the range of the variables by adding constraints to the problem or to reformulating the mathematical model.

(2) There may be poor *scaling* among the variables and constraints. Scaling difficulties can occur, for instance, when one of the terms in the criterion is of a much different order of magnitude than another in view of the significant figures in each term. Then the criterion is insensitive to changes in the values of the variables in the small term. For example, the value of an objective function

$$y = 100x_1^2 - 0.010x_2^2$$

would be unaffected by changes in x_2 unless x_2, because of its physical units, was of much greater magnitude than x_1. If x_2 were of the same magnitude as x_1, one or both variables could be multiplied by scaling factors which converted the two terms on the right-hand side of the equation to roughly equal magnitude. Let

$$\tilde{x}_1 = 10x_1 \qquad\qquad x_1^2 = 10^{-2}\tilde{x}_1^2$$
$$\tilde{x}_2 = 10^{-1}x_2 \qquad\qquad x_2^2 = 10^2\tilde{x}_2^2$$

Then the terms in the objective function become of the same order of magnitude. After the extremum was found for

$$y = \tilde{x}_1^2 - \tilde{x}_2^2$$

the values of x_1 and x_2 could be determined from the values of $\tilde{x}_1$ and $\tilde{x}_2$. Of course, it is not always so easy to rescale the functions in a mathematical model as just illustrated. Scaling of the constraints is equally important because the optimization codes essentially try to drive the right hand sides of the equations to zero--with some tolerance. If because of the scale of the variables in the terms in one equation, its deviation from zero is several orders of magnitude greater than that in another equation, the former equation will overshadow the latter in influencing the optimal solution. Consequently, it is essential to scale the equality constraints to yield deviations from zero of about the same order of magnitude. The same comment applies to the active $(g_j(x) = 0)$ inequality constraints.

(3) There may be *interaction* among the variables in a poorly designed mathematical model. Parameter interaction can be illustrated by examining an extremely simple criterion in which two variables are multiplied by each other:

$$y = 2x_1x_2 + 10$$

The individual values of x_1 and x_2 can range over any series of values for a given value of the product x_1x_2. Scaling is more difficult if interactions exists. A more subtle example, but one just as vulnerable to interaction among the variables, involves a model such as

$$y = x_1^2 + 2x_1 x_2 + x_2^2 + 2$$
$$= (x_2 + x_2)^2 + 2$$

After the transformation $x_1 + x_2 = \tilde{x}_1$ is made, we find

$$y = \tilde{x}_1^2 + 2$$

Observe that only one variable is left, $\tilde{x}_1$, that need to be varied to find the extremum of y.

(4) Another feature of chemical plant models that requires some attention is: how can the stochastic (random) nature of real variables be handled? Can we ignore the very real possibility that the coefficients and variables

in the mathematical model may be random variables with substantial uncertainly? We have to bypass considerations of uncertainty here because of lack of space, but the existence of uncertainty often causes people to wonder whether precise optimization is warranted.

(5) Process measurements are often unavailable for significant periods of time because of equipment problems, sensor degradation, or other factors. If the optimization loop is disabled every time a single input is off line, then on line optimization will not be feasible. What is worse is the possibility that the optimizer uses the invalid data as a basis and calculates a supposedly valid answer that is actually infeasible or definitely suboptimal.

To sum up, like all mathematical tools, nonlinear programming techniques cannot be blindly applied to a given problem without some forethought. Like all artisans' tools, the use of nonlinear programming requires some skill on the part of the user. There is no avoidance of the requirements that the mathematical model be carefully designed, and that appropriate numerical procedures be employed.

III. TECHNIQUES OF OPTIMIZATION FOR CHEMICAL PLANTS

In optimization of a chemical process, using equation based process simulators, for design (synthesis) or operations, you usually must handle a large set (perhaps 5,000 to 50,000) of equality constraints, many of which are nonlinear. The model representation may include an almost equal number of inequality constraints which, with the use of slack variables, can be converted to equality constraints with bounds on the slack variables. Three main generic categories of non-linear programming techniques have been applied to solving such large scale steady state problems (when they can be solved):

1. Successive (sequential) linearization (iterative LP, gradient projection, generalized reduced gradient, etc.)
2. Successive (sequential) quadratic programming
3. Random search.

No codes will solve all problems, and some codes will solve only special structures. Programs for both parallel and sequential computations exist, but as of this date almost all commercial codes are implemented on sequential machines. Table I lists a number of commercial codes used for plant optimization. We consider here the two main nonlinear programming procedures that are used in practical, namely (1) the generalized reduced gradient method [9,10,11] and (2) successive

Table I Process Simulators That Include Optimization

Name	Ref.	Source
ASPENPLUS	[1]	Aspen Technology Corp., Cambridge, Massachusetts 02139: original ASPEN from Nat. Tech. Info. Service, Springfield, Virginia
FLOWTRAN	[2]	Monsanto Co., St. Louis, Missouri
DESIGN/2000	[3]	Chem Share, Houston, Texas 77251-1885
PROCESS	[4]	Simulation Sciences, Fullerton, California 92633
SPEEDUP	[5]	Prosys Technology, Sheraton House, Castle Park, Cambridge, England

1. L. B. Evans, et al., "ASPEN: An Advanced System for Process Engineering," *Comput. Chem. Eng., 3*, 319 (1979).

2. E. M. Rose, and A. C. Pauls, "Computer Aided Chemical Process Design: The FLOWTRAN System," *Comp. & Chem. Eng., 1*, 11 (1977).

3. Chem. Share Corp., "Guide to Solving Process Engineering Problems by Simulation (Users Manual)", Chem. Share Corp., Houston, Texas, 1979

4. N. F. Brannock, V. S. Verneuil, and Y. L. Wang, "PROCESS Simulation Program--A Comprehensive Flowsheeting Tool for Chemical Engineers," *Comput. & Chem. Eng., 3*, 329 (1979).

5. J. D. Perkins, R. W. H. Sargent, and S. Thomas *in* "Computer Aided Process Plant Design", M. E. Leesley (Ed.), Gulf Publ., Houston, 1982.

quadratic programming [12,13] and then describe briefly mixed integer nonlinear programming. Fig. 4 illustrates the entire sequence of information flow needed in conjunction with the actual execution of the optimization phase. To be effective, the optimization phase must be suitably meshed with the process flowsheeting (simulation) code. Beigler [14] has surveyed existing process flowsheeting codes and describes their capabilities, characteristics, and distributors.

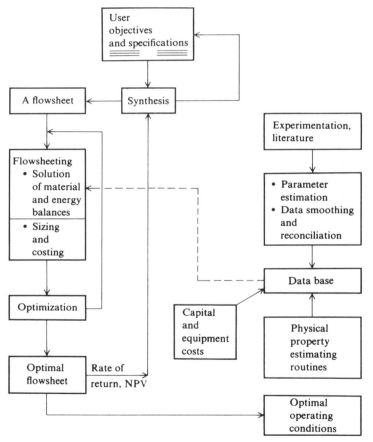

Fig. 4 Information flow for optimization in the chemical industry. The flowsheeting (process simulator) code can be comprised of equations or modules. Synthesis (design) may not yield *the* optimal flowsheet configuration, but optimal operating conditions apply to any selected flowsheet. [Taken from T. F. Edgar and D. M. Himmelblau, "Optimization of Chemical Process", McGraw-Hill Book Co., New York, 1988 by permission]

A.   GENERALIZED REDUCED GRADIENT METHOD

The generalized gradient algorithm is an extension of the Wolfe algorithm [15] to accommodate both a nonlinear objective function and nonlinear constraints. In essence the method employs linear, or linearized constraints, defines new variables that are normal to some of the constraints, and transforms the gradient to this new basis. (Wolfe describes the relation of the original reduced gradient method to the simplex method of linear programming.)   Although the problem solved by the generalized reduced method is

$$
\begin{array}{lll}
\text{Minimize:} & f(\mathbf{x}) & \mathbf{x} \varepsilon E^n \\
\text{Subject to:} & h_j(\mathbf{x}) = 0 & j = 1, \ldots, m \\
& L_i \leq x_i \leq U_i & i = 1, \ldots, n
\end{array} \tag{4}
$$

inequality constraints can be accommodated by subtracting nonnegative slack variables from the inequality constraints thus:

$$
h_j(\mathbf{x}) = g_i(\mathbf{x}) - v_j^2 = 0
$$

and permitting the bounds on the v_j's to be $-\infty \leq v_j \leq \infty$. (The v_j's are added to the set of n variables.)   The reduced gradient procedure is well suited to plant models involving a substantial number of equality constraints because the method essentially uses the equality constraints to reduce the dimensionality of the problem to a set of truely independent variables. The primary disadvantage is that the entire set of equality constraints must converge at each iteration before a new iteration commences.

To illustrate the basic idea, consider the problem of minimizing an objective function of just two variables subject to one equality constraint:

$$
\begin{array}{ll}
\text{Minimize:} & f(x_1, x_2) \\
\text{Subject to:} & h(x_1, x_2) = 0
\end{array}
$$

For differential displacement in x_1 and x_2,

$$d\,f(\mathbf{x}) = \frac{\partial f(\mathbf{x})}{\partial x_1}\,d\,x_1 + \frac{\partial f(\mathbf{x})}{\partial x_2}\,d\,x_2$$

Furthermore,

$$dh\,(\mathbf{x}) = \frac{\partial h\,(\mathbf{x})}{\partial x_1}\,dx_1 + \frac{\partial h\,(\mathbf{x})}{\partial x_2}\,dx_2 = 0$$

These equations are linear in the differential displacement, so that the selected differential dependent variable can be eliminated from the differential objective function. The only admissible displacements are those along the linearized constraint.

Solve $dh\,(\mathbf{x}) = 0$ for dx_2:

$$d\,x_2 = \frac{\partial h\,(\mathbf{x})/\partial x_1}{\partial h\,(\mathbf{x})/\partial x_2}\,dx_1$$

and introduce dx_2 into the differential objective function

$$d\,f(\mathbf{x}) = \left(\frac{\partial f(\mathbf{x})}{\partial x_1} - \frac{\partial f(\mathbf{x})}{\partial x_2}\frac{\partial f(\mathbf{x})/\partial x_1}{\partial h\,(\mathbf{x})/\partial x_2} \right)dx_1$$

to yield the *reduced gradient*

$$\frac{d\,f(\mathbf{x})}{dx_1} = \frac{\partial f(\mathbf{x})}{\partial x_1} - \frac{\partial f(\mathbf{x})}{\partial x_2}\left[\frac{\partial h\,(\mathbf{x})}{\partial x_2} \right]^{-1}\frac{\partial h\,(\mathbf{x})}{\partial x_1}$$

One necessary condition for $f(\mathbf{x})$ to be a minimum is that $d\,f(\mathbf{x}) = 0$, or by analogy to the condition for an unconstrained minimum that

$$\frac{df\,(\mathbf{x})}{dx_1} = 0$$

Thus, the search for a minimum can be carried out in a space of reduced dimensions (equal in size to the degrees of freedom).

In the generalized reduced gradient method with x split into two components, one containing m dependent variables (*basic*) variables x_V and the other containing (n-m) independent (*superbasic*) variables x_U, the null space search direction s_Z is the one used in adjusting the values of the independent variables. Suppose that A, the Jacobian matrix of the (linearized) equality constraints in Eq. (4) is partitioned as follows

$$A = [V \vdots U]$$

Let the set of m columns of A form V, an m x m nonsingular matrix, and the balance of the columns of A form U. Partition X so that

$$x = \left[\begin{array}{c} x_V \\ x_U \end{array} \right]$$

Then the linearized equality constraints are

$$[V \vdots U] \left[\begin{array}{c} x_V \\ x_U \end{array} \right] = b \tag{5}$$

The basis in the null space Z is orthogonal to all the vectors in the range space Y so that

$$Z = \left[\begin{array}{cc} -V^{-1} & U \\ I & \end{array} \right]$$

as can be demonstrated by direct multiplication of A times Z: $AZ = O$.

In the generalized reduced gradient method, the null space search direction s_Z is just s_U with (n-m) elements and the reduced gradient is

$$s = Z^T \nabla f(x) = \left[\begin{array}{cc} -V^{-1} & U \\ I & \end{array} \right] \left[\begin{array}{c} \nabla_{x_V} f(x) \\ \nabla_{x_U} f(x) \end{array} \right]$$

Once the search direction is obtained in the null space, a secant or quasi-Newton unconstrained search algorithm is used to take one or more steps in the s direction. On completion of a   step Δx in the direction s from a feasible point, the new x will ordinarily be a nonfeasible point, and then the basic (dependent) variables (only) are modified to obtain a feasible point. Let $x_V^{(k+1)}$ be the vector of dependent variables at stage $(k + 1)$, and $\tilde{x}_U^{(k+1)}$ be the temporary vector of independent variables at the same stage which causes the point $x^{(k+1)}$ to be infeasible, i.e., $h(x_V^{(k+1)}, \tilde{x}_U^{(k+1)}) \neq 0$; $x_U^{(k+1)}$ is the feasible component of $x^{(k+1)}$. If the constraints are linearized by a truncated Taylor series, we can find the $x_U^{(k+1)}$ that causes $h(x_V^{(k+1)}, x_U^{(k+1)})$ to vanish:

$$h(x_V^{(k+1)}, x_U^{(k+1)}) \approx h(x_V^{(k+1)}, \tilde{x}_U^{(k+1)}) + \frac{\partial h(x_V^{(k+1)}, \tilde{x}_U^{(k+1)})}{\partial x_U}(x_U^{(k+1)} - \tilde{x}_U^{(k+1)}) = 0$$

or
$$x_U^{(k+1)} - \tilde{x}_U^{(k+1)} = -\left(\frac{\partial h}{\partial x_U}\right)^{-1} h(x_V^{(k+1)}, \tilde{x}_U^{(k+1)}) \qquad (6)$$

Application of (6) is termed an "iteration by Newton's method" and is continued until one of the following outcomes is obtained.

If $(x_V^{(k+1)}, \tilde{x}_U^{(k+1)})$ becomes feasible (to within a selected tolerance), then the final $\tilde{x}_U^{(k+1)}$ becomes $x_U^{(k+1)}$. If $f(x_V^{(k+1)}, \tilde{x}_U^{(k+1)}) < f(x_V^{(i)}, x_U^{(i)})$, where the superscript i designates the most recent feasible x vector, the iteration by Newton's method is terminated, and the search starts over again. If $\tilde{x}_U^{(k+1)}$ is an interior or boundary point but $f(x_V^{(k+1)}, \tilde{x}_U^{(k+1)}) > f(x_V^{(i)}, x_U^{(i)})$, or if the iteration by Eq. (6) fails to converge in a fixed number of iterations, say 20, then the step length in the search direction is reduced by some fraction (such a 1/2 or 1/10), and the iteration by Eq. (6) is repeated. If neither of the first two outcomes is achieved, and if the last point obtained by Eq. (6) is not an interior or boundary point, a change in basis is carried out as described in the literature.

B.   SUCCESSIVE QUADRATIC PROGRAMMING

As mentioned above, most chemical process simulators (flowsheeting codes) are composed of modules (portable subroutines). An executive controls the flow of information among modules, reads data, stores data, and generates reports or graphs. Considerable work has focused on how to optimize plants without having to satisfy the entire set of equations, or the entire set of modules, on each cycle of iteration.

Powell (12) and Han (13) described an optimization algorithm which approximates the initial nonlinear programming problem with a sequence of quadratic programming problems. By incorporating the equality constraints into the quadratic programming subproblem, they were able to eliminate the inner convergence loop entirely and converge to the optimum value for the objective function and while satisfying the equality constraints simultaneously. The original Han-Powell method is not well suited for large problems because it requires the use of a non-sparse Hessian matrix containing all the variables in the problem. However modification of the method [16,17,18,19] have proved to be quite effective in practice, especially for modular based plant representations. In particular, repetitive and time consuming procedures for the convergence of nested calculational loops involving recycle streams have been replaced by simultaneous methods.

Successive quadratic program actually refers to the subproblem of determining the search direction to take for the problem outlined by Eqs. (4). Movement in the search direction can take place via many different techniques. The first order necessary conditions for optimality of Eqs. (4) can be obtained from the Kuhn-Tucker conditions applied to Lagrange function

$$\nabla L \equiv \nabla f(\mathbf{x}) + \Sigma \omega_j \nabla h_j(\mathbf{x}) = 0$$

$$h_j(\mathbf{x}) = 0 \tag{7}$$

If Newton's method is applied to solve the above two equations, you get

$$\begin{bmatrix} \nabla_x^2 L & -\mathbf{J} \\ -\mathbf{J}^T & 0 \end{bmatrix} \begin{bmatrix} \Delta \mathbf{x} \\ \Delta \omega \end{bmatrix} = \begin{bmatrix} -\nabla_x L \\ h(\mathbf{x}) \end{bmatrix} \tag{8}$$

where $\mathbf{J}$ is the Jacobian matrix of the (linearized) equality constraints. At the optimum, $(\mathbf{x}^*, \omega^*)$, the second order necessary conditions require that the projection of $\nabla_x^2 L(x^*, \omega^*)$ into the null space of $J(x^*)$ be positive semi-definite. Therefore, for some local region about (x^*, ω^*), a Newton step is equivalent to solving the following quadratic programming problem:

$$\text{Minimize: } F(s) = s^T \nabla f(x) = \frac{1}{2} s^T B s \tag{9}$$

$$\text{Subject to: } h_j(x) + s^T \nabla h_j(x) = 0 \quad j = 1, \ldots, m$$
$$x_L \leq x \leq x_U$$

where s is the search direction and B is a positive definite approximation of $\nabla_x^2 L(x)$ usually evaluated from first derivatives, or their finite difference substitutes. Inequality constraints can also be included in Eqs. (9) analogous to the linearized equality constraints.

For plants represented solely by equations and inequalities, the use of successive quadratic programming plus a search routine to minimize in the s direction is fairly straightforward. However, how to apply the concept to plants represented by modules requires a bit of explanation. Examine the vastly simplified set of modules illustrated in Fig. 5. Two classes of optimization strategies can exist if a sequential solution of the modules is to be used in the flowsheeting code:

(1) feasible path strategies
(2) infeasible path strategies

In the general nonlinear programming problem only two sets of constraints were listed, equality and inequality constraints. Now we must consider three classes of constraints:

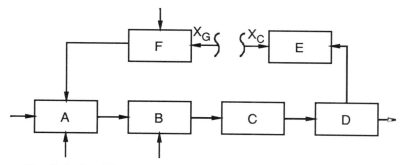

Fig. 5 A simplified configuration of modules that represents a chemical plant

(a) inequality constraints
(b) equality constraints that involve the design variables
(c) equality constraints that represent the stream interconnections between modules

In addition, the vector **x** is split into vectors for the

(1) design variables (independent variables)
(2) stream variables, i.e., those in the interconnections
(3) dependent variables, i.e., those calculated from (a) and (b) above only

Feasible path strategies, as the name implies, on each iteration solve the equality constraints, that is seek convergence for each module, for fixed values of the design variables, and then adjust the design variables via the optimization procedure. Infeasible path strategies (which we will describe), on the other hand, do not require exact solution of the modules on each pass through the simulator. Thus, if an infeasible path method fails, the concluding solution is of little value.

To effectively solve the set of equations that would be represented if each module could be converted into equations, a procedure caller *tearing* is employed. Briefly, tearing means selecting certain output variables from a module involved in a recycle loop of information so that the remaining values of variables in the loop can be solved serially. Tearing decouples the interconnections between modules. Given an initial guess for values of the torn variables, the module outputs are calculated and the values for the torn variables compared with the guesses. If the deviations are too big, the calculated values form new guesses, and so on, until the deviations are sufficiently small. Infeasible path methods do not require precise convergence for the set of modules (analogous to inexact solution of a set of equations) until the optimum is reached.

We can reformulate the problem given by Eqs. (1)-(3) as follows:

Minimize: $F(x_I, x_D, x_G)$ (10a)
x_I

Subject to: $g(x_I, x_D, x_G) \geq 0$ inequality constraints   (10b)

$h(x_I, x_D, x_G) = 0$ design constraints   (10c)

$\overline{h}(x_I, x_D, x_G, x_C) = 0$ stream interconnection constraints for

recycle
                                                    tear equations   (10d)

$x_L \leq x \leq x_L$ (10e)

where x_I as the vector of independent variables (decision variables), x_D is the vector of dependent variables (those used to directly calculate f, g, and h), x_G is the vector of guessed stream variables in torn streams between modules, and x_C is the vector of calculated stream variables for the torn streams. The set of constraints involving the torn variables represents the difference between the guessed values of the torn variables and the respective calculated values:

$$\overline{h} = x_G - x_C(x_I, x_D, x_G) = 0 \qquad (11)$$

Convergence of Eq. (11) does not take place on every pass through the simulator. Instead $\overline{h} \to 0$ as the objective function and other constraints converge to their limits during the optimization. Table II lists the steps in the algorithm described by Lang and Beigler [20] in which optimization was added to the flowsheeting code FLOWTRAN.

Table II   Steps for the Infeasible Path Sequential Modular Optimization
Algorithm [20]

Step 0   Select the design variables x_I, the dependent variables x_D, and identify the torn variables x_G that must be returned from the simulator modules to calculate values of $f(x)$ and the constraints. Choose a calculational sequence that includes all the x_I, x_D, and x_G elements in a single calculational loop (so that the perturbations can be accomplished in the proper sequence). Initialize the SQP counter to 0 and let $x_I = x_I^{(0)}$, $x_G = x_G^{(0)}$; $x_G^{(0)}$ can be calculated by partially converging the set of modules. Set the convergence criteria.

Step 1   Calculate the approximate gradients for f, g, h, and $\overline{h}$ with respect to x_I and x_G using forward finite difference substitutes by chainruling or direct perturbation of the modules.

Step 2   Solve the quadratic programming problem, Eq (9), (including inequality constraints) in which x is split into x_I and x_G to get to get the next search direction $s^{(1)}$ for x_I and x_G.

Step 3   Evaluate the convergence criteria which are described in detail in the reference, and if satisfied stop. Otherwise

Step 4   Carry out a line search in the s direction. A given step size λ along the search direction is accepted if a "sufficient" decrease is observed in some merit function. In the algorithm of Han and Powell, this function was the exact penalty function; in the reference algorithm an augmented Lagrange function was used.

Step 5   Calculate $x_I^{(k+1)} = x_I^{(k)} + \lambda s_{x_I}^{(k)}$ and $\widetilde{x}_G^{(k+1)} = x_G^{(k)} + \lambda s_{x_G}^{(k)}$

Step 6   Set the iteration counter for the Broyden update of **A** to 0, and calculate all the variables in the flowsheet given $x_I^{(k+1)}$ and $\widetilde{x}_G^{(k+1)}$. If certain criteria developed in the article are satisfied, set $x_I^{(k+1)} = \widetilde{x}_G^{(k+1)}$ and go to step 10. Otherwise continue.

Step 7   Set $x_G^{(0)} = \widetilde{x}_G^{(k+1)}$ and apply a modified Broyden relation (given in the reference) to update $H = [\nabla_{x_I} \overline{h}^T \nabla_{x_G} h^T]$ p times until certain convergence criteria in the reference are satisfied. Then set $x_G^{(k+1)} = \widetilde{x}_G^{(k+1)}(p)$. If the criteria cannot be satisfied with $p \leq 5$, set $x_G^{(k+1)} = \widetilde{x}_G^{(k)}(0)$.

Step 8   Calculate the gradients of $f^{(k+1)}, g^{(k+1)}, h^{(k+1)}$, and $\overline{h}^{(k+1)}$ at step $(k+1)$ as in step 1.

Step 9   Update **B** in the quadratic programming subproblem.

Step 10 let $k = k + 1$ and return to step 1.

In the subproblem given by Eqs. (9), the Hessian matrix approximation **B** is approximated as a dense matrix via quasi-Newton (secant) updating formulas such a BFGS. For chemical plant representations it makes more sense to approximate **B** in some sparse form, and adapt the quadratic programming subroutine to take advantage of sparsity in **B** and the constraint gradients. This step will lead to more efficient QP solutions as well as to better approximations of **B** and fewer SQP iterations.

Various ways exist to provide the values needed to calculate finite difference substitutes for derivatives. One way to calculate the Jacobian matrix of the constraints is to perturb each module in sequence with respect to the torn variables. Another way to calculate the partial derivatives is to simulate each module individually (rather than in sequence) after each unknown input variable is perturbed by a small amount. This method of calculation of the Jacobian matrix is usually referred to as full-block perturbation.

Because of the large number of modular simulations required for full-block perturbation, Mahelec, et al. [21] suggested using a diagonal approximation to the Jacobian matrix in which only the jth element of the input stream vector affected the corresponding element of the output stream vector. As might be expected, this approach sometimes leads to an exceedingly fast solution of a problem and other times leads to failure and is not recommended. Chen and Stadtherr [16] evaluated the relative effectiveness of the sequential perturbation method versus full-block perturbation in calculating the Jacobian matrix. When full-block perturbations are used, the internal variables for the module can be temporarily saved, thus reducing the CPU time. Sequential calculations lose this advantage. One can calculate the first element of the gradient of h_i with respect to the independent variable x_{I1} from the chain rule with full-block perturbation thus

$$\left(\frac{\partial h_i}{\partial x_{I1}}\right) = \left(\frac{\partial h_i}{\partial x_{D1}}\right)\left(\frac{\partial x_{D1}}{\partial x_{I1}}\right) + \left(\frac{\partial h_i}{\partial x_{D2}}\right)\left(\frac{\partial x_{D2}}{\partial x_{I1}}\right) + \cdots$$

and other elements similarly. Since $(\partial h_i/\partial x_{Dk})$ often can be evaluated explicitly, only $(\partial h_{Dk}/\partial x_{Ij})$ needs to be evaluated by perturbation.

Because optimization of chemical processes involves relatively few degrees of freedom, if successive quadratic programming can be applied in the null space of the equality constraints, a much smaller quadratic programming problem is involved. Suppose that an optimization problem has only equality constraints. An orthonormal null space to $\nabla_z h$, Z, and a range space, Y, satisfy the following relations:

$$\nabla h^T Z = 0$$
$$Z^T Z = I$$
$$YS = \nabla_z h$$
$$Y^T Z = 0$$
$$Y^T Y = I$$

where $\nabla_Z h \in R^{n \times m}$, $Y \in R^{n \times m}$, $Z \in R^{n \times (n-m)}$, and $S \in R^{n \times m}$ is some nonsingular matrix. Then the successor to Eq. (8) becomes

$$
\begin{bmatrix}
Y^TBY & Y^TBZ & S \\
Z^TBY & Z^TBZ & 0 \\
S^T & 0 & 0
\end{bmatrix}
\begin{bmatrix}
s_Y \\
s_Z \\
\omega^{(k+1)}
\end{bmatrix}
= -
\begin{bmatrix}
Y^T\nabla_Z f \\
Z^T\nabla_Z f \\
h
\end{bmatrix}
\tag{12}
$$

From the second and third rows of Eq. (12) one can observe that search direction, s_Z, is independent of the Lagrange multipliers and the search direction in the range space, s_Y, does not depend on B. Moreover, Gill, Murray and Wright [22] showed that by eliminating Y^TBY, Y^TBZ and Z^TBY (or setting them to zero) and using only the projected Hessian Z^TBZ, one can solve a much smaller problem:

$$
\begin{aligned}
Z^TBZs_Z &= -s^T\nabla f \\
S^Td_Y &= -h \\
Sv^{(k+1)} &= -Y^T\nabla_Z f
\end{aligned}
\tag{13}
$$

and still obtain a good estimate of the multipliers.

C. MIXED INTER NONLINEAR PROGRAMMING (MINLP)

An *integer linear programming problem* is a linear programming problem in which all the variables are constrained to take non-negative integer values. In a *mixed integer linear programming problem* some of the variables are constrained to be integer while the rest are continuous. Analogously, mixed integer nonlinear programming (MINLP) problems are nonlinear programming problems in which some of the variables are restricted to integer values. Such problems have the common feature that at least one of the variables is discrete. Problems with discrete variables are important in the chemical industry because in many processes some of the variables must be discrete. For instance, problems dealing with the number of pumps, tanks, reactors, and so on include variables with physically indivisible units. If a problem of this type is attached by solving the NLP problem as non-discrete and by adjusting or rounding the solution thus obtained so that it will fulfill the discreteness requirements, the adjusted solution may prove to be far from optimal for the discrete problem. In many calculations it will be more effective to use MINLP methods so that an optimal solution can be found.

Another reason why a discrete treatment is important is because logical relations may be expressed by meas of Boolean variables, and thus some logical constraints can be incorporated within a problem. Many combinatorial problems exist that can be formulated as discrete problems by using integer and Boolean variables.

Techniques of solving MINLP problems, not surprisingly, are in a much worse state than for nonlinear programming problems themselves [23]. Three general types of algorithms are used to solve the MINLP problem are:

(1) Branch and bound [24]
(2) Generalized Benders decomposition [25]
(3) Outer-approximation [26]

Some simplifications can occur if the problem has (1) separable objective function and constraints, (2) separable objective function and linear constraints, or (3) special structure.

A common problem in the chemical industries is the synthesis problem, that the problem of design such that optimal configuration of equipment can be selected for a given goal (objective function).

Such a problem is posed as follows [27]

$$
\begin{aligned}
&\text{Minimize:} && \mathbf{c}^T\mathbf{y} + f(\mathbf{x}) \\
&\quad \mathbf{x},\mathbf{y} \\
&\text{Subject to:} && \mathbf{h}(\mathbf{x}) = \mathbf{0} \\
& && \mathbf{g}(\mathbf{x}) \le \mathbf{0} \\
& && \mathbf{A}\mathbf{x} = \mathbf{a} \\
& && \mathbf{B}\mathbf{y} + \mathbf{C}\mathbf{x} \le \mathbf{d} && (14)\\
& && \mathbf{x}\varepsilon X = \{\mathbf{x} \mid \mathbf{x}\varepsilon R^n,\ \mathbf{x}^L \le \mathbf{x} \le \mathbf{x}^U\} \\
& && \mathbf{y}\varepsilon Y = \{\mathbf{y} \mid \mathbf{y}\varepsilon\{0,1\}^m,\ \mathbf{E}\mathbf{y} \le \mathbf{e}\}
\end{aligned}
$$

where $\mathbf{c}^T\mathbf{y}$ represents fixed investment costs, $f(\mathbf{x})$ includes operating costs and costs a function of process unit size, $\mathbf{x}$ is the vector of continuous variables representing flows, pressures, temperatures and unit sizes, $\mathbf{Y}$ is a vector of 0-1 binary variables representing the potential existence of unit processes that might be present in alternative flowsheet designs, $\mathbf{h(x)} = \mathbf{0}$ and $\mathbf{Ax} = \mathbf{a}$ are equations corresponding to the material and energy balances and design equations (models) in the process. Various specifications and constraints are represented by $\mathbf{g(x)} \leq \mathbf{0}$. Logical constraints and linear specifications that must hold for a flowsheet configuration to be selected from within the superstructure of process units are represented by the linear $\mathbf{By + Cx} \leq \mathbf{d}$ and $\mathbf{Ey} \leq \mathbf{e}$. The variables $\mathbf{x}$ are specified to lie within the compact set X consisting of lower and upper bounds.

Several algorithms have been proposed to solve problems such as Eqs. (14), but we do not have the space to describe them here. Grossmann [28] has reviewed the existing computational experience and concluded

(1) As expected, proper formulation of the problem statement is important and influences the selected solution technique. By tightly bounding the continuous variables, specifying the smallest possible upper bounds on logical constraints, and replacing constraints such as $\sum_{i=1}^{m} y_i - mz \leq 0$ with $y_i - z \leq 0$, $i = 1, \ldots$ m, the number of branches or major iterations in a generalized Benders decomposition or outer-approximation can be reduced. Also, one should avoid in so far as possible the use of products of Boolean variables with continuous variables or functions that introduce nonconvexities.

(2) As in NLP, no algorithm is best to solve all problems. Branch and bound is superior if the relaxed nonlinear programming problem exhibits integer solutions. But if this is not the outcome, the generalized Bender's decomposition and variations of the outer approximation algorithm do best, particularly in problems that involve the solution of hundreds or thousands of nonlinear programming subproblems.

(3) Nonconvexities cause two types of difficulties in the solution. First, the solution to the NLP subproblems may not be unique, and second, the optimal solution may be cut off in the generalized Benders and outer-approximation algorithms. Some variations of the outer approximation

algorithm by Grossman and coworkers seems to be fairly robust toward nonconvexities relative to the generalized Benders decomposition.

III.   DISTRIBUTED VERSUS CENTRALIZED OPTIMIZATION

Two general views of optimization, particular for control, are of interest. One is top-down, or centralized or global optimization, and the other is bottom-up, or distributed or hierarchial optimization.   Fig. 6 illustrates each concept. Implementation of these methods for control and scheduling is discussed by Lasdon [29].

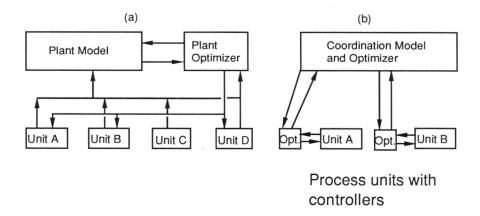

Fig. 6   Information flow for (a) centralized and (b) distributed optimization. In the latter each unit has its own optimizer.

In distributed optimization, the over-all problem is broken up into a number of local optimization problems that operate on individual pieces of equipment or sections of a process. Distributed optimization represents the operator's concept about the process, and control systems built on such a foundation are more likely to be utilized. An overall optimizer still exists, but its role becomes one of coordinating the local optimizers and determining the optimum for a few selected key variables, such as the overall feed rate, feed type selection or reactor conversion.

Distributed optimization also permits the use of more complex, and hopefully more representative, models for individual unit modules. In addition to more accurate modeling, more constraints can be accommodated, and different models can be used for different optimization levels. Global on-line optimizers are limited in the variables that can be considered because of the limits of the computational requirements. Because an optimum operating point often lies on or near several constraints, improved modeling of the constraints can yield an improved optimum.

The frequency of cycles of on-line optimization is also more flexible with distributed optimizers. In the steady state, each cycle of optimization must be less than the settling time of the system plus the time needed to make the estimates of the current state. Using local optimizers, you can set the frequency to match the individual characteristics of the process units whereas with centralized optimization the execution frequency must conform to the response time of the entire plant. Also the unit models can be updated, or disconnected, independently of the other units.

IV.   OPTIMIZATION OF DYNAMIC PROCESSES

A vast literature exists [30] describing in general and in particular the optimization of dynamic processes, particularly for process control. Unless a very good dynamic model of the process exists, process identification and optimization must occur jointly. Thus, the coupling between these two activities is inherent in optimization. From the viewpoint of nonlinear programming, the model serves as the constraints and some criterion is the objective function. Any dynamic mathematical model consists of unknown variables and parameters that have to be estimated. In general, these parameters are not estimated exactly, whether in the time or frequency domains, but rather are estimated or determined under non-optimal conditions. Accordingly, the solution generated from such system model may be suboptimal. Identification may be either static or dynamic but is usually dynamic, especially if the parameters of the process model are time varying or even functions of the state variables of the process.

Five typical of dynamic system optimization problems are: (1) the optimal control problem, (2) parameter estimation, (3) process identification, (4) two point-boundary value problems, and (5) data reconciliation. The problems can be posed as follows in the time domain.

A. PROCESS CONTROL

The optimal control problem is concerned with finding $\mathbf{u}$ (t), the (m x 1) vector of unknown control functions, such that the response of the process minimizes (maximizes) a specified performance criterion. Under certain restrictions, this problem can be converted into the form of a parameter estimation problem or a two-point-boundary-value problem. The problem to be solved may be posed as a nonlinear programming problem

$$
\begin{aligned}
&\text{Minimize:} && \mathbf{F}\ (\mathbf{x}\ (t),\ \mathbf{u}\ (t);\ \mathbf{p}) \\
&\ \ \mathbf{u}\ (t) && \\
&\text{Subject to:}\ \mathbf{f}\ (d\mathbf{x}\ (t)\ /\ dt,\ \mathbf{x}\ (t),\ \mathbf{u}\ (t);\ \mathbf{p}) = \mathbf{0} \\
&\qquad\qquad \mathbf{h}\ (\mathbf{x}\ (t),\ \mathbf{u}\ (t);\ \mathbf{p}) = \mathbf{0} \\
&\qquad\qquad \mathbf{g}\ (\mathbf{x}\ (t),\ \mathbf{u}\ (t);\ \mathbf{p}) \geq \mathbf{0} \\
&\qquad\qquad \mathbf{x}\ (t_0) = \mathbf{x}_0
\end{aligned}
\tag{15}
$$

where $\mathbf{x}$ (t) and $\mathbf{u}$ (t) are vector valued state and control (manipulated) variables, respectively, and $\mathbf{p}$ is a vector of model parameters, which may include disturbances. Note that the objective function $\mathbf{F}$ is not required to take any special form, although various choices of the form will lead to simpler solution techniques. The differential-algebraic equations, $\mathbf{f}$, are also not required to take any special form and may include integral and/or spatial derivative terms as well as time delays.

Quite a few methods for solving the nonlinear programming problem (15) have appeared in the open literature. The standard techniques of optimal control ensue by adding the constraint equations (with unknown penalty parameters) to the performance objective and deriving the necessary conditions for an optimum [31,32] leading a two point boundary value problem that must be solved to give the solution of the original problem. Many methods exist for the solving the boundary value problem, including finite element and collocation methods, though much of the literature seems to be centered around shooting methods. A method which is equivalent to solving the necessary conditions via a weighted residual technique can be obtained by writing the necessary conditions for the optimum of the original problem. The differential equations are first approximated by a set of algebraic equations using a weighted residual method. The resulting nonlinear program is solved using a Lagrange-Newton method.

Other methods for solving nonlinearly constrained optimization problems may be used, but the advantage of using an infeasible path method is that the model equations are solved simultaneously with the other constraints [33]. Using this approach, the differential equations are treated in the same manner as the other constraints and are not solved accurately until the final iterations of the optimization.

Orthogonal collocation on finite elements may be used in connection with successive quadratic programming to solve Eqs. (15). Eqs. (15) restated becomes

$$
\begin{aligned}
\text{Minimize:} \quad & F(\mathbf{x},\mathbf{u};\mathbf{p}) \\
\mathbf{x},\mathbf{u} \quad & \\
\text{Subject to:} \quad & \mathbf{f}(d\mathbf{x}/dt,\mathbf{x},\mathbf{u};\mathbf{p}) = \mathbf{0} \\
& \mathbf{h}(\mathbf{x},\mathbf{u};\mathbf{p}) = \mathbf{0} \\
& \mathbf{g}(\mathbf{x},\mathbf{u};\mathbf{p}) = \mathbf{0} \\
& \mathbf{x}(t_0) = \mathbf{x}_0
\end{aligned}
\tag{16}
$$

where $d\mathbf{x}/dt$ is approximated by $\mathbf{Ax}$, where $\mathbf{A}$ is a matrix of collocation weights [34]. Any derivatives with respect to spatial coordinated maybe handled in a similar manner, and integral terms may be included efficiently by using appropriate quadrature formulae. A wide range of constraints may be easily incorporated, the manipulated variable profile may be either continuous or discrete, and the solution of the optimization problem provides useful sensitivity information at little additional cost. The primary disadvantage of this approach is that it is difficult to handle discontinuities in the state variable profiles.

Two techniques which are commonly used that requires a feasible set of inputs and solve the model accurately at each iteration are a feasible path nonlinear programming algorithm such as the generalized reduced gradient method, or a feasible path successive quadratic programming algorithm and (2) the solution of the differential equations each time an objective function evaluation is required. For most nonlinear chemical process models of interest, analytical solutions for the differential equation do not exist and numerical solution techniques are required.

These methods are generally thought to be inefficient because they require that the model equations be solved accurately at each iteration within the optimization, even when the iterates are far from the final optimal solution. But integrating the differential equations at each iteration offers at least two important advantages over other methods in that it is possible (1) to accommodate discontinuities in the state profiles, and (2) to solve stiff systems of equations. One serious disadvantage of this approach is that variations in the accuracy of the solution of the differential equations from one iteration to the next may make it impossible to obtain reliable gradient information. Although in some cases it may be possible to avoid this difficulty by decreasing the value of the desired accuracy, it is also possible that such action will cause the ODE solver to fail, or to increase the time required to solve the system of equations beyond a reasonable limit.

B. PARAMETER ESTIMATION

The parameter estimation problem is to determine the m-dimensional parameter vector $\mathbf{p}$ and any free initial conditions such that the response of the system minimizes (or maximizes) a specific performance criterion

$$
\begin{aligned}
&\text{Minimize:} && F(\mathbf{x}(t),\mathbf{p};\mathbf{u}(t)) \\
&\quad\mathbf{p} && \\
&\text{Subject to:} && \mathbf{f}(d\mathbf{x}(t)/dt,\mathbf{x}(t),\mathbf{p};\mathbf{u};(t)) \;=\; 0 \\
& && \mathbf{h}(\mathbf{x}(t),\mathbf{p};\mathbf{u}(t)) \;=\; 0 \\
& && \mathbf{g}(\mathbf{x}(t),\mathbf{p};\mathbf{u};(t)) \;\geq\; 0 \\
& && \mathbf{x}(t_0) \;=\; \mathbf{x}_0
\end{aligned}
\tag{17}
$$

Eqs. (17) have been posed as a nonlinear programming problem except that now the degrees of freedom are the same as the number of parameters to be estimation. Although F can be of any form, it most likely will be a least squares or maximum likelihood function. Any of the techniques mentioned in Secion A can be employed to solve Eqs. (17).

C   SYSTEM IDENTIFICATION PROBLEM

The mathematical form of the problem is equivalent to Eqs. (17) above except that the precise form of F as well as the parameters are unknown. This is an old problem [35] which we do not have the space to discuss, but is especially important in the area of chemical kinetics and reaction engineering.

D.   TWO-POINT BOUNDARY VALUE PROBLEM

In this class of problems, the objective is to find the m-vector of unknown initial conditions so that the response of the process model satisfies the boundary conditions

$$M_i(x(t_j),t_j) = 0 \qquad \begin{aligned} i &= 1, 2, \ldots, MZ \\ j &= 0, \ldots, f \end{aligned}$$

Each of the above problems can be expressed to a form in which a scalar function dependent explicitly on the system response $x(t)$ is to be minimized (or maximized). Because $x(t)$ depends implicitly on a set of parameters and initial condition, these parameters and initial conditions may be considered as the decision variables in an equivalent mathematical programming problem [30].

E.   RECONCILIATION OF DATA FROM A DYNAMIC PROCESS

Reconciliation of data taken from nonlinear dynamic process (after removal of gross errors) involves the removal of measurement noise. The data reconciliation problem can be formulated as

$$\begin{aligned} \text{Minimize:} \quad & F(\mathbf{y}_t,\widehat{\mathbf{y}}(t),\mathbf{u}(t);\mathbf{p}) \\ \widehat{\mathbf{y}}(t),\mathbf{x}(t) \end{aligned} \tag{18}$$

$$\begin{aligned} \text{Subject to:} \quad & \frac{d\mathbf{x}}{dt} = \mathbf{f}(\mathbf{x},\mathbf{u};\mathbf{p}) \\ & \widehat{\mathbf{y}} = \mathbf{h}(\mathbf{x},\mathbf{u};\mathbf{p}) \end{aligned}$$

where

y_t is the vector of the measured values of the process variables

$\hat{y}$ is the vector of the estimates of the measured quantities

x is the vector of the state variables

u is the vector of the manipulated variables

p is the vector of the model parameters

Often, the objective function is in the form of a weighted-least squares sum of the differences between the measurements and their corresponding estimates. The weights correspond to the reproducibility of the measurement devices.

The estimation of consistent measurement data constrained by ordinary differential equations is currently being solved using orthogonal collocation on finite elements. Orthogonal collocation [34] consists of approximating the solution as a linear combination of orthogonal basis functions over the interval of interest. The method of weighted residuals is used to reduce the data reconciliation problem to an algebraically-constrained nonlinear programming problem. The residuals are evaluated at the discrete roots of the collocation functions.

Unfortunately, the resulting problem contains a relatively large number of added variables and constraints. To keep the size of the problem manageable, a *moving window* is used on the data stream. As each new measurement set is obtained, all measurements within the time window are reconciled using the redundant information contained within the measurement set and the model equations. Then, only the most recent set of estimates is saved, and the procedure is repeated at the next time step.

References

1. H. P. Hutchinson, "The Development of An Equation-Oriented Flowsheet Simulation and Optimization Package - I. The Quasilin Program", *Comput. Chem. Eng. 10*, 19-29 (1986).

2. N. Ganesh and L. T. Biegler, "A Robust Technique for Process Flowsheet Optimization Using Simplified Model Approximations", *Comput. Chem. Eng. 11*, 553-565 (1987).

3. L. T. Biegler and I. E. Grossmann, "Strategies for the Optimization of Chemical Processes", *Chem. Rev. 3*, 2-47 (1985).

4. T. P. Kisala, R. A. Trevino-Lozano, J. F. Boston, and I. H. Britt, "Sequential Modular and Simultaneous Modular Strategies for Process Flowsheet Optimization", *Comput. Chem. Eng. 11*, 567-579 (1987).

5. L. B. Evans, "Flowsheeting, A State of the Art Review", *in* "Proceed. Chemcomp. 1982", (G. F. Froment, ed.) KVI, Antwerp, Belgium, 1982.

6. E. Rosen, "Steady State Chemical Simulation--State of the Art Review" *in* "Computer Applications to Chemical Engineering" (R. G. Squires and G. V. Reklaitis, eds.) ACS Symposium Series No. 124, Amer. Chem. Soc., Wash. D.C., 1980.

7. T. L. Wayburn and J. D. Seader, "Homotopy Continuation Methods for Computer-Aided Process Design", *Comput. Chem. Eng. 11*, 7-25 (1987).

8. M. Fukushima, "Solving Inequality Constrained Optimization Problems by Differential Homotopy Continuation Methods", *J. Math. Anal. Applns. 133*, 109-121 (1988).

9. J. Abadie and J. Carpentier, "Generalization of the Wolfe Reduced Gradient Method to the Case of Noninear Constraints", *in* "Optimization," (R. Fletcher, ed) Academic Press, London, 1969.

10. J. Abadie, "Application of the GRG Algorithm to Optimal Control Problems", *in* "Integer and Nonlinear Programming" (J. Abadie, ed.) North Holland, Amsterdam, 1970.

11. J. Liebman, L. Lasdon, L. Schrage, and A. Waren, "Modeling and Optimization with GINO", Scientific Press, Palo Alto, CA, 1986.

12. M. J. D. Powell, "The Convergence of Variable Metric Methods for Nonlinearly Constrained Optimization Calculations" *in* "Nonlinear

Programming 3", O. L. Mangasarian, et. al. (eds.), Academic Press, New York, 1978.

13. S. P. Han, "Superlinearly Convergent Variable-metric Algorithms for General Nonlinear Programming Problems", *Math. Programming 11*, 263-282 (1976).

14. L. T. Biegler, "Chemical Process Simulation", *Chem. Eng. Progress*, 50-61 (Oct. 1989).

15. P. Wolfe, "Methods of Nonlinear Programming" *in* "Recent Advances in Mathematical Programming," R. O. Graves and P. Wolfe (eds.) McGraw-Hill Book Company, New York, 1963.

16. H. S. Chen and M. A. Stadtherr, "A Simultaneous Modular Approach to Process Flowsheeting and Optimization, Part I: Theory and Implementation", *AIChE J. 31*, 1843- 1856 (1985).

17. L. T. Biegler and R. R. Hughes, "Infeasible Path Optimization of Sequential Modular Simulators", *AIChE J. 28*, 994-1002 (1982).

18. L. T. Biegler and R. R. Hughes, "Feasible Path Optimization for Sequential Modular Simulators", *Comp. Chem. Eng. 9*, 379-387 (1985).

19. J. D. Perkins, "Efficient Solution of Design Problems Using a Sequential Modular Flowsheeting Programme", *Comp. Chem. Eng. 3*, 375-382 (1979).

20. Y. D. Lang and L. T. Biegler, "A Unified Algorithm for Flowsheet Optimization", *Comp. Chem. Eng. 11*, 143-158 (1987).

21. V. Mahelec, H. Kjuzik, and L. B. Evans, "Simultaneous Modular Algorithm for Steady State Flowsheet Simulation and Design," paper presented at the 12th European Symposium on Computer in Chemical Engineering, Montreux, Switzerland (1979).

22. P. E. Gill, W. Murray, and M. H. Wright, "Practical Optimization", Academic Press, New York, 1981.

23. L. Wolsey, "Strong Formulations for Mixed Integer Programming: A Survey", *Math. Programm. 45*, 173-191 (1989).

24. E. M. L. Beale, "Integer Programming", in "The State of the Art in Numerical Analysis" (D. Jacobs, ed.), Academic Press, London, 1977.

25. J. F. Benders, "Partitioning Procedures for Solving Mixed-Variables Programming Problems", *Numer. Math. 4*, 238-252 (1962).

26. M. A. Duran and I. E. Grossman, "An Outer-Approximation Algorithm for a Class of Mixed-Integer Nonlinear Programs." *Math. Programm. 36*, 307-339 (1986).

27. G. R. Kocis and I. G. Grossmann, "Relaxation Stretegy for the Structural Optimization of Flowsheets", *Ind. Eng. Chem. Res. 26*, 1869-1880 (1987).

28. I. E. Grossman, "MINLP Optimization Strategies and Algorithms for Process Synthesis", *in* "FOCAPD 89", (J. J. Siirola and I. E. Grossmann, eds.) Elsevier Publ. Co., Amsterdam, 1990.

29. L. S. Lasdon, "The Integration of Planning, Scheduling, and Process Control", *in* "Chemical Process Control-CPC III", (M. Morari and T. J. McAvoy, eds.), Elsevier, Amsterdam, 1986.

30. L. C. W. Dixon and G. P. Szego, "Numerical Optimization of Dynamic Systems: A Survey", *in* "Numerical Optimization of Dynamic Systems", North-Holland Publ. Co., Amsterdam, 1980.

31. A. E. Bryson and Y. Ho, "Applied Optimal Control", Hemisphere Publishing, New York, 1975.

32. W. H. Ray, "Advanced Process Control," McGraw-Hill, New York, 1981.

33. L. T. Biegler, "Strategies for Simultaneous Solution and Optimization of Differential-Algebraic Systems", *in* "FOCAPD 89", Elsevier, Amsterdam, 1990.

34. J. V. Villadsen and W. E. Stewart, "Solution of Boundary-Value Problems by Orthogonal Collocation," *Chem. Eng. Sci. 22*, 1483-1501 (1967).

35. A. P. Sage and J. L. Melsa, "System Identification", Academic Press, New York, 1971.

INDEX

407